T0183106

Lecture Notes in Artificial Intelligence 9449

Subseries of Lecture Notes in Computer Science

More information about this series at http://www.springer.com/series/1244

Adrian-Horia Dediu · Carlos Martín-Vide
Klára Vicsi (Eds.)

Statistical Language and Speech Processing

Third International Conference, SLSP 2015
Budapest, Hungary, November 24–26, 2015
Proceedings

 Springer

Editors
Adrian-Horia Dediu
Research Group on Mathematical
   Linguistics
Rovira i Virgili University
Tarragona
Spain

Carlos Martín-Vide
Research Group on Mathematical
   Linguistics
Rovira i Virgili University
Tarragona
Spain

Klára Vicsi
Department of Telecommunications
   and Media Informatics
Budapest University of Technology
   and Economics
Budapest
Hungary

ISSN 0302-9743   ISSN 1611-3349 (electronic)
Lecture Notes in Artificial Intelligence
ISBN 978-3-319-25788-4   ISBN 978-3-319-25789-1 (eBook)
DOI 10.1007/978-3-319-25789-1

Library of Congress Control Number: 2015952770

LNCS Sublibrary: SL7 – Artificial Intelligence

Springer Cham Heidelberg New York Dordrecht London

Printed on acid-free paper

Springer International Publishing AG Switzerland is part of Springer Science+Business Media
(www.springer.com)

Preface

This volume contains the papers presented at the Third International Conference on Statistical Language and Speech Processing (SLSP 2015), held in Budapest, Hungary, during November 24–26, 2015.

SLSP 2015 was the third event in a series to host and promote research on the wide spectrum of statistical methods that are currently in use in computational language or speech processing; it aims to attract contributions from both fields. The conference encourages discussion on the employment of statistical methods (including machine learning) within language and speech processing. The scope of the SLSP series is rather broad, and includes the following areas: anaphora and coreference resolution; authorship identification, plagiarism, and spam filtering; computer-aided translation; corpora and language resources; data mining and the Semantic Web; information extraction; information retrieval; knowledge representation and ontologies; lexicons and dictionaries; machine translation; multimodal technologies; natural language understanding; neural representation of speech and language; opinion mining and sentiment analysis; parsing; part-of-speech tagging; question-answering systems; semantic role labeling; speaker identification and verification; speech and language generation; speech recognition; speech synthesis; speech transcription; spelling correction; spoken dialogue systems; term extraction; text categorization; text summarization; user modeling.

SLSP 2015 received 71 submissions, which were reviewed by the Program Committee members, some of whom consulted with external referees as well. Most of the papers received three reviews, and several even four reviews. After a thorough and lively discussion, the committee decided to accept 26 papers (representing an acceptance rate of 37 %). The program also included three invited talks.

Part of the success in managing such a high number of submissions is due to the excellent facilities provided by the EasyChair conference management system. We would like to thank all invited speakers and authors for their contributions, the Program Committee and the reviewers for their cooperation, and Springer for its very professional publishing work.

August 2015

Adrian-Horia Dediu
Carlos Martín-Vide
Klára Vicsi

Organization

SLSP 2015 was organized by the Laboratory of Speech Acoustics (LSA), Department of Telecommunications and Telematics, Budapest University of Technology and Economics, Hungary, and the Research Group on Mathematical Linguistics (GRLMC), Rovira i Virgili University, Tarragona, Spain.

Program Committee

Steven Abney	University of Michigan, Ann Arbor, USA
Roberto Basili	University of Rome Tor Vergata, Italy
Jean-François Bonastre	University of Avignon, France
Jill Burstein	Educational Testing Service, Princeton, USA
Nicoletta Calzolari	National Research Council, Pisa, Italy
Kevin Bretonnel Cohen	University of Colorado, Denver, USA
W. Bruce Croft	University of Massachusetts, Amherst, USA
Marc Dymetman	Xerox Research Centre Europe, Meylan, France
Guillaume Gravier	IRISA, Rennes, France
Kadri Hacioglu	Sensory Inc., Santa Clara, USA
Udo Hahn	University of Jena, Germany
Thomas Hain	University of Sheffield, UK
Mark Hasegawa-Johnson	University of Illinois, Urbana, USA
Jing Jiang	Singapore Management University, Singapore
Tracy Holloway King	A9.com, Palo Alto, USA
Sadao Kurohashi	Kyoto University, Japan
Claudia Leacock	McGraw-Hill Education CTB, Monterey, USA
Mark Liberman	University of Pennsylvania, Philadelphia, USA
Carlos Martín-Vide (Chair)	Rovira i Virgili University, Tarragona, Spain
Alessandro Moschitti	University of Trento, Italy
Jian-Yun Nie	University of Montréal, Canada
Maria Teresa Pazienza	University of Rome Tor Vergata, Italy
Adam Pease	IPsoft Inc., New York, USA
Fuchun Peng	Google Inc., Mountain View, USA
Bhiksha Raj	Carnegie Mellon University, Pittsburgh, USA
Javier Ramírez	University of Granada, Spain
Paul Rayson	Lancaster University, UK
Dietrich Rebholz-Schuhmann	University of Zurich, Switzerland
Douglas A. Reynolds	Massachusetts Institute of Technology, Lexington, USA
Michael Riley	Google Inc., Mountain View, USA
Horacio Saggion	Pompeu Fabra University, Barcelona, Spain
David Sánchez	Rovira i Virgili University, Tarragona, Spain

Roser Saurí	Oxford University Press, Oxford, UK
Stefan Schulz	Medical University of Graz, Austria
Efstathios Stamatatos	University of the Aegean, Karlovassi, Greece
Yannis Stylianou	Toshiba Research Europe Ltd., Cambridge, UK
Maosong Sun	Tsinghua University, Beijing, China
Tomoki Toda	Nara Institute of Science and Technology, Japan
Yoshimasa Tsuruoka	University of Tokyo, Japan
Klára Vicsi	Budapest University of Technology and Economics, Hungary
Enrique Vidal	Technical University of Valencia, Spain
Atro Voutilainen	University of Helsinki, Finland
Andy Way	Dublin City University, Ireland
Junichi Yamagishi	University of Edinburgh, UK
Luke Zettlemoyer	University of Washington, Seattle, USA
Pierre Zweigenbaum	LIMSI-CNRS, Orsay, France

Additional Reviewers

Chenhui Chu	Sergio Martínez
Ying Ding	Jianfei Yu
Mónica Domínguez Bajo	

Organizing Committee

Adrian-Horia Dediu	Rovira i Virgili University, Tarragona, Spain
Carlos Martín-Vide (Co-chair)	Rovira i Virgili University, Tarragona, Spain
György Szaszák	Budapest University of Technology and Economics, Hungary
Klára Vicsi (Co-chair)	Budapest University of Technology and Economics, Hungary
Florentina-Lilica Voicu	Rovira i Virgili University, Tarragona, Spain

Invited Talks (Abstracts)

Low-Rank Matrix Learning for Compositional Objects, Strings and Trees

Xavier Carreras

Xerox Research Centre Europe, 38240 Meylan, France
xavier.carreras@xrce.xerox.com

Many prediction tasks in Natural Language Processing deal with structured objects that exhibit a compositional nature. Consider this sentence in the context of a Named Entity Recognition task:

$$\overbrace{\text{Loc}}$$

A shipload of 12 tonnes of rice arrives in $\overbrace{\text{Umm Qasr}}^{\text{Loc}}$ port in the Gulf.

In this example we can distinguish three main parts: the entity mention Umm Qasr annotated with a location tag (Loc), a left context, and a right context. We could generate other examples by substituting one of these parts with other values. We would like to develop statistical models that correctly capture the compositional properties of our structured objects (i.e. what parts can be combined together to form valid objects), and exploit these properties to make predictions.

For instance, we could design a model that computes predictions using this expresssion:

$$\alpha(\text{A shipload of 12 tonnes of rice arrives in})^\top \gamma_{\text{Loc}} (\text{Umm Qasr}) \, \beta(\text{port in the Gulf.})$$

where:

- The model induces latent spaces of n dimensions
- α: is a function that embeds left contexts to vectors in $\mathbb{R}^n$
- β: is a function that embeds right contexts into vectors in $\mathbb{R}^n$
- γ_{Loc}: is a function that embeds a an entity candidate of type location into a matrix in $\mathbb{R}^{n\times n}$

For named entity prediction, we would like the α and β embeddings to capture the relevant properties of the left and right entity contexts, and γ_t to capture the relevant properties of entities of type t. We say that the model above is compositional in the sense that it mimics the compositional for of the structures. For instance, if we are given an entity candidate and want to predict its type, we will evaluate the model expression for every possible type: in these evaluations, the computations of α and β are shared, and only the computation of γ needs to be done for each type. Similarly we could share computation among structures that share left or right contexts.

The challenge we focus on is how to learn compositional models like the one above that are compact (in terms of embedding parts of a structure in low-dimensional spaces) and accurate (in terms of achieving low error on one or several predefined tasks).

To solve this, we explore a framework based on formulating the learning problem as learning low-rank matrices, or in general, learning low-rank tensors. One key idea is to use tensors of m orders to represent compositions of m elements: each of the axis of the tensor represents on part of the structure. The rank of the tensor corresponds to the intrinsic dimension of the model. Thus, it is natural to use the rank as penalty for controlling the capacity of the model, and thus we aim at obtaining models that have low rank. In our learning framework, we formulate this using the nuclear norm regularizer, which corresponds to a convex relaxation of a low-rank penalty [7]. Finally, we use a combined objective that includes a standard discriminative loss (in our case the negative log-likelihood) and a nuclear-norm regularizer. Our optimization is convex.

This invited talk overviews some instantiations of this framework. The first, described by Madhyastha, Carreras and Quattoni [4], is a simple application to model bilexical relations, or linguistic relations between two lexical items. For example, for a noun-adjective relation, we are given a query noun such as cat and we want to retrieve adjectives that can modify the noun, such as small, hungry, ... In a series of experiments we show that our framework can be used to induce word embeddings that are tailored for the relation; and, as a specific case, we show that our method can be used to refine existing word embeddings. In general, the task-related embeddings we obtain are very compact and accurate.

The second application is a model for Named Entity Recognition (NER), described by Primadhanty, Carreras and Quattoni [5]. State-of-the-art discriminative models for NER make use of features of an entity mention, its left and right contexts, and specially conjunctions of all these features. This results in high-dimensional feature spaces, and for standard datasets only a relatively small number of conjunctions are observed during training. Standard learning methods can only put weight on observed conjunctions, thus, the feature space of NER models is not exploited at its full potential because most conjunctive parameters remain zero after training. In contrast, our low-rank method can propagate weights from seen to unseen conjunctions. In experiments we observe large improvements over standard models when classifying unseen entities.

In the last part, I will focus on compositional structures of arbitrary size, such as sequences and trees, and weighted finite automata to model these structures. In this setting a key object is the Hankel matrix, a matrix that collects statistics over decomposed structures, and whose factorization reveals the state space of a weighted automata. Following the framework by Balle and Mohri [3] on learning low-rank Hankel matrices, I will illustrate two applications: a hidden-state sequence tagger [6], and an unsupervised method to learn transducers and grammars in an unsupervised fashion [1, 2].

References

1. Bailly, R., Carreras, X., Luque, F.M., Quattoni, A.: Unsupervised spectral learning of WCFG as low-rank matrix completion. In: Proceedings of the 2013 Conference on Empirical Methods in Natural Language Processing, pp. 624–635. Association for Computational Linguistics, Seattle, Washington, October 2013. http://www.aclweb.org/anthology/D13-1059

2. Bailly, R., Carreras, X., Quattoni, A.: Unsupervised spectral learning of finite state transducers. In: Burges, C., Bottou, L., Welling, M., Ghahramani, Z., Weinberger, K. (eds.) Advances in Neural Information Processing Systems, vol. 26, pp. 800–808. Curran Associates Inc. (2013). http://papers.nips.cc/paper/4862-unsupervised-spectral-learning-of-finite-state-transducers.pdf

3. Balle, B., Mohri, M.: Spectral learning of general weighted automata via constrained matrix completion. In: Pereira, F., Burges, C., Bottou, L., Weinberger, K. (eds.) Advances in Neural Information Processing Systems, vol. 25, pp. 2159–2167. Curran Associates Inc. (2012). http://papers.nips.cc/paper/4697-spectral-learning-of-general-weighted-automata-via-constrained-matrix-completion.pdf

4. Madhyastha, P.S., Carreras, X., Quattoni, A.: Learning task-specific bilexical embeddings. In: Proceedings of COLING 2014, the 25th International Conference on Computational Linguistics: Technical Papers, pp. 161–171. Dublin City University and Association for Computational Linguistics, Dublin, August 2014. http://www.aclweb.org/anthology/C14-1017

5. Primadhanty, A., Carreras, X., Quattoni, A.: Low-rank regularization for sparse conjunctive feature spaces: an application to named entity classification. In: Proceedings of the 53rd Annual Meeting of the Association for Computational Linguistics and the 7th International Joint Conference on Natural Language Processing (Volume 1: Long Papers), pp. 126–135. Association for Computational Linguistics, Beijing, July 2015. http://www.aclweb.org/anthology/P15-1013

6. Quattoni, A., Balle, B., Carreras, X., Globerson, A.: Spectral regularization for max-margin sequence tagging. In: Jebara, T., Xing, E.P. (eds.) Proceedings of the 31st International Conference on Machine Learning (ICML 2014), pp. 1710–1718. JMLR Workshop and Conference Proceedings (2014). http://jmlr.org/proceedings/papers/v32/quattoni14.pdf

7. Srebro, N., Shraibman, A.: Rank, trace-norm and max-norm. In: Auer, P., Meir, R. (eds.) COLT 2005. LNCS (LNAI), vol. 3559, pp. 545–560. Springer, Heidelberg (2005)

Towards Two-Way Interaction
with Reading Machines

Sebastian Riedel[1], Sameer Singh[2], Guillaume Bouchard[1],
Tim Rocktäschel[1], and Ivan Sanchez[1]

[1] Department of Computer Science, University College London, London, UK
{s.riedel, g.bouchard, t.rocktaschel,
i.sanchezg}@cs.ucl.ac.uk
[2] University of Washington, Computer Science and Engineering,
Seattle, WA, USA
sameer@cs.washington.edu

Abstract. As machine learning models that underlie machine reading systems
are becoming more complex, latent, and end-to-end, they are also becoming less
interpretable and controllable. In times of rule-based systems users could *interact* with a system in a two-way fashion: injecting their own background
knowledge *into* the system through explanations in the form of rules, and
extracting explanations *from* the system in the form of derivations. It is not clear
how this type of communication can be realized within more modern architectures. In this position paper we propose a research agenda that will (re-)
enable this two-way communication with machine readers while maintaining the
benefits of the models of today. In fact, we argue for a paradigm in which the
machine reading system is an agent that communicates with us, learning from
our examples and *explanations*, and providing us with explanations for its
decisions we can use to debug and improve the agent further.

Contents

Towards Two-Way Interaction with Reading Machines

Sebastian Riedel[1]([✉]), Sameer Singh[2], Guillaume Bouchard[1],
Tim Rocktäschel[1], and Ivan Sanchez[1]

[1] Department of Computer Science, University College London, London, UK
{s.riedel,g.bouchard,t.rocktaschel,i.sanchez}@cs.ucl.ac.uk
[2] Computer Science & Engineering, University of Washington, Seattle, WA, USA
sameer@cs.washington.edu

Abstract. As machine learning models that underlie machine reading systems are becoming more complex, latent, and end-to-end, they are also becoming less interpretable and controllable. In times of rule-based systems users could *interact* with a system in a two-way fashion: injecting their own background knowledge *into* the system through explanations in the form of rules, and extracting explanations *from* the system in the form of derivations. It is not clear how this type of communication can be realized within more modern architectures. In this position paper we propose a research agenda that will (re-)enable this two-way communication with machine readers while maintaining the benefits of the models of today. In fact, we argue for a paradigm in which the machine reading system is an agent that communicates with us, learning from our examples and *explanations*, and providing us with explanations for its decisions we can use to debug and improve the agent further.

1 Introduction

Machine Reading is the task of extracting machine-interpretable representations of knowledge from text. Often this involves the creation of knowledge bases (KBs): large graphs of entities and their relations [16]. In turn, such KBs can be used downstream for humans to answer questions, or by other machines to further analyse the data in the spirit of the semantic web.

While in the early days of machine reading and information extraction, most approaches were rule-based [2], nowadays machine learning-based approaches are dominating the academic field [4,8,11]. Machine learning-based methods can often be trained to achieve higher coverage and deal with noise and uncertainty in data.

However, rule-based approaches have several benefits over their machine learning-based counterparts, and these relate to the direction of human-computer communication:

- *Computer-to-User*: any output from a rule-based system can be backtracked to identify the rules that were triggered in the form of derivations. This enables the user to understand what goes wrong in the system.

© Springer International Publishing Switzerland 2015
A.-H. Dediu et al. (Eds.): SLSP 2015, LNAI 9449, pp. 1–7, 2015.
DOI: 10.1007/978-3-319-25789-1_1

- *User-to-Computer*: it is simple to incorporate prior knowledge: users can explain the machine their own understanding through the incorporation of new rules. Users can also ask the machine for explanations.

In other words, rule-based systems allow users to two-way communicate with the reading machine, which makes them apparently easier to maintain, comprehend and debug.

However, rule-based approaches have several benefits over their machine learning-based counterparts: they tend to be easier to maintain, comprehend and debug, and it is simple to incorporate prior knowledge. In other words, rule-based systems allow users to *two-way communicate* with the reading machine. They can ask the machine for explanations in the form of derivations, and they can explain the machine their own understanding through the incorporation of new rules. In fact, it is due to these features that rule-based systems still dominate the commercial sector [2].

In recent years machine reading systems based on machine learning have become more powerful, but also evermore complex. While earlier models were essentially classifiers, many current models rely on latent feature representations [4,9,11,15,18], often learned in an unsupervised setting, as well as deeper architectures. The two-way communication we illustrated above is difficult in this setting. The only explanation we can usually give for a prediction is: If the machine reader connects one entity to another using some relation, it will be because the latent representations of these entities and relations are somewhat compatible. And to incorporate background knowledge we still need to find ways to alter these latent representations effectively.

Our goal is to build robust machine learning-based reading machines that support this type of two-way communication. In fact, in order to make this communication as natural and effective as possible, we envision reading machines that can communicate in natural language. To make this more concrete, let us consider the following dialogue between a human user and the reading machine.

HUMAN: Read http://www.riedelcastro.org/.
MACHINE: Done.
HUMAN: Where does Sebastian work?
MACHINE: Sebastian is a Reader at UCL.
HUMAN: Does Sebastian live in London?
MACHINE: I don't know.
HUMAN: Why?
MACHINE: Sebastian works at UCL, which is in London, but its not clear what this means for whether Sebastian lives in London.
HUMAN: If someone works in an organisation based in some place, then he is likely to live in that place.
MACHINE: Okay.
HUMAN: Does Sebastian live in London?
MACHINE: Most likely.

We believe that this type of dialogue can enable non-experts to rapidly develop high-coverage and robust reading machines. If we are okay with communicating directly using rules, it is relatively straight-forward to support dialogues as the above using rule-based systems. How to enable this type of dialogue in modern machine-learning systems, and using natural language instead of rules, remains an open problem. In the following we describe some of the challenges that this problem involves.

2   Challenges

Above all, a reading machine needs to be able to learn lexical and syntactic transformations in order to go from a statement such as "Sebastian is a Reader at UCL" to "Sebastian works at UCL", and to support inference patterns such as "if X works for Y and Y is in Z then X likely lives in Z". In this paper we mention this problem only in passing: it needs to be solved in any case, and much recent work has made substantial progress in this regard [6,9]. One of our core hypotheses is that we will see further progress, and that this progress will stem from more latent and deeper models in which transformations and inferences are becoming increasingly latent. Notice that in the following we will mostly remain agnostic with respect to the specific type of models used. Sometimes we will refer to our own model in which a knowledge graph is modelled as a matrix of entity pairs as columns, and relations as rows, and in which we learn a low-rank factorization of this matrix [11].

Based on the example dialogue above, and in addition to the general requirement of we identify two core challenges: (1) Injecting explanations and (2) Extracting explanations. We briefly discuss these challenges below.

2.1   Injecting Explanations

How can we enable the reading machine to understand and follow a statement such as the one below?

HUMAN: If someone works in an organisation based in some place, then he is likely to live in that place.

We first notice that with a logic and rule-based system the above statement amounts to adding the rule $worksFor(X,Y) \land contains(Z,Y) \implies livesIn(X,Z)$. Hence we can conceptually divide our problem into: how to map natural language to such rules, and how to inject these rules into models that do not use such rules *explicitly*?

For the latter question of injecting rules into machine learning-based models there is a large body of previous work [5,7,14]. One option is to add logical rules as terms to the training objective. We show that by simply moving along their gradient in stochastic gradient descent leads to strong results compared to a range of baselines [12] when applied to our matrix factorization models [11]. However, fundamentally these methods require a full *grounding* of rules: a rule such as *worksFor*(X,Y) $\land$ *contains*(Z,Y) $\implies$ *livesIn*(X,Z) is instantiated for a large number for triples (X, Y, Z), and these become terms in the objective. In contrast, logical reasoners usually do not require this type of preemptive grounding, and hence scale much better with the number of elements in the domain. It is currently unclear how this behaviour can be implemented when adding rules in the learning objective.

Conversing in logical rules is unnatural, and likely difficult for the non-expert user that knows her domain, but not as much about rule-based formalisms. How can we enable natural language interactions? We discuss two options: the use of a semantic parser, and the direct injection of statements into the model.

- Employing a semantic parser that maps statements to their logical form. There has been remarkable progress in this field, but most work focuses on parsing questions to a database [1, 10, 20]. Extracting semantic representations of rules is a different matter. In any case, adding a semantic parser to the pipeline introduces another source of errors, generally more brittleness, and requires the costly development of this component for new problems and domains. Can this be avoided? We could turn the semantic parser into a differentiable component that is trained together with the reading machine on unlabelled data.
- A less obvious avenue could be to consider to drop the two-stage architecture altogether, and directly inject natural language statements into the model. One could also imagine learning from explanations itself as a supervised learning problem in which the user provides examples of natural language explanations and the effect these explanations should have, as has been proposed for toy tasks in Facebook bAbI project [19].

2.2   Extracting Explanations

How can the machine return explanations such as the one of our example?

MACHINE: Sebastian works at UCL, which is in London, but its not clear what this means for whether Sebastian lives in London.

The above explanation works in terms of reasoning *steps*: the machine knows how to go from Sebastian to London via UCL, but it misses a step from there to the answer of the target question. Thinking in terms of these steps seems useful from a user perspective because she can add missing steps in form of explanations. However, these steps may not correspond to how inference in the model actually proceeds. For example, in our relation extraction models [11] the answer to a question such as "Does Sebastian live in London?" is determined by some algebraic operations on low-dimensional *embeddings* for Sebastian, UCL and the relation "lives in". Showing the user these embeddings will not be very helpful in terms of fixing a problem. The question then becomes: can we explain the predictions in a stepwise fashion that makes it easy to fix problems, even though the underlying model may operate in a different manner?

This problem has been discussed in the past—primarily during the first advent of neural networks [3,17]. Several papers discussed how interpretable proxy models could be extracted from neural networks. For example, one can use decision trees to simulate the behaviour of neural networks with high fidelity. These decision trees then can be inspected and used for explaining predictions. Unfortunately previous research has mostly focused on *local* classifiers that make independent decisions based on a set of observed features. In our reading machines predictions are correlated, and these correlations are important when explaining a prediction. For example, the predictions of *livesIn*(X,Z), *worksFor*(X,Y) and *contains*(Z,Y) will be correlated, even though the inferences that are made are not happening in a step-wise fashion.

We have recently begun to extract interpretable models from the matrix factorization models that underlie our reading machines [13]. Instead of using classifiers, we opted for tree-structured Bayesian Networks. These networks define *joint* distributions, are very scalable, easy to inspect and provide explanations in forms of steps: as paths from observed variables to predicted ones. One can use such explanations to find wrong edges, or missing hops. We showed that these networks reach high fidelity with respect to the predictions of the matrix factorization donor model.

While there has been progress in this regard, we are still far away from giving explanations as the one in our running example. There are two reasons for this. First, we have yet to translate Bayesian Network-based explanations into the natural language representations. This could be tackled through the use of specific grammars and templates, or maybe in a learning-based fashion, for example by using RNNs as concept-to-string generators. Second, it is difficult to reliably and efficiently evaluate the quality of an explanation, and hence to make measurable progress in this direction. An explanation is as good as it helps the user to fix problems, and ultimately evaluation hence requires humans in the loop. While this is possible, there is a need for automatic evaluations that can speed up development in the way that BLEU scores have worked for machine translation.

3 Conclusion

We propose a research agenda around two-way communication with reading machines that features both injection and extraction of explanations. We identify several challenges involved in this type of communication that stem from the fact that modern reading machines do not explicitly operate in term of rules.

Acknowledgments. This work was supported in part by Microsoft Research through its PhD Scholarship Programme, in part by CONAYCT, in part by the TerraSwarm Research Center, one of six centers supported by the STARnet phase of the Focus Center Research Program (FCRP) a Semiconductor Research Corporation program sponsored by MARCO and DARPA, in part by an ARO grant number W911NF-13-1-0246, and in part by the Paul Allen Foundation through an Allen Distinguished Investigator grant.

References

1. Berant, J., Liang, P.: Semantic parsing via paraphrasing. In: Association for Computational Linguistics (ACL) (2014)
2. Chiticariu, L., Li, Y., Reiss, F.R.: Rule-based information extraction is dead! long live rule-based information extraction systems! In: EMNLP, pp. 827–832, ACL (2013). http://dblp.uni-trier.de/db/conf/emnlp/emnlp2013.html#ChiticariuLR13
3. Craven, M.W., Shavlik, J.W.: Extracting tree-structured representations of trained networks. In: Advances in Neural Information Processing Systems (NIPS-8), pp. 24–30 (1996)
4. Culotta, A., Sorensen, J.: Dependency tree kernels for relation extraction. In: Proceedings of the 42nd Annual Meeting of the Association for Computational Linguistics (ACL 2004), Barcelona, Spain (2004). http://www.cs.umass.edu/culotta/pubs/tkernel.pdf
5. Ganchev, K., Graca, J., Gillenwater, J., Taskar, B.: Posterior regularization for structured latent variable models. Technical report MS-CIS-09-16, University of Pennsylvania Department of Computer and Information Science (2009)
6. Gardner, M., Talukdar, P., Krishnamurthy, J., Mitchell, T.: Incorporating vector space similarity in random walk inference over knowledge bases. In: Proceedings of EMNLP (2014)
7. Mann, G.S., McCallum, A.: Generalized expectation criteria for semi-supervised learning of conditional random fields. In: Annual Meeting of the Association for Computational Linguistics (ACL), pp. 870–878 (2008)
8. Mintz, M., Bills, S., Snow, R., Jurafsky, D.: Distant supervision for relation extraction without labeled data. In: Proceedings of the Joint Conference of the 47th Annual Meeting of the ACL and the 4th International Joint Conference on Natural Language Processing of the AFNLP (ACL 2009), pp. 1003–1011. Association for Computational Linguistics (2009)
9. Neelakantan, A., Roth, B., McCallum, A.: Compositional vector space models for knowledge base completion. In: Association for Computational Linguistics (ACL) (2015)
10. Reddy, S., Lapata, M., Steedman, M.: Large-scale semantic parsing without question-answer pairs. Trans. Assoc. Comput. Linguist. (TACL) **2**, 377–392 (2014)
11. Riedel, S., Yao, L., Marlin, B.M., McCallum, A.: Relation extraction with matrix factorization and universal schemas. In: Joint Human Language Technology Conference/Annual Meeting of the North American Chapter of the Association for Computational Linguistics (HLT-NAACL 2013), June 2013
12. Rocktäschel, T., Singh, S., Riedel, S.: Injecting logical background knowledge into embeddings for relation extraction. In: Proceedings of the 2015 Human Language Technology Conference of the North American Chapter of the Association of Computational Linguistics (2015)
13. Sanchez, I., Rocktaschel, T., Riedel, S., Singh, S.: Towards extracting faithful and descriptive representations of latent variable models. In: AAAI Spring Symposium on Knowledge Representation and Reasoning, March 2015
14. Singh, S., Hillard, D., Leggetter, C.: Minimally-supervised extraction of entities from text advertisements. In: North American Chapter of the Association for Computational Linguistics - Human Language Technologies (NAACL HLT) (2010)
15. Socher, R., Chen, D., Manning, C.D., Ng, A.: Reasoning with neural tensor networks for knowledge base completion. In: Advances in Neural Information Processing Systems, pp. 926–934 (2013)

16. Suchanek, F., Fan, J., Hoffmann, R., Riedel, S., Talukdar, P.P.: Advances in auto-
    mated knowledge base construction. In: SIGMOD Records Journal, March 2013.
    http://suchanek.name/work/publications/sigmodrec2013akbc
17. Thrun, S.: Extracting rules from artificial neural networks with distributed repre-
    sentations. In: Advances in Neural Information Processing Systems, pp. 505–512
    (1995)
18. Toutanova, K., Chen, D., Pantel, P., Poon, H., Choudhury, P., Gamon, M.: Rep-
    resenting text for joint embedding of text and knowledge bases. In: ACL Associ-
    ation for Computational Linguistics, September 2015. http://research.microsoft.
    com/apps/pubs/default.aspx?id=254916
19. Weston, J., Chopra, S., Bordes, A.: Memory networks. CoRR abs/1410.3916
    (2014). http://arxiv.org/abs/1410.3916
20. Zettlemoyer, L.S., Collins, M.: Learning to map sentences to logical form: Struc-
    tured classification with probabilistic categorial grammars. In: Uncertainty in Artif-
    ical Intelligence (UAI) (2005)

The Prediction of Fatigue
Using Speech as a Biosignal

Khan Baykaner[1], Mark Huckvale[1(✉)], Iya Whiteley[2], Oleg Ryumin[3],
and Svetlana Andreeva[3]

[1] Speech Hearing and Phonetic Sciences, UCL,
London, UK
{k.baykaner,m.huckvale}@ucl.ac.uk
[2] Centre for Space Medicine, UCL,
Dorking, Surrey, UK
i.whiteley@ucl.ac.uk
[3] Gagarin Cosmonaut Training Centre,
Star City, Russia

Abstract. Automatic systems for estimating operator fatigue have application in safety-critical environments. We develop and evaluate a system to detect fatigue from speech recordings collected from speakers kept awake over a 60-hour period. A binary classification system (fatigued/not-fatigued) based on time spent awake showed good discrimination, with 80 % unweighted accuracy using raw features, and 90 % with speaker-normalized features. We describe the data collection, feature analysis, machine learning and cross-validation used in the study. Results are promising for real-world applications in domains such as aerospace, transportation and mining where operators are in regular verbal communication as part of their normal working activities.

Keywords: Fatigue · Speech · Computational paralinguistics

1   Introduction

There are a variety of safety-critical environments for which operator fatigue is a significant risk factor, including aerospace, transportation and mining. In response to this risk, a variety of systems have been developed to detect or estimate fatigue. Some of these are accurate but are based on measurements which require expensive or intrusive equipment. However, in some safety-critical environments operators are engaged in regular or constant verbal communication, and in such environments a fatigue monitoring system based on analyzing speech might provide a cheaper and less-intrusive solution.

Existing models for estimating fatigue from speech have tended to focus on predictions of subjective ratings of sleepiness given by the speakers themselves. This paper describes a corpus of speech data from subjects kept awake over a three day period in which they became demonstrably fatigued, and the model training procedures used to classify fatigue based on the objective property of time awake.

© Springer International Publishing Switzerland 2015
A.-H. Dediu et al. (Eds.): SLSP 2015, LNAI 9449, pp. 8–17, 2015.
DOI: 10.1007/978-3-319-25789-1_2

2 Background

In safety critical-environments fatigue is a significant risk factor. One of the clearest examples of this is in transportation, where driver fatigue is widely considered to be an important contributory factor in fatal and serious accidents [1, 2]. It is difficult to pinpoint the exact proportion of accidents caused by fatigue, but the consensus of scientists studying safety and accident prevention is that fatigue is the largest identifiable and preventable cause of accidents in transportation, accounting for around 15-20 % of all accidents [3]. Fatigue is just as significant a risk in other safety critical settings where vigilance is important, such as aerospace [4].

Models have been developed in the domain of computer-vision to predict and monitor fatigue based on video recordings of operators, and these can be quite accurate (see [5] for a review). Even more accurate ways of monitoring fatigue are possible using intrusive physiological measurements (see [6] for a review of the capacity of electroencephalography, electrocardiography, elektro-okulogram, and pulse oximetry measurements to assess fatigue).

While the vision-based and physiological approaches may be accurate, measuring these features presents a significant challenge to user acceptance in many practical applications because additional, expensive or intrusive equipment is required. By contrast, a cheap and non-intrusive fatigue monitoring system could be implemented if it were possible to predict fatigue by analyzing the voice. This would be particularly useful in those situations requiring drivers or operators to regularly communicate by speaking, (e.g. in aviation, spaceflight, or mining transportation industries).

Existing research has identified vocal correlates with measures of fatigue. For example in [7] it was demonstrated that when subjects were kept awake for a period of 24 h, the duration of their pauses gradually increased for read speech, and the variation in the 4th formant decreased for sustained vowel sounds.

A variety of models were generated and tested for the Interspeech 2011 speaker state challenge [8] aimed at predicting subjective ratings of sleepiness. For the training and testing data the Sleepy Language Corpus (SLC) was developed, consisting of a mixture of isolated vowels, sustained vowels, commands, and natural speech for 99 speakers. For measures of fatigue, the Karolinska Sleepiness Scale (KSS) was used, which is a subjective scale ranging from 1 (extremely alert) to 10 (extremely sleepy, cannot stay awake). The data was divided into two sets for classification, with the non-sleepy group being all ratings from 1 to 7 (sleepy, but no effort to stay awake), and the sleepy group being all ratings from 8 (sleepy, some effort to stay awake) to 10. The optimal proposed baseline model was able to achieve an unweighted accuracy (UA) of 70.3 %, and the winner of the challenge achieved an UA of 71.7 % [9]. A higher UA of 82.8 % has also been reported in another study based on a subset of the SLC [10].

Since there is no easily accessible ground truth for fatigue it is sometimes unclear what ought to be predicted. The KSS, used in the Interspeech 2011 speaker state challenge, has been validated against performance and EEG measures [11], and although significant correlations were found for most of the measures the highest correlation found (that between KSS and reaction time) had only $r = 0.57$ (standard deviation = 0.25). The authors point out that subjective sleepiness cannot be regarded as a substitute for performance measurements, and similarly that one performance measure cannot usually be substituted for another. Further evidence of the imperfect relationship between subjective scores and performance can be seen in other studies where the correlation between reaction time and KSS scores has been moderate but highly variable ($r = 0.49\text{-}0.71$ depending on the subject) [12], or non-existent [13].

In this work models are produced aiming to predict whether or not a subject is fatigued based on sleep latency (i.e. the time the subjects have been kept awake) rather than subjective ratings of sleepiness.

3   Corpus Collection and Labelling

The goal of corpus collection was to collect speech samples from speakers over an extended period in which they became demonstrably fatigued. The subjects were seven native Russian speakers (six male, one female) who were taking part in a psychological study of the effects of isolation and sleep deprivation. In this study the subjects were isolated and asked to keep awake for over 60 h. All subjects began the study at 10am on day one and finished the study at 9 pm on day three. They were given a range of tasks to occupy their time, including physiological and psychological tests. They were continuously monitored from outside the isolation chamber, but could not communicate with the experimenters.

Speech was collected from the subjects at regular intervals of approximately 6 h. The subjects were asked to read prose from a computer monitor into a Roland R-05 digital recorder sitting on the desk in front of them. The selected prose came from a Russian novel, and was chosen to create a simple and unstressful task for the subjects. The subjects were able to decide themselves how much to read, so recording durations varied between 105 and 495 s. Recordings were collected at 24-bit resolution at 44100 samples/sec.

The recordings were post-processed by editing out any speech that was not part of the reading task, by normalizing the signal level and by conversion to 16-bit PCM audio files. In total, 74 speech recordings were made, labelled by speaker and sleep latency (i.e. time since the start of the experiment).

4   Feature Extraction

Previous work on predicting sleepiness from speech [e.g. 8] has demonstrated how high-dimensionality feature vectors are useful to capture and represent variation in the signal across conditions. In this work we use a similar approach but using our own feature analysis tools to extract from the whole of each recording a representation of the variation of the speech signal in the frequency domain, the time domain and the modulation domain. The recordings were analyzed and summarized to produce fixed-length feature vectors as follows:

1. The waveform is pre-emphasized and divided into 50 ms Hamming-windowed sections overlapping by 10 ms.

2. An FFT is applied to each window and a bank of triangular filters is used to calculate a smoothed spectrum on a non-linear frequency scale. The filters are 200mel wide and spaced by 100mel.
3. A cosine-transform of the log-compressed smoothed spectrum is taken to generate 19 MFCC parameters per frame.
4. The first and second temporal differences of the MFCC parameters are computed.
5. The autocorrelation of each window is also computed and interpolated onto a log delay axis.
6. A cosine transform of the log delay autocorrelation function is taken to generate 19 autocorrelation shape parameters per frame.
7. The first and second temporal differences of the autocorrelation shape parameters are computed.
8. The energy of each window is calculated, and the first and second temporal difference is computed.
9. The distributions of the MFCC, autocorrelation and energy parameters are collected over the whole file.
10. The distribution of each parameter was then summarized using the quantile values at 5 %, 10 %, 25 %, 50 %, 75 %, 90 % and 95 % together with robust measures of skewness and kurtosis.
11. The audio file was then band-pass filtered between 300 and 3500 Hz, rectified and low-pass filtered at 80 Hz to generate a temporal envelope trace. The modulation spectrum of the temporal envelope was calculated using 40 band-pass filters logarithmically-spaced between 0.1 and 50 Hz. These parameters were added to the summary statistics parameters generated from the MFCC, autocorrelation and energy analysis.

Ultimately, each file was described by a feature vector containing 1093 parameters.

5 Model Construction Procedure

In our first study, all recordings prior to 10am on day two were labelled as non-fatigued and the remaining sessions were labelled as fatigued for binary classification. Setting any particular threshold for classification is arbitrary; but setting a threshold of 10am has two distinct benefits. Firstly, it seems reasonable to suggest that any subject would be fatigued after a full 24 h of wakefulness. Secondly, selecting this threshold results in a corpus with 31 non-fatigued cases and 43 fatigued cases; giving roughly balanced classes with 41.9 % and 58.1 % of the corpus in each class. Having well balanced classes is important for training a classifier because large imbalances tend to result in models which preferentially predict the majority class.

As in [10], the relative data sparsity makes a speaker-dependent multiple hold-out cross validation approach most appropriate. Specifically a 'leave one sample out' cross validation procedure was implemented, where in each iteration a model was trained on data from all subjects with a single sample withheld for validation. The final classification error is calculated by averaging over all 74 classifiers.

For each classifier a support vector machine (SVM) model using a linear kernel was trained, with a margin constraint of 1. Ideally, with a larger corpus available, the approach discussed in [15] would be utilized, and an isolated development set would be used to train SVMs with radial basis function (RBF) kernels, identifying the optimal margin constraint and sigma parameter. With the relatively small corpus, however, it was considered fairer to only use the linear kernel SVM with a fixed margin constraint. Models were trained using the Weka machine-learning toolkit (http://www.cs.waikato. ac.nz/ml/weka/).

Table 1 shows the model performance alongside the performance of two simpler models for comparison: the ZeroR and the OneR [16]. The ZeroR model ignores any features and simply predicts the more common label for every sample. In this case the ZeroR model predicts every sample to be fatigued, and therefore will always have an unweighted accuracy (UA) of 50 %. The OneR model utilizes 1-level decision trees based on the single best feature and offers a useful point of comparison since the use of relatively complex machine learning approaches should be justified by showing an improvement over simpler approaches. The performance of the speech model can therefore be considered with reference to the improvement in UA over these simpler models.

The results show that in this case the linear kernel SVM performed well with a substantially higher unweighted accuracy than either of the simpler models.

Table 1. Comparison of the SVM, ZeroR, and OneR fatigue prediction models using speech features on the 24-hour fatigue task. Positive corresponds to the fatigued class and negative corresponding to the non-fatigued class.

Measure	SVM	ZeroR	OneR
True Positive	36	43	26
False Positive	6	31	12
True Negative	25	0	19
False Negative	7	0	17
Precision	85.7 %	58.1 %	68.4 %
Recall	83.7 %	100 %	60.4 %
Unweighted Accuracy	**82.2 %**	50.0 %	60.8 %

6   Using Gaussianized Features

Any feature normalization applied by the SVM models trained in Sect. 5 is performed over all recordings despite the fact that they came from different speakers. Improvements in performance should be possible by explicit normalization of feature distributions for each speaker prior to training. The goals of feature normalization are to remove speaker-specific differences in terms of mean, range and distribution shape.

A normalization process called "Gaussianization" [18] was used to transform the speech feature values in such a way that ensures a normal distribution of each feature for each individual speaker. This process maps the empirical feature distributions to a normal distribution by mapping empirical percentiles to values drawn from the inverse cumulative normal distribution.

Table 2. Prediction performance on 24 h fatigue task using models trained with gaussianized features.

Measure	SVM	OneR
True Positive	38	34
False Positive	5	15
True Negative	26	16
False Negative	5	9
Precision	88.4 %	69.4 %
Recall	88.4 %	79.1 %
Unweighted Accuracy	**86.1 %**	65.3 %

The use of speaker-specific normalization has important consequences for practical applications of this fatigue-detection system. Any implementation would now require an enrolment process for each new speaker to establish their personal distribution of feature values. Although the models themselves would be trained from multiple speakers, this limitation effectively renders such systems speaker dependent.

Table 2 shows the results for the model which utilizes Gaussianized speech features. The equivalent performance measures are also shown for the OneR model using the same feature set.

The results show an improved UA of 86.1 % when using Gaussianized features for the SVM models, compared with 82.2 % on raw features. This resulted from increases in both the true positive rate and true negative rate. The simpler OneR model also showed a small increase in UA mostly from an increase in true positives and a decrease in false negatives.

7 Shifting the Classification Threshold

The choice of 10am as the classification threshold between fatigued and non-fatigued was, as previously noted, an arbitrary one. Here we investigate how well the classification system would operate if the threshold were set earlier to 2am of day two, based on a wakefulness of 16 h.

Redrawing the classes in this way produces a non-fatigued class size of 24, and a fatigued glass size of 50. This is a more serious imbalance than was the case for the 10am boundary and needed to be addressed. It was decided to use the same strategy as in [9] and utilize the Synthetic Minority Over-sampling Technique (SMOTE) algorithm [14] for generating additional synthetic non-fatigued samples. Using the SMOTE algorithm encourages the classifier training to produce a more generalizable model because the synthetic samples have feature values randomly generated between nearest neighbors; this is particularly important for relatively small corpora such as that considered here. After applying SMOTE, there were 48 non-fatigued samples, giving 98 samples in total and a more balanced corpus with the non-fatigued class accounting for 49.0 % of samples.

Table 3 shows the results of carrying out a 98 fold leave-one-out cross validation scheme using the modified boundary.

Table 3. Cross validation fatigue prediction performance with a classification threshold set to 16 h awake

Measure	SVM Raw	OneR Raw	SVM Gauss	OneR Gauss
True Positive	42	38	46	39
False Positive	9	18	2	14
True Negative	39	30	46	34
False Negative	8	12	4	11
Precision	82.3 %	67.9 %	95.8 %	73.6 %
Recall	84.0 %	76.0 %	92.0 %	78.0 %
Unweighted Accuracy	**82.6 %**	69.3 %	**93.9 %**	74.4 %

The cross-validation performance of the SVM models using the 2am threshold is similar to those obtained using the 10am threshold, with 82.6 % and 82.2 % unweighted accuracy achieved respectively when using raw features, and 93.9 % and 86.1 % when using gaussianized features. The performance cannot be directly compared since the former model was trained without synthetic data being added. In both cases the simpler OneR models performed more poorly.

8 Split Validation

Although good performance was shown by the models in the previous sections, there is still uncertainty about whether the choice of leave-one-out cross-validation is giving unfair advantage to the classifier. This is particularly relevant for SVM models which operate by selecting and retaining specific training vectors. It may be that good performance came from retaining vectors which happen to be close to each test vector.

For more robust assessment we would ideally use a validation set of new speakers independent from those used in training, but our small corpus size makes this difficult. Instead we have implemented a split validation procedure to give an indication of the expected accuracy when making predictions on new data. If performance after split-validation was significantly worse than leave-one out cross-validation then we would need to be concerned about the robustness of our approach.

The split validation test was first performed on the 10am threshold data set. The data were split into a 2/3 training (49 samples) and 1/3 validation (25) set with samples selected at random. A linear kernel SVM and a OneR model was generated based on the training data for the raw and Gaussianized feature sets, and performance was measured by testing predictions on the validation set. Table 4 shows the performance for these models.

The split validation performance indicated that the SVM linear kernel model based on gaussianized features produced the highest UA of 93.8 % (cf. 86.1 % for leave-one-out), with a lower UA of 79.5 % (cf. 82.2 % for leave-one-out) achieved using raw features.

The split validation procedure was then repeated on the 2am threshold corpus described in Sect. 7. Table 5 shows the performance.

Table 4. Prediction performance on the 24 h fatigue task using separated training and validation data sets.

Model	Precision	Recall	Unweighted Accuracy
Raw features			
ZeroR	64.0 %	100.0 %	50.0 %
OneR	72.7 %	50.0 %	58.3 %
SVM	86.7 %	81.3 %	**79.5 %**
Gaussianized features			
OneR	70.0 %	43.8 %	55.2 %
SVM	100 %	87.5 %	**93.8 %**

Table 5. Prediction performance on the 16 h fatigue task using separated training and validation data sets.

Model	Precision	Recall	Unweighted Accuracy
Raw features			
ZeroR	56.0 %	100.0 %	50.0 %
OneR	54.5 %	42.9 %	44.2 %
SVM	85.7 %	85.7 %	**83.8 %**
Gaussianized features			
OneR	62.5 %	71.4 %	58.4 %
SVM	92.3 %	85.7 %	**88.3 %**

Using raw features, the UA was 83.8 % (cf. 82.6 % for leave-one-out). Using Gaussianized features, the UA was 88.3 % (cf. 93.9 % for leave-one-out).

Generally model performance held-up well under split-validation, with performance varying from 7.7 % better to 5.1 % worse compared to cross-validation. As with the cross-validation, the split validation performance of the models with the 2am threshold were similar to those obtained using the 10am threshold.

It is likely that the variation in unweighted accuracy from the split validation procedure observed here is a result of the relatively small corpus size, which in turn produces a higher variability in performance measures. In general, this suggests that performance differences on the order of 8 % are possible when applying this modelling approach to new data.

9 Conclusion

Speech recordings were gathered from subjects undergoing three days without sleep. Over this period the subjects became increasingly fatigued. Features were generated by analyzing speech recordings collected at regular intervals and these features were used to train SVM classifiers predicting whether the subject was fatigued or non-fatigued based on sleep latency. The model training results show that the use of

Gaussianized features significantly improves prediction accuracy; but it should be noted that practical implementations using Gaussianized features require a speaker enrolment stage for each new speaker. On separated validation data the unweighted accuracy of the SVM classifiers was around 80 % for raw features and 90 % for Gaussianized features. Good performance was obtained for fatigue thresholds set at either 16 h or 24 h of sleep deprivation.

It may be inconvenient in some applications to require several minutes of speech in order to predict fatigue. However, the generated features can be produced based on much shorter recordings, and similar results have been obtained utilizing speech excerpts of 20 s. Further work should aim to determine whether tasks based on spontaneous speech would be more revealing of fatigue compared to the read speech used here.

This work shows that using features generated from recordings of speech, meaningful predictions can be made of sleep latency as an objective measure of fatigue. This is promising for the development of decision-making aids applied in safety-critical environments where the fatigue level of an operator is an important risk factor.

Acknowledgements. The authors would like to acknowledge the European Space Agency and University College of London who are jointly responsible for funding this work.

References

1. Dobbie, K.: Fatigue-related crashes: an analysis of fatigue-related crashes on Australian roads using an operational definition of fatigue. Australian transport safety bureau (OR23), (2002)
2. FMCSA: Regulatory impact analysis – hours of service final rule. Federal motor carrier safety administration, December 2011
3. Åkerstedt, T.: Consensus statement: Fatigue and accidents in transport operations. J. Sleep Res. **9**, 395 (2000)
4. Rosekind, M., Gander, P., Miller, D., Gregory, K., Smith, R., Weldon, K., Co, E., McNally, K., Lebacqz, J.: Fatigue in operational settings: examples from the aviation environment. J. Hum. Factors. **36**, 327–338 (1994)
5. Barr, L., Howarth, H., Popkin, S., Carroll, R.: A review and evaluation of emerging driver fatigue detection measures and technologies. John A. Volpe National Transportation Systems Center (2005)
6. Begum, S.: Intelligent driver monitoring systems based on physiological sensor signals: a review. In: IEEE Annual Conference on Intelligent Transportation Systems (ITSC) (2013)
7. Vogel, A., Fletcher, J., Maruff, P.: Acoustic analysis of the effects of sustained wakefulness on speech. J. Acoust. Soc. Am. **128**, 3747–3756 (2010)
8. Schuller, B., Batliner, A., Steidl, S., Schiel, F., Zrajewski, F.: The Interspeech 2011 Speaker state challenge. In: Proceedings of Interspeech 2011, pp. 2301–2304 (2011)
9. Huang, D., Ge, S., Zhang, Z.: Speaker State Classification Based on Fusion of Asymmetric SIMPLS and Support Vector Machines. In: Proceedings of the Interspeech 2011, pp. 3301–3304 (2011)

10. Krajewski, J., Batliner, A., Golz, M.: Acoustic sleepiness detection: Framework and validation of a speech-adapted pattern recognition approach. Behav. Res. Methods **41**, 795–804 (2009)
11. Kaida, K., Takahashi, M., Akerstedt, T., Nakata, A., Otsuka, Y., Haratani, T., Fukasawa, K.: Validation of the Karolinska sleepiness scale against performance and EEG variables. Clin. Neurophysiol. **117**, 1574–1581 (2006)
12. Gillberg, M., Kecklund, G., Akerstedt, T.: Relations between performance and subjective ratings of sleepiness during a night awake. J. Sleep **17**, 236–241 (1994)
13. Åhsberg, G., Kecklund, G., Åkerstedt, T., Gamberale, F.: Shiftwork and different dimensions of fatigue. Int. J. Ind. Ergon. **26**, 457–465 (2000)
14. Chawla, N., Bowyer, K., Hall, L., Kegelmeyer, W.: SMOTE: synthetic minority over-sampling technique. J. Artif. Intell. Res. **16**, 321–357 (2002)
15. Hsu, C., Chang, C., Lin, C.: A practical guide to support vector classification. Department of computer science technical report, National Taiwan University (2010)
16. Holte, R.: Very simple classification rules perform well. J. Mach. Learn. **11**, 63–91 (1993)
17. Chen, S., Gopinath, R.: Gaussianization. In: Proceedings of the NIPS 2000, Denver Colorado (2000)

Supertagging for a Statistical HPSG Parser for Spanish

Luis Chiruzzo[✉] and Dina Wonsever

Instituto de Computación, Facultad de Ingeniería,
Universidad de la República, Montevideo, Uruguay
{luischir,wonsever}@fing.edu.uy

Abstract. We created a supertagger for the Spanish language aimed at disambiguating the HPSG lexical frames for the verbs in a sentence. The supertagger uses a CRF model and achieves an accuracy of 83.58 % for the verb classes on the test set. The tagset contains 92 verb classes, extracted from a Spanish HPSG-compatible annotated corpus that was created by automatically transforming the Ancora Spanish corpus. The verb tags include information about the arguments structure and syntactic categories of the arguments, so they can be easily translated into HPSG lexical entries.

Keywords: HPSG · Supertagging · Parsing · Spanish

1 Introduction

This paper describes a partial result of an ongoing project for creating a statistical HPSG parser for Spanish. In particular, we describe the creation of a supertagger, a tool that is used prior to the parse process in order to narrow the search space. Our supertagger is trained specifically to deal with the verb structures in Spanish.

Head-driven Phrase Structure Grammars (HPSG) [1] are a lexicalized grammars formalism. HPSG grammars are very expressive and can model many linguistic phenomena capturing both syntactic and semantic notions at the same time. The rules used in these grammars are very generic, inspired in X' theory, generally indicating how to combine expressions with their complements, modifiers and specifier. The syntactic categories are organized in a type hierarchy, and the parse trees use feature structures as nodes.

In order to build a statistical HPSG parser we need to define a HPSG grammar (a feature structure for words and phrases, and a set of rules determining how to combine them) and a corpus of parse trees that use this grammar.

From this corpus, we can extract a set of lexical entries, the lexicon, which comprises all the words used in the corpus annotated with their appropriate feature structure. Lexical entries can be clustered in sets that we call *lexical frames*, which represent classes of words that share combinatorial properties, i.e. they can be used in the same way. For example, a simple way of separating the verbs in classes could be using the number of arguments they take (e.g. intransitive, transitive, ditransitive, etc.).

Given an unannotated sentence, a parser should be able to assign the appropriate lexical frame for each word, as well as arrange all the words into a suitable parse tree.

© Springer International Publishing Switzerland 2015
A.-H. Dediu et al. (Eds.): SLSP 2015, LNAI 9449, pp. 18–26, 2015.
DOI: 10.1007/978-3-319-25789-1_3

The number of different lexical frames a word could appear in is large; for example, common verbs like *"ir"* (*"to go"*) or *"hacer"* (*"to do/make"*) could have dozens of possible lexical frames; and the parser would need to check against every lexical frame for every word in the sentence, which yields a combinatorial explosion of possibilities.

One way of making this problem tractable is by using a supertagger. A supertagger is a module that decides the most likely lexical frames for a given word in the context of a sentence, so that the parser does not need to test against all possible combinations. The current paper describes the construction of a supertagger for a statistical HPSG parser for Spanish that aims to assign the correct lexical frame for the verbs in a sentence. We focus on the tagging of verbs because it is the category with the most combinatorial complexity in Spanish, and thus the most difficult to deal with. Our hypothesis is that after solving the problem for verbs the rest of the categories should be easier.

2 Related Work

The term supertagging is used for the first time in [2] to describe a way of prepro-cessing words, assigning appropriate classes to the tokens that are more specific than POS and closer to the grammar categories, in the context of a Lexicalized Tree Adjoining Grammar (LTAG) parser. They report 68 % accuracy using a trigram model and they say that the result could be boosted to 77 % by incorporating long-range dependencies information.

After this, the technique has been applied successfully to other lexicalized gram-mars. For example, [3] explores the possibility of allowing the supertagger to output multiple tags in order to improve the performance of a Combinatory Categorial Grammar (CCG) parser. Their method implies getting the tag with the highest prob-ability, and all the tags whose probabilities are up to a value β apart from it. They report an accuracy of 97.7 % using an ambiguity level of 1.4 categories per word. More recently, [4] explores the possibility of using word embeddings to improve the per-formance of a supertagger. This allows them to use millions of unsupervised words to train the model, which boosts the performance of the supertagger and the CCG parser.

There has been work on supertagging for HPSG as well. For example [5] describes several experiments to improve the performance of the PET parser for English using a supertagger. The supertagger achieves a 90.2 % of accuracy when selecting one tag, and 97.2 % when selecting two tags. Reference [6] describes the supertagger used for the Enju parser. They use GFG-filtering by creating a CFG that approximates their HPSG grammar, and incorporate this information during the training of the supertagger (a perceptron), achieving a 93.98 % accuracy. Reference [7] experiments training three types of supertaggers for the LXGram HPSG grammar for Portuguese. Their best tagger achieves an accuracy of 92.63 % for verbs and 95.18 % for nouns training over a corpus of 21,000 sentences.

Although a working parser for HPSG in Spanish has been created [8], the methodology used is different as it does not train a supertagger. The only report of a supertagger for Spanish that we know of is shown in [9]. They extract a LTAG

grammar from the Ancora Spanish corpus and use it to train a supertagger with a
Maxent model, reporting an accuracy of 79.64 %.

3   Lexical Frames

In this section we describe the creation of the corpus we use and the way we automatically extract the classes of lexical frames for verbs.

3.1   Corpus

The Ancora corpus [10] is a corpus in Spanish and Catalan containing 17,000 sentences
(half a million words). All the sentences are annotated with syntactic information in a
CFG style, but also each word and phrase are enriched with useful information such as
the syntactic function and the arguments structure.

In a previous work [11] a set of manually crafted rules was used to convert the CFG
structure of the Ancora corpus into a HPSG compatible style. In order to do this, all
phrases in the corpus were binarized, and for each expression the HPSG rule and the
syntactic head was identified.

When identifying the rule that is applied to a phrase, we need a way of deciding
whether an element that accompanies a syntactic head is acting as a specifier, a complement or a modifier (adjunct). This is also decided using a set of hand crafted rules.
The arguments of a verb are the specifier and the complements, but not the modifiers.

3.2   Verb Classes

HPSG grammars are lexicalized grammars; this means that the lexical entries (and not
the syntactic rules) are in charge of defining the combinatorial properties of the words.
In HPSG, the lexical entries define a rich set of features that control how the word can
be used in the context of a sentence. In our work, we include morphological, syntactic
and semantic information into the lexical entries. Figure 1 shows the basic lexical entry
and the features we use.

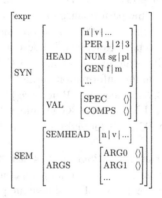

Fig. 1. Feature structure for the basic lexical entry

Although many HPSG grammars employ minimal recursion semantics [12] to model semantic information (a rich semantics framework which handles quantifiers), our approach to semantics is simpler, similar to the approach used in Enju [13]: we are only interested in modeling the arguments structure of the verbs in a format akin to PropBank [14]. Ancora includes information about the arguments structure into its nodes, this information is similar to thematic roles. However, these annotations are not always complete in the corpus. There are almost 90,000 elements in the corpus acting as arguments of a verb, but only 75 % of those elements are annotated with the appropriate argument structure attribute.

In particular, whenever there is an auxiliary (*"ser"*/*"to be"*), modal (*"poder"*/ *"can"*) or aspectual (*"empezar"*/*"to start"*) verb, the complement verb or phrase is not denoted with any attribute. This lack of information is problematic if we want to learn from these data, so we annotated the complements of these verbs with a new argument value *"argC"*.

We focus our analysis on classes of verbs, as they are the most complex categories in terms of possible lexical frames: in our corpus there are more than 450 lexical frames for verbs. The verb classes we consider can be built from the verb's argument structure. Consider the following example:

$$ofrecer/v-arg0s\_sn-arg1\_sn-arg2\_sp\_a$$

The tag contains the following information: the word is a verb, and it has three arguments. The first argument is marked as *arg0* (agent) in the lexical argument structure, it is a noun phrase and it is also the subject of the verb. The second argument is *arg1* (theme), it is also a noun phrase. The third argument is *arg2* (beneficiary) and it is a prepositional phrase with head *"a"*. This is a possible subclass for the verb *"ofrecer"* (*"to offer"*) as seen in the corpus.

To sum up, the information included in the verb class tag is:

- Argument structure values (*arg0*, *arg1*, *arg2*, *arg3*, *arg4*, *argM*, *argL* or our *argC*).
- Syntactic category for each argument (*sn*, *S*, *sa*, *sadv*, *sp* or *v*). In the case of a preprositional phrase, we include the preposition: *sp_a*, *sp_de*, *sp_con*, *sp_en*, or just *sp* for the rest.
- Whether the argument is the specifier (subject) of the verb.

The reason we use this complex class scheme for verbs is because from the tag we are able to create the correct lexical frame for the verb. No extra information is needed. In the previous example, the lexical frame is the one shown in Fig. 2.

4 Experiments

A series of experiments were run using variations of the corpus and different sets of features in order to find the best configuration for the supertagger. In all the experiments we train a conditional random fields (CRF) [15] model using the CRF++ tool [16]. The whole Ancora corpus contains 517,000 tokens in 17,000 sentences and there are 61,000 instances of verbs. We consider only the verb classes that appear at least 30

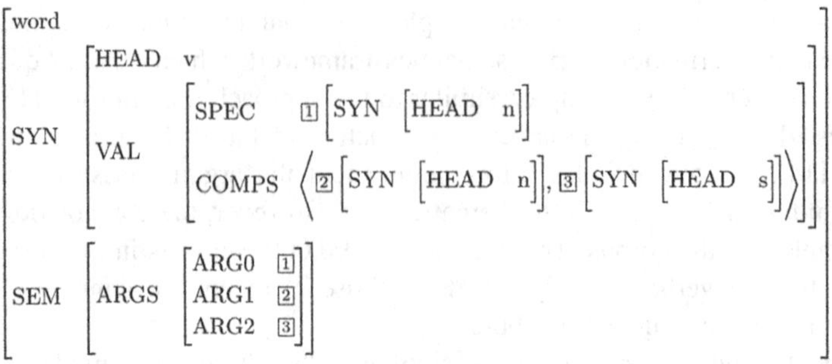

$$
\left[
\begin{array}{l}
\text{word} \\
\text{SYN}
\left[
\begin{array}{ll}
\text{HEAD} & \text{v} \\
\text{VAL} &
\left[
\begin{array}{ll}
\text{SPEC} & \boxed{1}\left[\text{SYN} \left[\text{HEAD} \quad \text{n}\right]\right] \\
\text{COMPS} & \left\langle \boxed{2}\left[\text{SYN} \left[\text{HEAD} \quad \text{n}\right]\right], \boxed{3}\left[\text{SYN} \left[\text{HEAD} \quad \text{s}\right]\right] \right\rangle
\end{array}
\right]
\end{array}
\right] \\
\text{SEM} \quad
\left[
\text{ARGS}
\left[
\begin{array}{ll}
\text{ARG0} & \boxed{1} \\
\text{ARG1} & \boxed{2} \\
\text{ARG2} & \boxed{3}
\end{array}
\right]
\right]
\end{array}
\right]
$$

Fig. 2. Lexical frame for the tag

times in the corpus, the rest of the verbs are marked with a generic tag *"v"*; this reduces the tagset from 453 to 92 possible values.

Only the accuracy for the verb classes is reported. In all cases we consider a match if any of the three best (most likely) values returned by the tagger is the correct one. The number of candidate tags was chosen so as to have a fixed way to compare the experiments. Choosing only the first candidate seemed too restrictive and generally had poor accuracy values given the high number of categories the classifier has to choose from, but allowing any number of candidates would also yield artificially high accuracies. We decided to settle in a constant number of candidates that made the different experiments comparable, although in a real scenario it would be more useful if the supertagger could return a variable number of candidates depending on what the parser needs [3].

4.1 Lemmas

The first set of experiments use only the lemmas of the words as features, the variations between experiments are:

- *context*: the number of words in context (5, 7, 9 or 11)
- *threshold:* the threshold of infrequent words. Words that appear fewer times than this value in the corpus are substituted with an "unknown" token (threshold values are 20, 30 or 50). Some preliminary experiments were run to check which thresholds were more likely to yield the best results. We tried using values of 10, 15, 20, 30 and 50, but the first two underperformed so we focused on the others.

The corpus was separated in a training set of 470,000 tokens (15,600 sentences), and both a development and a test set of roughly 23,000 tokens each (around 800 sentences). We first calculated a baseline for this data: assign to the test data the most common tags seen in the training data for each word. This yields an accuracy of 58.59 % for verbs (considering the top three tags).

Using the CRF tagger, the best accuracy for the development corpus (78.1 %) was found using context 5 and threshold 30. In the test corpus this yields an accuracy of 78.7 %.

4.2   Lemmas and POS

The second series of experiments use both lemmas and the POS tag included in Ancora as features. The rationale behind the use of a precalculated POS tag is that it might help focus the tagger on the categories we want: all the categories except *"v"* should be tagged exactly as the POS tag feature indicates; the only categories with new information are subclasses of *"v"*. This could help the tagger not to mix the *"v"* tokens with other categories.

In the previews experiments we saw that the training of taggers for such a number of different tags might be quite slow (the longest experiment took around 36 h), so we decided to break the training corpus into smaller corpora. We trained on the smaller corpora first, and then we could train on the big corpus focusing only in the configurations that yielded the most promising results.

The variations between experiments are *context* and *threshold* (the same as before) and also the size of the smaller corpus (50,000; 106,000 or 218,000 tokens). Table 1 summarizes the results for this series of experiments.

Table 1. Accuracies over the development set for the lemma/POS experiments. The left table uses threshold 30 and the right table uses threshold 50. The columns indicate the size of the training set and the rows indicate the number of words in context.

	50k	106k	218k		50k	106k	218k
5	72.70 %	75.90 %		5	71.88 %	75.20 %	
7	72.29 %	76.43 %	80.11 %	7	72.64 %	75.85 %	79.70 %
9	71.88 %	77.07 %	80.11 %	9	71.41 %	75.73 %	79.23 %
11	70.48 %	74.97 %		11	69.72 %	74.91 %	

Considering these results, we trained taggers using context 5 and 7, and threshold 30, for the whole training corpus. The tagger for context 5 yielded an accuracy of 81.51 % on the development corpus, while the tagger for context 7 yielded an accuracy of 81.21 %. We tested the first of these taggers using the test corpus, and it had an accuracy of 81.35 %.

For comparison, we also trained maximum entropy classifiers for the same data and features using the Stanford Classifier [17]. The best results were achieved using context 5 and threshold 30, yielding an accuracy of 78.90 %, a few points below the best CRF classifier.

4.3   Only Verbs with Arguments

Examining the results of the tagger, we found out that a large number of the errors occurred because the tagger assigned a generic *"v"* frame to tokens that should have a proper verb class. This could happen because there were a lot of tokens annotated with a generic *"v"*. This tag is applied in one of two cases:

- The appropriate class for the verb is not very frequent (i.e. it appears less than 30 times in the corpus).
- There is no information in the corpus about the arguments of the verb, thus we cannot infer a subclass.

This second case is the most common one: out of 61,000 instances of verbs in the corpus, more than 11,000 are not annotated with any argument. This missing information is a problem about our corpus. It is important to notice that there is a category of verbs in Spanish that do not take any argument (impersonal verbs such as *"llover"/"to rain"*), but this is not the case for most of this 11,000 examples, which include instances of *"ganar"* (*"to win"*) and *"encontrar"* (*"to find"*).

We created a new corpus to fix this issue. The aim was to remove from the corpus all the verbs that do not have information about their arguments (no argument structure is annotated). Besides filtering out the problematic examples, the process would prune all the (valid) instances of impersonal verbs. However, we found these verbs are very infrequent in Ancora, so we considered them negligible.

The main drawback of this approach is that we had to prune all the sentences that contain at least one verb without arguments, removing a great number of good training examples that did have arguments, but it was not possible to find a way of extracting the verbs and their contexts without manual intervention. In the end we were left with a new corpus that was about half the size of the original Ancora: 260,000 tokens in almost 10,000 sentences.

We separated this corpus into a training set of 230,000 tokens (8600 sentences), and both a development and a test set of 13,000 tokens each (around 600 sentences). As this corpus is different from the original, we calculated a new baseline by assigning the most common tags to the test data. This yields an accuracy of 59.84 % for verbs (considering the top three tags).

We also created smaller training sets of 60,000 and 120,000 tokens in order to quickly create and evaluate models. The features we use for these experiments are the lemma and POS tag of each word, and the variations between experiments are *context* (5, 7, 9 or 11) and *threshold* (20 or 30). The results for this series of experiments are shown in Table 2.

Table 2. Accuracies over the development set for the experiments using only verbs with arguments. The left table uses threshold 20 and the right table uses threshold 30. The columns indicate the size of the training set and the rows indicate the number of words in context.

	60k	120k		60k	120k
5	76.71 %	78.82 %	5	76.39 %	78.59 %
7	76.71 %	80.39 %	7	76.16 %	80.23 %
9	75.76 %	79.06 %	9	76.47 %	79.29 %
11	74.35 %	78.51 %	11	74.04 %	78.43 %

The configurations using 7 words of context of both sets of experiments seemed promising, so we trained both taggers on the whole corpus. For threshold 20 the accuracy of the new tagger was 82.35 %, and for threshold 30 it was 81.88 %. We tested only the first one using the test corpus, and obtained an accuracy of 83.58 %.

We also trained maximum entropy classifiers for this data. The best classifier uses context 5 and threshold 30 and has an accuracy of 82.35 % over the development corpus, the same as our best CRF tagger. However, the performance of this classifier against the test corpus dropped to 80.71 %. On the other hand, it is important to notice that the training time was a lot shorter for the maximum entropy models.

5   Conclusions

We extracted a set of verb classes from a Spanish corpus annotated in a HPSG compatible format. There are 453 verb classes in the corpus. A class includes the verbal arguments structure, their corresponding syntactic categories, and the distinction between specifier and complements. With this fine grained information we can create the correct HPSG lexical entry for a verb from its class.

We trained a series of supertaggers to disambiguate between these classes. In the experiments, only the 92 categories that appear 30 times or more in the corpus are used. The supertagger that achieved the best results was a CRF model trained using lemma and POS features, 7 words of context, and considering the words that appear less than 20 times in the corpus as unknown words. This supertagger achieves an accuracy of 83.58 % over the verb classes (it achieves an accuracy of 95.40 % over all the tags). This result is comparable to [8], which reports an accuracy of 79.64 % for a supertagger over the same base corpus, but with a much larger tagset in a different grammar framework. However, the corpus we used was only a subset of the original corpus because we had to prune many sentences that did not provide appropriate information about the verbs arguments.

As the performance of the supertagger has not plateaued, we consider that the results could be further improved by adding more training data. One way of doing this would be annotating the arguments that are missing in the corpus, or trying to leverage the information contained other Spanish corpora annotated in different formalisms.

References

1. Pollard, C., Sag, I.A.: Head-Driven Phrase Structure Grammar. University of Chicago Press/CSLI Publications, Chicago/Stanford (1994)
2. Joshi, A.K., Srinivas, B.: Disambiguation of super parts of speech (or supertags): almost parsing. In: Proceedings of the 15th Conference on Computational Linguistics, vol. 1, pp. 154–160. Association for Computational Linguistics (1994)
3. Curran, J.R., Clark, S., Vadas, D.: Multi-tagging for lexicalized-grammar parsing. In: Proceedings of the 21st International Conference on Computational Linguistics and the 44th Annual Meeting of the Association for Computational Linguistics, pp. 697–704. Association for Computational Linguistics (2006)
4. Lewis, M., Steedman, M.: Improved CCG parsing with semi-supervised supertagging. Trans. Assoc. Comput. Linguist. 2, 327–338 (2014)
5. Dridan, R.: Using lexical statistics to improve HPSG parsing. Doctoral dissertation, University of Saarland (2009)

6. Zhang, Y.Z., Matsuzaki, T., Tsujii, J.I.: Forest-guided supertagger training. In: Proceedings of the 23rd International Conference on Computational Linguistics, pp. 1281–1289. Association for Computational Linguistics (2010)
7. Silva, J., Branco, A.: Assigning deep lexical types. In: Sojka, P., Horák, A., Kopeček, I., Pala, K. (eds.) TSD 2012. LNCS, vol. 7499, pp. 240–247. Springer, Heidelberg (2012)
8. Marimon, M., Bel, N., Espeja, S., Seghezzi, N.: The spanish resource grammar: pre-processing strategy and lexical acquisition. In: Proceedings of the Workshop on Deep Linguistic Processing, pp. 105–111. Association for Computational Linguistics (2007)
9. Kolachina, P., Bangalore, S., Kolachina, S. Extracting LTAG grammars from a Spanish treebank. In: Proceedings of ICON-2011: 9th International Conference on Natural Language Processing. Macmillan Publishers, India (2011)
10. Taulé, M., Martí, M.A., Recasens, M.: Ancora: multilevel annotated corpora for catalan and Spanish. In: Proceedings of 6th International Conference on Language Resources and Evaluation, Marrakesh, Morocco (2008)
11. Chiruzzo, L., Wonsever, D.: Desarrollo de un parser HPSG Estadístico para el Español. In: Proceedings of I Workshop on Tools and Resources for Automatically Processing Portuguese and Spanish, São Carlos, SP, Brazil (2014)
12. Copestake, A., Flickinger, D., Pollard, C., Sag, I.A.: Minimal recursion semantics: an introduction. Res. Lang. Comput. 3(2–3), 281–332 (2005)
13. Miyao, Y., Ninomiya, T., Tsujii, J.: Corpus-oriented grammar development for acquiring a head-driven phrase structure grammar from the penn treebank. In: Su, K.-Y., Tsujii, J., Lee, J.-H., Kwong, O.Y. (eds.) IJCNLP 2004. LNCS (LNAI), vol. 3248, pp. 684–693. Springer, Heidelberg (2005)
14. Babko-Malaya, O.: PropBank annotation guidelines (2005)
15. Lafferty, J., McCallum, A., Pereira, F.C.: Conditional random fields: probabilistic models for segmenting and labeling sequence data (2001)
16. Kudo, T.: CRF++: yet another CRF toolkit (2005). Software available at http://crfpp. sourceforge.net
17. Manning, C., Klein, D.: Stanford classifier. The Stanford Natural Language Processing Group (2003). Software available at http://nlp.stanford.edu/software/classifier.shtml

Residual-Based Excitation with Continuous F0 Modeling in HMM-Based Speech Synthesis

Tamás Gábor Csapó[1](✉), Géza Németh[1], and Milos Cernak[2]

[1] Department of Telecommunications and Media Informatics,
Budapest University of Technology and Economics,
Magyar Tudósok körútja 2, Budapest, Hungary
{csapot,nemeth}@tmit.bme.hu
[2] Idiap Research Institute, Rue Marconi 19, Martigny, Switzerland
Milos.Cernak@idiap.ch

Abstract. In statistical parametric speech synthesis, creaky voice can cause disturbing artifacts. The reason is that standard pitch tracking algorithms tend to erroneously measure F0 in regions of creaky voice. This pattern is learned during training of hidden Markov-models (HMMs). In the synthesis phase, false voiced/unvoiced decision caused by creaky voice results in audible quality degradation. In order to eliminate this phenomena, we use a simple continuous F0 tracker which does not apply a strict voiced/unvoiced decision. In the proposed residual-based vocoder, Maximum Voiced Frequency is used for mixed voiced and unvoiced excitation. As all parameters of the vocoder are continuous, Multi-Space Distribution is not necessary during training the HMMs, which has been shown to be advantageous. Artifacts caused by creaky voice are eliminated with this speech synthesis system. A subjective listening test of English utterances has shown improvement over the traditional excitation.

Keywords: Speech synthesis · HMM · Creaky voice · Vocoder · Pitch tracking

1 Introduction

State-of-the-art text-to-speech synthesis is either based on unit selection or statistical parametric methods. Recently, particular attention has been paid to hidden Markov-model (HMM) based text-to-speech (TTS) synthesis [29], which has gained much popularity due to its flexibility, smoothness and small footprint. In this speech synthesis technique, the speech signal is decomposed to parameters representing excitation and spectrum of speech, and are fed to a machine learning system. After the training data is learned, during synthesis, the parameter sequences are converted back to speech signal with reconstructing methods (e.g. speech vocoders, excitation models).

There are three main factors in statistical parametric speech synthesis that are needed to deal with in order to achieve as high quality synthesized speech

© Springer International Publishing Switzerland 2015
A.-H. Dediu et al. (Eds.): SLSP 2015, LNAI 9449, pp. 27–38, 2015.
DOI: 10.1007/978-3-319-25789-1_4

as unit selection: vocoder techniques, acoustic modeling accuracy and over-smoothing during parameter generation [31]. In this paper, we investigate the first factor. A large number of improved excitation models have been proposed recently (for a comparison, see [18]). Statistical parametric speech synthesis and most of these excitation models are optimized for regular, modal voices (with quasi-periodic vibration of the vocal folds in voiced regions) and may not produce high quality with voices having frequent non-modal sections.

Irregular voice is such a non-modal phonation mode, which can cause disturbing artefacts in hidden Markov-model based text-to-speech synthesis. During regular phonation (modal voice) in human speech, the vocal cords are vibrating quasi-periodically. For shorter or longer periods of time instability may occur in the larynx causing irregular vibration of the vocal folds, which is referred to as irregular phonation, creaky voice, glottalization, vocal fry and laryngealization [4,5]. It leads to abrupt changes in the fundamental frequency (F0), amplitude of the pitch periods or both. Irregular phonation is a frequent phenomenon in both healthy speakers and people having voice disorders. It is often accompanied by extremely low pitch and the quick attenuation of glottal pulses. Glottalization can cause problems for standard speech analysis methods (e.g. F0 tracking and spectral analysis) and it is often disturbing in speech technologies [8].

In this paper we propose an attempt to eliminate the phenomena of non-modal phonation in HMM-based speech synthesis. More specifically, we hypothesize that a continuous F0 tracker, which does not apply a strict voiced/unvoiced decision caused by creaky voice, can 'smooth' the voice irregularities that further improves modeling capabilities of the HMM-based training framework.

2   Related Work

In our earlier studies, we modeled the creaky voice in HMM-TTS explicitly using a rule-based [7] and a data-driven irregular voice model [8]. We used a residual codebook based excitation model [6,9]. We also created an irregular-to-regular transformation method to smooth irregularities in speech databases [10]. Another alternative for overcoming the issues caused by creaky voice is to eliminate miscalculation of pitch tracking by using a more accurate fundamental frequency (F0) estimation method.

It has been shown recently that continuous F0 has advantages in statistical parametric speech synthesis [17]. For example, it was found that using a continuous F0, more expressive F0 contours can be generated [26–28] than using Multi-Space Distribution (MSD) [25] for handling discontinuous F0. Another important observation is that the voiced/unvoiced decision can be left up to the aperiodicity features in a mixed excitation vocoder [22]. This decision can also be modeled using a dynamic voiced frequency [13].

Accurate modeling of the residual has been shown to improve the synthesis quality [11,18]. Using a Principal Component Analysis-based (PCA) 'Eigen-Residual' results in significantly more natural speech (in terms of artificiality, buzziness) than the traditional pulse-noise excitation [16].

In the following, we introduce a new combination of continuous F0 (based on pitch tracking with Kalman-smoother based interpolation, [17]), excitation modeling (PCA-based residual, [11]) and aperiodicity features (based on Maximum Voiced Frequency, MVF, [13]).

3   Methods

We trained three HMM-TTS systems with various parameter streams using two voices. In the following the used databases, the methods for analysis, training of HMMs and synthesis are presented for a baseline and two improved systems.

3.1   Data

Two English speakers were chosen from the CMU-ARCTIC database [21], denoted EN-M-AWB (Scottish English, male) and EN-F-SLT (American English, female). Both of them produced irregular phonation frequently, mostly at the end of sentences. For speaker EN-M-AWB, the ratio of the voiced frames produced with irregular phonation vs. all voiced frames is 2.25 %, whereas for speaker EN-F-SLT, this ratio is 1.88 %, measured on the full database using automatic creaky voice detection [12].

3.2   Analysis

A. HTS-F0std. In the baseline system, the input is a speech waveform low-pass filtered at 7.6 kHz with 16 kHz sampling rate and 16 bit linear PCM quantization. The fundamental frequency (F0) parameters are calculated by a standard pitch tracker, the RAPT algorithm [23] implemented in Snack [2]. We denote this as 'F0std'. 25 ms frame size and 5 ms frame shift are used. In the next step 34-order Mel-Generalized Cepstral analysis (MGC) [24] is performed on the speech signal with $\alpha = 0.42$ and $\gamma = -1/3$. The results are the F0std and the MGC parameter streams.

As PCA-based residual has been shown to produce better speech quality than pulse-noise excitation [16], we perform residual analysis in the baseline system. The residual signal (excitation) is obtained by MGLSA inverse filtering [19]. The SEDREAMS Glottal Closure Instant (GCI) detection algorithm [15] is used to find the glottal period boundaries in the voiced parts of the residual signal. Pitch synchronous, two period long frames are used according to the GCI locations and they are Hann-windowed. The pitch-synchronous residual frames are resampled to twice the average pitch period of the speaker (e.g. for EN-M-AWB, $Fs = 16$ kHz, $F0_{avg} = 123$ Hz, $T0_{avg} = 130 samples$, $framelen_{resampled} = 260 samples$). Finally, Principal Component Analysis is applied on these frames, and the first principal component is calculated. Figure 1 shows examples for the PCA residual. Instead of impulses, this PCA residual will be used for the synthesis of the voiced frames.

For text processing (e.g. phonetic transcription, labeling, etc.), the Festival TTS front-end is used [1].

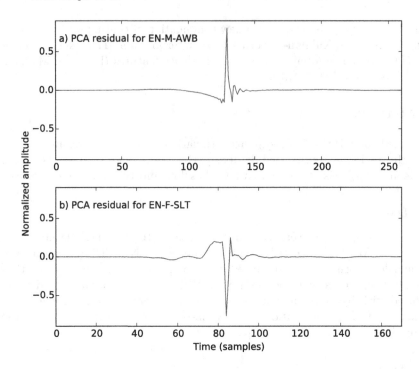

Fig. 1. PCA residuals obtained from two period long pitch-synchronous residuals.

B. HTS-F0std+MVF. In the 2nd system, the analysis of the speech wave-form and residual is similar to HTS-F0std, resulting in the MGC and F0std parameters. After these steps, Maximum Voiced Frequency is calculated from the speech signal using the MVF_Toolkit [14] with 5 ms frame shift, resulting in the MVF parameter.

C. HTS-F0cont+MVF. In the third system, we use the same MGC parameter stream as in the baseline. For the calculation of the fundamental frequency, the open-source implementation [3] of a simple continuous pitch tracker [17], denoted as 'F0cont', is used. In regions of creaky voice, this pitch tracker interpolates F0 based on a linear dynamic system and Kalman smoothing. Similarly to the 2nd system, MVF is also calculated here. That is, in the 3rd system we use the F0cont, MGC and MVF parameter streams.

Figure 2 (above the dashed line) shows all the steps applied in the analysis part of the HTS-F0cont+MVF system.

3.3   HMM Training

For training, the various parameters are calculated from each frame to describe the residual (F0std/F0cont/MVF) and the spectrum (MGC). During training,

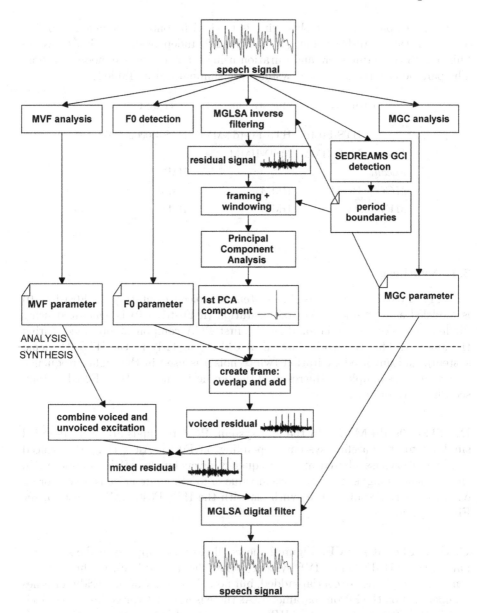

Fig. 2. Analysis (above the dashed line) and synthesis (below the dashed line) with the HTS-F0cont+MVF system.

logarithmic values are used as they were found to be more suitable in our experiments. F0std is modeled with MSD-HMMs because this stream does not have values in unvoiced regions. All the other parameter streams (F0cont, MVF, and MGC) are modeled as simple HMMs. The first and second derivatives of all of the parameters are also stored in the parameter files and used in the training phase.

Decision tree-based context clustering is used with context dependent labeling applied in the English version [29] of HTS 2.2. Independent decision trees are built for all the parameters and duration using a maximum likelihood criterion. The parameter streams for the systems are summarized in Table 1.

Table 1. Parameter streams of the three systems.

	HTS-F0std	HTS-F0std+MVF	HTS-F0cont+MVF
F0std	MSD-HMM	MSD-HMM	-
F0cont	-	-	HMM
MGC	HMM	HMM	HMM
MVF	-	HMM	HMM

3.4   Synthesis

A. HTS-F0std. In the baseline system, unvoiced excitation (if F0std = 0) is modeled as white noise. Voiced excitation (if F0std > 0) is generated using pitch-synchronously overlap-adding the first PCA component obtained during the analysis. This is lowpass filtered at 6 kHz (similarly to the HTS 2.2 demo system), and unvoiced excitation (white noise) is used in the higher frequency bands. For an example for the result of the synthesis with the HTS-F0std system, see Fig. 3 (a) and (b).

B. HTS-F0std+MVF. In the 2nd system, PCA residuals are overlap-added similarly to the baseline system, depending on F0std. After that, this voiced residual is lowpass filtered at the frequency given by the MVF parameter. In the frequencies higher than the actual value of MVF, white noise is used. For an example for the result of the synthesis with the HTS-F0std+MVF system, see Fig. 3 (c) and (d).

C. HTS-F0cont+MVF. Figure 2 shows all the steps applied in the synthesis part of the HTS-F0cont+MVF system (below the dashed line). In this 3rd system, PCA residuals are overlap-added, but now the density of the residual frames is dependent on the F0cont parameter. As there is no strict voiced/unvoiced decision in the F0cont stream, the MVF parameter models the voicing information: for unvoiced sounds, the MVF is low (around 1 kHz), for voiced sounds, the MVF is high (typically above 4 kHz), whereas for mixed excitation sounds, the MVF is in between (e.g. for voiced fricatives, MVF is around 2–3 kHz). Voiced and unvoiced excitation is added together similarly to the 2nd system, depending on the MVF parameter stream (see 'mixed residual signal' in Fig. 2).

   Figure 3 (e) and (f) shows an example for the result of the synthesis with the HTS-F0cont+MVF system. By comparing all three sentence variants, it can be seen that in the baseline and 2nd systems (subfigures (a) and (c)), F0std

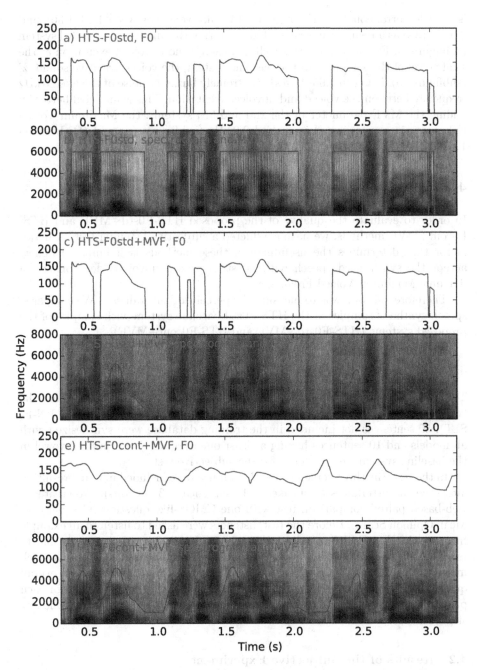

Fig. 3. Synthesis examples for the sentence: *'Please Mom, is this New Zealand, or Australia?'*, by speaker EN-M-AWB. Subfigures (a), (c) and (e) show the generated F0std/F0cont parameters; subfigure (b) shows a spectrogram and a fixed 6 kHz MVF; subfigures (d) and (e) show the spectrograms and the generated MVF parameter.

is modeled erroneously at the regions of creaky voice (between 1.23–1.36 s and 2.98–3.13 s) as a result of miscalculated F0 during the analysis. In the 3rd system (subfigure e), F0cont models this well and there is no unvoiced excitation at the final vowel of the sentence. In the baseline system, the voiced fricative sound 'z' (subfigure (b), between 1.65–1.74 s) is extremely buzzy because of the fixed 6 kHz frequency between the voiced and unvoiced excitation. This is modeled better by adding the MVF parameter in the 2nd and 3rd systems: the Maximum Voiced Frequency in the 'z' sound is around 2.2 kHz (subfigures (d) and (f), between 1.65–1.74 s).

4 Evaluation

In order to evaluate the quality of the proposed HTS-F0std+MVF and HTS-F0cont+MVF methods, we have conducted a subjective listening test. A major factor that determines the usefulness of these methods is if human listeners accept the synthesized speech with no strict voiced/unvoiced decision and a dynamic Maximum Voiced Frequency.

Therefore, our aim was to measure the perceived 'naturalness'. We compared speech synthesis samples of the HTS-F0std baseline system with samples of the proposed systems (HTS-F0std+MVF and HTS-F0cont+MVF).

4.1 Methods of the Subjective Experiment

To obtain the speech stimuli, we created six models with the baseline and the two proposed systems and the two English speakers (EN-M-AWB and EN-F-SLT). 50 sentences not included in the training database were synthesized with all models and 10 sentences having at least one irregularly synthesized vowel in the baseline system were selected for the subjective test.

In the test, the three versions of each sentence were included in pairs, resulting altogether 60 utterances (2 speakers · 10 sentences · 3 versions). We created a web-based paired comparison test with one CMOS-like question (Comparative Mean Opinion Score). Before the test, listeners were asked to listen to an example from speaker EN-F-SLT to adjust the volume. In the test, the listeners had to rate the naturalness ('Which of the sentences is more natural?', '1 – 1st is much more natural' ... '5 – 2nd is much more natural') as a question for overall quality. The utterances were presented in a randomized order (different for each participant). The listening test can be found at http://leszped.tmit.bme.hu/slsp2015_en/.

4.2 Results of the Subjective Experiment

Altogether 8 listeners participated in the test (3 females, 5 males). They were all students or speech experts, between 24–45 years (mean: 37 year). They were not native speakers of English. On average the whole test took 18 min to complete. Two listeners noted that some of the sentences were too long to evaluate properly.

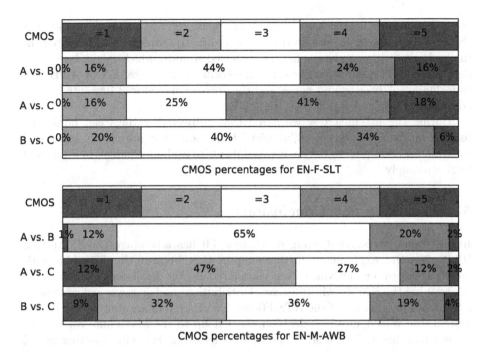

Fig. 4. Results of the listening test for the 'naturalness' question (top: speaker EN-F-SLT, bottom: speaker EN–M-AWB). A: HTS-F0std, B: HTS-F0std+MVF, C: HTS-F0cont+MVF. Average CMOS values can be found in the text of Sect. 4.2.

The results of the listening test are presented in Fig. 4 for the two speakers and three systems. The figure provides a comparison between the baseline A: HTS-F0std system and the two proposed systems (B: HTS-F0std+MVF, C: HTS-F0cont+MVF) pair by pair. The answers of the listeners for the 'naturalness' question were pooled together for the visualization (levels 1-2-3-4-5).

The ratings of the listeners were compared by t-tests as well, with a 95 % confidence level. For speaker EN-F-SLT, HTS-F0std+MVF was significantly preferred over HTS-F0std (average CMOS for A vs. B: 3.40) and HTS-F0cont+MVF was significantly preferred over both HTS-F0std and HTS-F0std+MVF (average CMOS for A vs. C: 3.60 and for B vs. C: 3.26). This result means that for the female voice, listeners evaluated the proposed systems as being significantly more natural than the baseline system. Figure 4 also shows that for speaker EN-F-SLT (top subfigure), there was no '=1' answer from the listeners.

For speaker EN-M-AWB, system B was slightly preferred over system A, although this difference is not significant (average CMOS for A vs. B: 3.10). However, both system A and system B reached significantly higher CMOS scores than system C (average CMOS for A vs. C: 2.46 and for B vs. C: 2.76). From this result we can conclude that adding the MVF parameter increased the naturalness, but combined with F0cont, this introduced audible vocoding artifacts.

By investigating the synthesis samples of speaker EN-M-AWB we found that the HTS-F0cont+MVF system often resulted in too strong voiced component in the lower frequency bands for the unvoiced sounds, which might have been disturbing for the listeners. The original recordings of speaker EN-M-AWB contain significant background noise, and the vocoder introduced unwanted buzzy components because of this.

During the listening test we noted that subjective ratings can hardly be focused on buzziness or voiced/unvoiced transitions, but are mostly influenced by overall speech quality. Hence, it was difficult to evaluate changes in segmental level separately.

5 Discussion and Conclusions

It was found earlier that using continuous F0 has advantages in HMM-TTS [17,26–28]. Our experiments further support this, because the disturbing artifacts caused by creaky voice were eliminated by the proposed vocoder. During training the HMMs, Multi-Stream Distribution modeling was not necessary, because all paramaters of the HTS-F0cont+MVF system are continouos. In this system, the voiced/unvoiced decision was left up to the Maximum Voiced Frequency parameter. This kind of aperiodicity modeling is similar to other mixed excitation vocoders [13,22], but our system is simpler, i.e. uses less parameters compared to STRAIGHT-based mixed excitation [20,30]. However, MVF does not always work well for voiced/unvoiced decision (e.g. in case of unvoiced stops there is a disturbing component in the samples of HTS-F0cont+MVF). In future work we will decrease the periodic component of unvoiced sounds.

In this paper we introduced a new vocoder, using (1) Principal Component Analysis-based residual frames, (2) continuous pitch tracking, and (3) Maximum Voiced Frequency. In a listening test of English speech synthesis samples, the proposed system with a female voice was evaluated as significantly more natural than a baseline system using only PCA-based residual excitation. The listening test results of the male voice have shown that there is room for improvement in modeling the unvoiced sounds with this continuous F0 model. MVF-based mixed voiced and unvoiced excitation was found to be extremely useful for modeling the voiced fricatives (e.g. 'z' in Fig. 3). However, in case of unvoiced sounds, the lowest MVF value was 1 kHz, which was disturbing for the male voice, but acceptable for the female voice. It is a question whether the buzziness caused by the combined F0cont and MVF modeling can be reduced. In the future, we plan to add a Harmonics-to-Noise Ratio parameter to both the analysis and synthesis steps in order to investigate this and to further reduce the buzziness caused by vocoding.

With the proposed methods we extend previous speech processing techniques dealing with irregular phonation. The above models and results might be useful for more natural, expressive, personalized speech synthesis.

Acknowledgments. We would like to thank the listeners for participating in the subjective test. We thank Philip N. Garner for providing the continuous pitch tracker open source and Bálint Pál Tóth for useful comments on this manuscript. This research is partially supported by the Swiss National Science Foundation via the joint research project (SCOPES scheme) SP2: SCOPES project on speech prosody (SNSF no IZ73Z0_152495-1) and by the EITKIC project (EITKIC_12-1-2012-001).

References

1. The Festival Speech Synthesis System [Computer program], Version 2.1 (2010). http://www.cstr.ed.ac.uk/projects/festival/
2. The Snack Sound Toolkit [Computer program], Version 2.2.10 (2012). http://www.speech.kth.se/snack/
3. Speech Signal Processing - a small collection of routines in Python to do signal processing [Computer program] (2015). https://github.com/idiap/ssp
4. Bohm, T., Audibert, N., Shattuck-Hufnagel, S., Németh, G., Aubergé, V.: Transforming modal voice into irregular voice by amplitude scaling of individual glottal cycles. In: Acoustics 2008, Paris, France, pp. 6141–6146 (2008)
5. Blomgren, M., Chen, Y., Ng, M.L., Gilbert, H.R.: Acoustic, aerodynamic, physiologic, and perceptual properties of modal and vocal fry registers. J. Acoust. Soc. Am. **103**(5), 2649–2658 (1998)
6. Csapó, T.G., Németh, G.: A novel codebook-based excitation model for use in speech synthesis. In: IEEE CogInfoCom, Kosice, Slovakia, pp. 661–665, December 2012
7. Csapó, T.G., Németh, G.: A novel irregular voice model for HMM-based speech synthesis. In: Proceedings of the ISCA SSW8, Barcelona, Spain, pp. 229–234 (2013)
8. Csapó, T.G., Németh, G.: Modeling irregular voice in statistical parametric speech synthesis with residual codebook based excitation. IEEE J. Sel. Top. Sig. Proc. **8**(2), 209–220 (2014)
9. Csapó, T.G., Németh, G.: Statistical parametric speech synthesis with a novel codebook-based excitation model. Intell. Decis. Technol. **8**(4), 289–299 (2014)
10. Csapó, T.G., Németh, G.: Automatic transformation of irregular to regular voice by residual analysis and synthesis. In: Proceedings of the Interspeech (2015). (accepted)
11. Drugman, T., Dutoit, T.: The deterministic plus stochastic model of the residual signal and its applications. IEEE Trans. Audio Speech Lang. Proc. **20**(3), 968–981 (2012)
12. Drugman, T., Kane, J., Gobl, C.: Data-driven detection and analysis of the patterns of creaky voice. Comput. Speech Lang. **28**(5), 1233–1253 (2014)
13. Drugman, T., Raitio, T.: Excitation modeling for HMM-based speech synthesis: breaking down the impact of periodic and aperiodic components. In: Proceedings of the ICASSP, Florence, Italy, pp. 260–264 (2014)
14. Drugman, T., Stylianou, Y.: Maximum voiced frequency estimation : exploiting amplitude and phase spectra. IEEE Sig. Proc. Lett. **21**(10), 1230–1234 (2014)
15. Drugman, T., Thomas, M.: Detection of glottal closure instants from speech signals: a quantitative review. IEEE Trans. Audio Speech Lang. Process. **20**(3), 994–1006 (2012)
16. Drugman, T., Wilfart, G., Dutoit, T.: Eigenresiduals for improved parametric speech synthesis. In: EUSIPCO09 (2009)

17. Garner, P.N., Cernak, M., Motlicek, P.: A simple continuous pitch estimation algorithm. IEEE Sig. Process. Lett. **20**(1), 102–105 (2013)
18. Hu, Q., Richmond, K., Yamagishi, J., Latorre, J.: An experimental comparison of multiple vocoder types. In: Proceedings of the ISCA SSW8, pp. 155–160 (2013)
19. Imai, S., Sumita, K., Furuichi, C.: Mel Log Spectrum Approximation (MLSA) filter for speech synthesis. Electron. Commun. Jpn. (Part I: Communications) **66**(2), 10–18 (1983)
20. Kawahara, H., Masuda-Katsuse, I., de Cheveigné, A.: Restructuring speech representations using a pitch-adaptive time-frequency smoothing and an instantaneous-frequency-based F0 extraction: Possible role of a repetitive structure in sounds. Speech Commun **27**(3), 187–207 (1999)
21. Kominek, J., Black, A.W.: CMU ARCTIC databases for speech synthesis. Tech. rep, Language Technologies Institute (2003)
22. Latorre, J., Gales, M.J.F., Buchholz, S., Knil, K., Tamura, M., Ohtani, Y., Akamine, M.: Continuous F0 in the source-excitation generation for HMM-based TTS: Do we need voiced/unvoiced classification? In: Proceedings of the ICASSP, Prague, Czech Republic, pp. 4724–4727 (2011)
23. Talkin, D.: A robust algorithm for pitch tracking (RAPT). In: Kleijn, W.B., Paliwal, K.K. (eds.) Speech Coding and Synthesis, pp. 495–518. Elsevier, Amsterdam (1995)
24. Tokuda, K., Kobayashi, T., Masuko, T., Imai, S.: Mel-generalized cepstral analysis - a unified approach to speech spectral estimation. In: Proceedings of the ICSLP, Yokohama, Japan, pp. 1043–1046 (1994)
25. Tokuda, K., Mausko, T., Miyazaki, N., Kobayashi, T.: Multi-space probability distribution HMM. IEICE Trans. Inf. Syst. **E85–D**(3), 455–464 (2002)
26. Yu, K., Thomson, B., Young, S., Street, T.: From discontinuous to continuous F0 modelling in HMM-based speech synthesis. In: Proceedings of the ISCA SSW7, Kyoto, Japan, pp. 94–99 (2010)
27. Yu, K., Young, S.: Continuous F0 modeling for HMM based statistical parametric speech synthesis. IEEE Trans. Audio Speech Lang. Process. **19**(5), 1071–1079 (2011)
28. Yu, K., Young, S.: Joint modelling of voicing label and continuous F0 for HMM based speech synthesis. In: Proceedings of the ICASSP, Prague, Czech Republic, pp. 4572–4575 (2011)
29. Zen, H., Nose, T., Yamagishi, J., Sako, S., Masuko, T., Black, A.: The HMM-based speech synthesis system version 2.0. In: Proceedings of the ISCA SSW6, Bonn, Germany, pp. 294–299 (2007)
30. Zen, H., Toda, T., Nakamura, M., Tokuda, K.: Details of the Nitech HMM-based speech synthesis system for the Blizzard Challenge 2005. IEICE Trans. Inf. Syst. **E90–D**(1), 325–333 (2007)
31. Zen, H., Tokuda, K., Black, A.W.: Statistical parametric speech synthesis. Speech Commun. **51**(11), 1039–1064 (2009)

Discourse Particles in French: Prosodic Parameters Extraction and Analysis

Mathilde Dargnat[1], Katarina Bartkova[1], and Denis Jouvet[2,3,4(✉)]

[1] Université de Lorraine & CNRS, ATILF, UMR 7118, 44, Avenue de la Libération,
30687, 54063 Nancy Cedex, France
{mathilde.dargnat,katarina.bartkova}@atilf.fr
[2] Speech Group, Inria, LORIA, 54600 Villers-lès-Nancy, France
denis.jouvet@inria.fr
[3] Université de Lorraine, LORIA, UMR 7503, 54600 Villers-lès-Nancy, France
[4] CNRS, LORIA, UMR 7503, 54600 Villers-lès-Nancy, France

Abstract. Detecting the correct syntactic function of a word is of great importance for language and speech processing. The semantic load of a word is different whether its function is a discourse particle or a preposition. Words having the function of a discourse particle (DP) are very frequent in spontaneous speech and their discursive function is often expressed only by prosodic means. Our study analyses some prosodic correlates of two French words (*quoi*, *voilà*), used as discourse particles or pronoun (*quoi*) or preposition (*voilà*). Our goal is to determine to what extent intrinsic and contextual prosodic properties characterize DP and non-DP functions. Prosodic parameters are analyzed with respect to the DP or non-DP function for these words extracted from large speech corpora. A preliminary test concerning the automatic detection of the word function is also carried out using prosodic parameters only, leading to an encouraging result of 70 % correct identification.

Keywords: Part-of-speech tagging · Prosody · Discourse particles · Discourse structure · Automatic prosodic annotation

1 Introduction

The correct category definition of homographic lexical units often determine their phonetic forms, stressed syllable or prosodic characteristics. But a word can also have different pragmatic properties which can impact the word semantic load. A word with low semantic content is often considered as a disturbance of the speech fluency and can then disappear in automatic processing (content retrieval, automatic translation ...). Therefore it becomes important to detect the syntactic and pragmatic properties of the words and, for a correct detection of these properties, the lexical context is sometimes not enough, prosodic information is also needed. The aim of this study is to investigate how the pragmatic properties of words can trigger different prosodic parameters. This study is part of a larger project on discourse particles (DP) in French [2]. It aims to correlate

© Springer International Publishing Switzerland 2015
A.-H. Dediu et al. (Eds.): SLSP 2015, LNAI 9449, pp. 39–49, 2015.
DOI: 10.1007/978-3-319-25789-1_5

DP's prosodic properties with their syntactical, semantic and pragmatic properties. In this paper, we focus here on the prosodic analysis of two words, **quoi** (*what*) and **voilà** (*here is, that's it, there*), frequently used as DP in formal and casual speech, and we mainly examine the relevance of some prosodic features in distinguishing between DP and non-DP use of these words.

Studies generally address DP through semantic and pragmatic descriptions, from synchronic or diachronic points of view (see [5,6,13,22,24,30]). Syntactic analysis is less frequent (see [10,17,29]), while prosodic considerations remain peripheral or too general (see [1,11,31]). The goal of our study is to construct a fine-grained corpus-based prosodic analysis, in order to identify possible correlations with other linguistic properties of DP. The main question addressed here concerns the correlation between syntactic properties (mainly position in the utterance) and discourse values (information structure) on the one hand, and prosodic features (pause, position in prosodic group, syllabic duration, tone, slope . . .) on the other hand. If such a correlation is confirmed, this could lead to an interesting tool for distinguishing different uses of the studied items (e.g. *quoi* as a pronoun or as a DP, and its different values as DP: closing, rhematic marker, reformulation, etc.).

2   Discourse Particles

DP convey information about utterance interpretation, epistemic state and affective mood of the speaker or the management of interaction [13]. DP do not form parts of speech like verbs or adjectives (contra Paillard [25]), but a 'functional category' [15,24] whose lexical members, in addition to being DP, have more traditional grammatical uses, like coordinating conjunctions, adverbs, verbs, pronouns, interjections, adjectives. We focus here on major grammatical and discursive uses of *quoi* (pronoun vs. DP) and *voilà* (preposition, introducer [18] vs. DP) in French. We do not propose for the moment a more fine-grained subcategorization.

2.1   Main Features

DP frequently exhibit phonetic attrition. They have prosodic autonomy and can be singled out by a pause or a specific prosodic pattern (see [1,14,20,21]). In French, they tend to be mono- or bisyllabic, but some of them are also 'complex', i.e., combinations like *'bon ben quoi voilà hein'*, *'mais enfin'* or *'écoutez donc'*. DP are neither argument nor circumstantial adjuncts. They are optional and their position in the utterance is neither fixed nor totally free (see [9,13,16,24,27]). DP do not contribute to the propositional content of the utterance. As a result, they do not affect its truth value. They have undergone a 'pragmaticalization' process whereby their initial meaning has given way to some pragmatic or procedural values [28] (for DP main feature descriptions, see also [2,6,12–14]).

2.2   Illustration: *quoi* and *voilà*

The words *quoi* and *voilà* have been chosen because they are very frequent in present-day French. They belong to different parts of speech, they occur in several positions and they play a role in discourse information structuring. Major DP values of *quoi* are closing, leftward focus marking (1), (re)phrasing signaling with a specificity to refer to the previous context (it has scope over the material to its left) (See [8,12,22,23]). Major DP values of *voilà*are closing, sometimes with agreement expression about the previous discourse, and stage marking in an ongoing non-linguistic activity (2). Its position depends on its pragmatic values and on the discourse type (monologue vs. dialogue) [6,7].

(1)   c'est un outil de travail mais c'est de l'abstrait quoi   c'est c'est
       it's   a   working tool   but   it's   abstract   PARTICLE, it's   it's
       c'est pas du concret quoi
       it's   not concrete   PARTICLE
       'It's a working tool, but an abstract one PARTICLE, it's it's it's not concrete PARTICLE'
       [22, p. 6]

(2)   c'est bon allez: ↓↓ on va   mouiller ↓ voilà: vous remuez
       it's   OK go,   we are gonna get wet   PARTICLE: you   stir
       'It's OK, we will get wet, that's it: move'
       [6, p. 366]

3   Methodology and Corpus

The study of the prosodic parameters of our DP words is corpus-based. The major part of our data processing is done automatically. An effort was made to build an automatic extraction and annotation procedure that will further allow enrichment of our DP database in a consistent way. However, manual intervention is still needed to distinguish between DP or non-DP uses. We kept only occurrences for which at least two expert annotators agreed.

3.1   Corpus Constitution and Extraction

All occurrences of *voilà* and *quoi* are extracted from the ESTER corpus (French broadcast news collected from various radio channels, about 200 hours of speech) and from the ETAPE corpus (debates collected from various French radio and TV channels, about 30 h of recordings). Their compositions slightly differ: ETAPE contains more spontaneous speech whereas ESTER is mainly constituted of broadcast news and includes relatively few interviews. An advantage in using such corpora is the rather good quality of the speech signal, which leads to a reliable acoustic analysis. Table 1 indicates the distribution of DP vs. non-DP uses of the two words, after manual annotation. As illustrated in Table 1, there is quite a difference between the DP uses of the two words studied here, in fact the DP use of *voilà* is almost twice as high as the DP use of *quoi*.

Table 1. Distribution of DP and non-DP uses (and number of occurrences) for the words *voilà* and *quoi* in ESTER and ETAPE corpora

	Number of occurrences	DP	Non-DP
quoi	1002	381 ⇨ 39 %	621 ⇨ 61 %
voilà	1407	971 ⇨ 69 %	436 ⇨ 31 %

3.2   Speech Data Pre-processing

ESTER and ETAPE contain manual transcriptions and information of different types (speakers, turn-takings, dysfluencies, noise, etc.). All the speech data processing is done automatically. First, grapheme-to-phoneme translation is carried out and the sound segmentation is achieved, using forced alignment (achieved with Sphinx tools). Subsequently, Prosotran annotator [3] is used, which, for each vowel, indicates the degree of its duration lengthening (compared to a mean vowel duration calculated on at least 5 adjacent syllables); its F0 slope, compared to the glissando threshold; its pitch level, quantized on a ten level scale calculated on the whole speaker data. Further prosodic information is provided by the detection of prosodic groups. Segmentation of the speech stream into prosodic groups is yielded by the ProsoTree software [4], which locates intonation group boundaries using information based on F0 slope values, pitch level and vowel duration (Fig. 1).

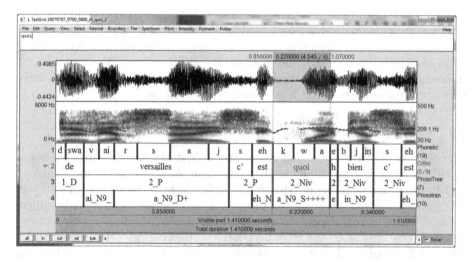

Fig. 1. Result of prosodic processing of speech data using Prosotran and Prosotree software

4 Analysis of Results

As mentioned before, we investigate here whether intrinsic and contextual prosodic properties are reliable cues to characterize DP and non-DP uses of *voilà* and *quoi*, either separately or jointly. In this section, we analyze and discuss for the studied words the role of pauses preceding or following them, their pitch level in isolation and in relation to their immediate preceding and following contexts, their vowel duration, whose lengthening may indicate a stressed position and their position in the intonation group.

4.1 Pauses

Information about pauses are collected automatically. The occurrences of pause contexts are presented in Table 2. Some pauses whose durations have not been measured (they occur before the first or after the last word of the speech segment) do not appear in Table 2.

Several differences in pause occurrences are noteworthy:

- **Non-DP uses**: *quoi* occurs predominantly without a pause; if *voilà* occurs with a pause, the pause is predominantly before, with a big proportion of long pauses (> 250 ms).
- **DP and non-DP uses**: when *quoi* occurs with a pause, this pause comes pre-dominantly after, with a large proportion of long pauses (> 250 ms) for DP uses.

Table 2. Number of pauses before and after *quoi* and *voilà*: number of occurrences (in parentheses) and percentage per category. (short $\leq$ 100 ms; mid $\leq$250 ms; long $\geq$ 250 ms).

		Pauses						No pause
		Before			After			
		short	mid	long	short	mid	long	
quoi	DP (399)	1.2% (5)	5.0% (20)	0.2% (1)	5.7% (23)	14% (55)	40.0% (160)	34.0% (135)
	Non-DP (630)	0.7% (3)	4.0% (25)	3.0% (19)	5.0% (32)	8.6% (54)	12.0% (78)	66.5% (419)
voilà	DP: (1019)	0.4% (4)	11.2% (114)	20.2% (206)	10.0% (103)	11.6% (118)	23.0% (232)	23.7% (242)
	Non-DP (416)	0.5% (2)	7.2% (30)	35.6% (148)	3.3% (14)	4.1% (17)	4.3% (18)	45.0% (187)

These pause occurrences correspond to syntactical and information structures of the non-DP and DP uses: non-DP *quoi* is often an argument of a verb, DP *quoi* is more often conclusive. Non-DP *voilà* begins an utterance and, as an introducer or a preposition, introduces the following discourse segment, which is syntactically dependent on it.

4.2   Position in the Intonation Group

The position of *quoi* and *voilà* in the intonation groups (IG) is analysed according to the intonation groups automatically detected using the Prosotree software. Results about their location in the intonation group are displayed in Fig. 2.

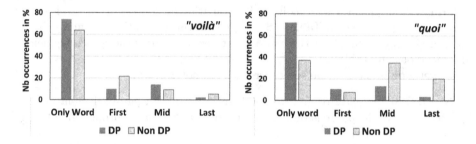

Fig. 2. Position of *voilà* and *quoi* in the intonation groups

According to our results, *quoi* and *voilà* more frequently occur as single words in their corresponding intonation group, but relevant distinctions can be made when they are integrated into a larger IG.

DP *quoi* occurs predominantly as a single word. Its prosodic detachment is coherent with its syntactical and semantical relative autonomy. Non-DP *quoi* occurs as a single word, but in equal proportion in middle position in IG. The intermediate position seems to be an indicator of its syntactic and semantic integration.

DP *voilà* occurs predominantly as a single word too, for the same reason as DP *quoi*. Non-DP *voilà*, in contrast to non-DP *quoi*, occurs preferentially at the beginning and not in an intermediate position. This corroborates it 'opening-introducing' function.

4.3   Pitch Level and F0 Slope

Pitch level values of the syllable nuclei are yielded by the Prosotran software, quantized on a ten degree scale. In order to compare pitch levels for *voilà* and *quoi* in our data, only measurements on the last syllable (last nuclei) are used.

As illustrated in Fig. 3, DP *quoi* is uttered very often at low pitch levels and very seldom at high pitch levels. It confirms its major conclusive function. On the other hand, *voilà* is often uttered at high pitch level.

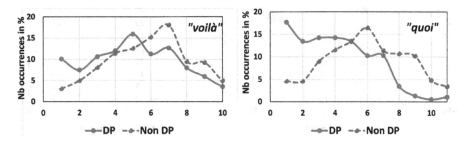

Fig. 3. Frequency of occurrences with respect to pitch level values (measured on the last syllable nuclei)

4.4   F0 Slopes Linking with Left and Right Contexts

The prosodic articulation between *voilà* and *quoi* and their neighbour words is measured by F0 slope values. The slope between *voilà* or *quoi* and its left context corresponds to the ΔF0 between the first (or unique) syllable of *quoi* or *voilà* and the last syllable of the previous word (a negative slope means that the *quoi* or *voilà* target syllable has a lower F0 value than the last syllable of the previous word). The slope with respect to the right context is obtained as the ΔF0 between the first syllable of the next word and the last or only syllable of *voilà* and *quoi* (a negative slope means that the *quoi* or *voilà* target syllable has a higher F0 value than the first syllable of the next word).

For *quoi* as DP (see Fig. 4), the F0 slope has often a falling pattern in relation to its left context. The DP is generally added at the end of a sentence as a kind of 'pronounced comma' and its F0 level is most of the time low or very low.

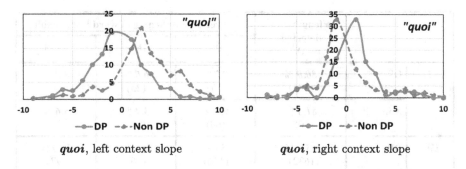

quoi, left context slope   *quoi*, right context slope

Fig. 4. Frequency of occurrences with respect to the ΔF0 value for *quoi*

With respect to the F0 slopes, the DP *voilà* has a similar behaviour as the DP *quoi*, since it is often used as a conclusive particle. However, its F0 pattern is different from the particle *quoi*. In fact, (see Fig. 5), this particle can be uttered with a conclusive (falling) but also with a continuative (rising) F0 pattern. This observation is corroborated also by its pitch level characteristics (see Fig. 3).

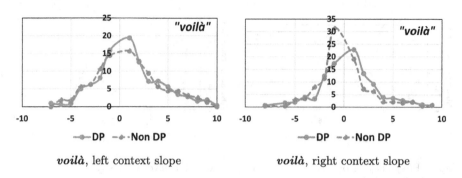

voilà, left context slope   voilà, right context slope

Fig. 5. Frequency of occurrences with respect to the ΔF0 value for *voilà*

4.5 Vowel Duration

Prosotran provides information about vowel duration, which includes three symbolic annotation degrees of vowel lengthening: very strong, strong and mild lengthening (see respectively D+++, D++ and D+ in Table 3 and Fig. 6). This annotation is used for comparing duration lengthening of the last (or unique) vowels of the words. As vowel duration lengthening is a strong parameter cue for stressed syllable in French, it seems worthwhile to analyse how vowel duration lengthening contributes to the prosodic characteristics of the DP. Table 3 and Fig. 6 show the number of occurrences and the percentage of lengthened vowels for both words, in their DP and non-DP uses.

Table 3. *quoi* and *voilà last* syllable duration, percentage of occurrences according to vowel lengthening, and number of occurrences in parentheses

		no lengthening detected D	Degree of lengthening of word last syllable		
			D+	D++	D+++
quoi	DP	80.0 % (305)	9.7 % (37)	6.0 % (23)	4.2 % (16)
	Non-DP	81.3 % (505)	9.7 % (60)	3.4 % (21)	5.6 % (35)
voilà	DP	65.0 % (632)	15.0 % (146)	9.9 % (96)	10.0 % (97)
	Non-DP	85.5 % (373)	8.9 % (39)	2.7 % (12)	2.7 % (12)

Statistics in Table 3 show that for each sub-category, the target vowel is predominantly non-lengthened. However, if we analyse only the cases with lengthening (see Fig. 6), we can observe that:

– The lengthening for *voilà* is more important when used as a DP than in a non-DP function;
– With respect to the strongest lengthening (D+++), *quoi* and *voilà* do not behave in the same way: DP *voilà* is more often markedly lengthened than non-DP *voilà*, contrary to DP *quoi*, which is less often markedly lengthened than non-DP *quoi*.

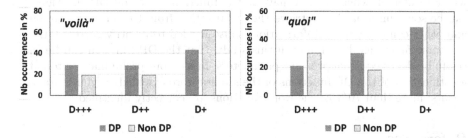

Fig. 6. *quoi* and *voilà*, last syllable vowel lengthening (percentage of the lengthened vowels)

4.6   Automatic DP Identification

An automatic identification of the DP function for the two words under study was carried out. The identification procedure relies only on the prosodic parameters described and analysed in the previous paragraphs.

We used the j48 decision tree [26] in the Weka toolkit [19]. The use of a decision tree is motivated by the adequacy of this technique for data which contain numeric and symbolic values. The decision tree is trained on 60 % of our data while the remaining 40 % is kept for evaluating the classifier (identification of the DP function).

The results obtained by the classifier (see Table 4) are very encouraging. In fact, in more than 70 % of the cases, the DP function is correctly identified using prosodic parameters only. Thus, one can reasonably expect a substantial improvement of these results when more linguistic information (part of speech ...) is introduced into the decision procedure.

Table 4. DP automatic identification scores in percentage

	Correct identification	Precision	Recall	F-Measure
quoi	<u>**71 %**</u>	71 %	71 %	71 %
voilà	<u>**73 %**</u>	73 %	73 %	73 %

5 Conclusion

Our study aims to identify pertinent prosodic parameters associated to two French words (*quoi* and *voilà*) when used as a discourse particle (DP function) or not (i.e., pronoun, preposition, ...).

The identification of the DP function of the words is very important for the automatic processing of speech data. For example, the translation of such words will be different whether they have a DP function or a non-DP function. It can be sometimes hard to retrieve the DP function from the written form only and other complementary information is also needed to identify correctly these discourse functions. It is found in this study, that the DP function can be successfully identified using prosodic parameters. In fact, our preliminary automatic identification of the DP functions of the studies words yielded very encouraging results: more than 70 % of the word functions are correctly identified.

References

1. Aijmer, K.: Understanding Pragmatic Markers: A Variational Pragmatic Approach. Edinburgh University Press, Edinburgh (2013)
2. Bartkova, K., Dargnat, M.: Pardi!: Particules discursives: sens et corrélats prosodiques, ATILF Research Project (2014–2015)
3. Bartkova, K., Delais-Roussarie, E., Santiago, F.: Prosotran: a tool to annotate prosodically non-standard data. In: Proceedings of Speech Prosody 2012, Shangaï. pp. 55–58 (2012)
4. Bartkova, K., Jouvet, D.: Automatic detection of the prosodic structures of speech utterances. In: Železný, M., Habernal, I., Ronzhin, A. (eds.) SPECOM 2013. LNCS, vol. 8113, pp. 1–8. Springer, Heidelberg (2013)
5. Beeching, K.: La co-variation des marqueurs discursifs bon, c'est-à-dire, enfin, hein, quand même, quoi et si vous voulez: une question d'identité ? Langue Française **154**, 78–93 (2007)
6. Brémond, C.: Les "petites marques du discours". Le cas du marqueur métadiscursif bon en français. Ph.D. thesis, Dpt. of Linguistics, Universitéde Provence (2002)
7. Bruxelles, S., Traverso, V.: Usages de la particule 'voilà' dans une réunion de travail: analyse multimodale. In: Drescher, M., Job, B. (eds.) Les marqueurs discursifs dans les langues romanes: approches théoriques et méthodologiques, pp. 71–92. Peter Lang, Bern (2006)
8. Chanet, C.: 1700 occurrences de la particule quoi en français parlé contemporain: approches de la "distribution" et des fonctions en discours. Marges Linguistiques **2**, 52–80 (2001)
9. Dargnat, M.: Marqueurs de discours et interjections. In: Abeillé, A., Godard, D. (eds.) La Grande Grammaire du Français. Actes Sud, Arles (2016), to appear
10. Degand, L., Fagard, B.: Alors between discourse and grammar: the role of syntactic position. Funct. Lang. **18**, 19–56 (2011)
11. Degand, L., Simon, A.: L'analyse des unités discursives de base: pourquoi et comment? Langue Française **170**, 45–59 (2011)
12. Denturck, E.: Étude des marqueurs discursifs, l'exemple de quoi. Master's thesis, Taal-en-Letterkunde, Universiteit Gent (2008)

13. Dostie, G.: Pragmaticalisation Et Marqueurs Discursifs: Analyse Sémantique Et Traitement Lexicographique. De Boeck-Duculot, Bruxelles (2004)
14. Dostie, G., Pusch, C.: Les marqueurs discursifs. présentation. Langue Française **154**, 3–12 (2001)
15. Fernandez-Vest, J.: Les particules énonciatives dans la construction du discours. Presses Universitaires de France, Paris (1994)
16. Fraser, B.: What are discourse markers? J. Pragmatics **37**, 931–952 (1999)
17. Gardent, C.: Syfrap project (syntaxe du français parlé), LORIA-ATILF-LLF Research Project (2011–2013)
18. Grevisse, M., Goose, A.: Le Bon Usage, 14th edn. De Boeck-Duculot, Bruxelles (2008)
19. Hall, M., Frank, E., Holmes, G., Pfahringer, B., Reutemann, P., Witten, I.H.: The weka data mining software: an update. SIGKDD Explorations 11(1) (2009). http://www.cs.waikato.ac.nz/ml/index.html
20. Hirschberg, J., Litman, D.: Empirical studies on desambiguisation of cue phrases. Comput. Linguist. **19**, 501–530 (1993)
21. Horne, M.P., Hansson, P., Bruce, G., Frid, J., Filipsson, M.: Cue words and the topic structure of spoken discourse: the case of Swedish men. J. Pragmatics **33**, 1061–1081 (2001)
22. Lefeubvre, F., Morel, M.A., Teston-Bonnard, S.: Valeurs prototypiques de quoi à travers ses usages en français oral, unpublished MS., Archives ouvertes HAL-SHS (2013). https://halshs.archives-ouvertes.fr/halshs-00840728
23. Martin, P.: Quoi peut être dedans ou dehors du noyau macro-syntaxique, quoi. Unpublished (2014), handout Journée Pardi!, ATILF, Nancy, 18–19 December 2014
24. Mosegaard Hansen, M.B.: The Function of Discourse Particles. A study with Special reference to Spoken Standard French. Benjamins, Amsterdam (1998)
25. Paillard, D.: Les mots du discours comme mots de langu. Le Gré des Langues **14**, 10–41 (1998)
26. Quinlan, J.R.: Programs for Machine Learning. Morgan Kaufmann Publishers, San Mateo (1993)
27. Shiffrin, D.: Discourse markers: Language, meaning, and context. In: Shiffrin, D., Tannen, D., Hamilton, H.E. (eds.) The Handbook of Discourse Analysis, pp. 54–75. Blackwell, Oxford (2001)
28. Sperber, D., Wilson, D.: Relevance: Communication and Cognition. Blackwell, Oxford (1986)
29. Teston-Bonnard, S.: Propriétés topologiques et distributionnelles des constituants non régis. Ph.D. thesis, Dpt. of Linguistics, Université de Provence (2006)
30. Vincent, D.: Les ponctuants de la langue et autres mots du discours. Nuits Blanches, Québec (1993)
31. Wichmann, A., Simon-Vandenbergen, A.A., Aijmer, K.: How prosody reflects semantic change: a synchronic case study of course. In: Davidse, K., Vandelanotte, L., Cuyckens, H. (eds.) Subjectification, Intersubjectification and Grammaticalization, pp. 103–154. Mouton de Gruyter, Berlin (2010)

Effects of Evolutionary Linguistics
in Text Classification

Julia Efremova[1]([✉]), Alejandro Montes García[1], Jianpeng Zhang[1],
and Toon Calders[1,2]

[1] Eindhoven University of Technology, Eindhoven, The Netherlands
{i.efremova,a.montes.garcia,j.zhang.4}@tue.nl
[2] Université Libre de Bruxelles, Brussels, Belgium
toon.calders@ulb.ac.be

Abstract. We perform an empirical study to explore the role of evolutionary linguistics on the text classification problem. We conduct experiments on a real-world collection with more than 100.000 Dutch historical notary acts. The document collection spans over six centuries. During such a large time period some lexical terms modified significantly. Person names, professions and other information changed over time as well. Standard text classification techniques which ignore temporal information of the documents might not produce the most optimal results in our case. Therefore, we analyse the temporal aspects of the corpus. We explore the effect of training and testing the model on different time periods. We use time periods that correspond to the main historical events and also apply clustering techniques in order to create time periods in a data driven way. All experiments show a strong time-dependency of our corpus. Exploiting this dependence, we extend standard classification techniques by combining different models trained on particular time periods and achieve overall accuracy above 90 % and macro-average indicators above 63 %.

1 Introduction

Text classification is a popular machine learning problem with many applications, such as: classification of news into groups, classification of fairy tales according to their genres, filtering emails into spam and not, mining opinions... [1,9].

Research on text classification has mainly focused on topic identification, keywords extraction, sparsity reduction and ignored the aspects of language evolution across different time periods. Research that investigates temporal characteristics of documents and their role in text classification has been scarce so far.

Evolutionary linguistics (EL) is the study of the origin and evolution of languages. It plays an important role in text classification. As a result of the evolution of language vocabulary changes. New words appear and other become outdated. Person names also vary. As an example, more than 100 variants of the first name *Jan* are known, (e. g. *Johan, Johannes, Janis...*) [7]. These modifications change the characteristics of the text, the weights of terms and therefore can

© Springer International Publishing Switzerland 2015
A.-H. Dediu et al. (Eds.): SLSP 2015, LNAI 9449, pp. 50–61, 2015.
DOI: 10.1007/978-3-319-25789-1_6

affect the classification results. Standard classification methods do not consider a time period of the documents to which they belong to [8,16]. They typically use supervised machine learning methods and compute weights of words in the collection.

In this paper, we investigate the role of EL from various perspectives. We make an extensive empirical study to identify robustness of a classifier in the case when the training and test data belong to different time periods. We analyse an impact of EL on the class distribution, correlation between term frequency and time periods, change in the vocabulary across several time periods. In the next part we design a simple framework that enhances existing techniques. The framework incorporates EL aspects and trains the model on relevant examples.

We carry out our experiments on the collection of Dutch historical notary acts from the 15th to the 20th century. Available data spans large time period and we analyse temporal aspects under the context of historical events. To identify main time periods in the history of the Netherlands we consider a time-line proposed by the Rijksmuseum in Amsterdam and split the data into several time periods. Moreover, we identify optimal time periods in a date-driven way and apply year clustering. We present results that show strong term-category-time period dependencies.

The contribution of this paper is summarised as follows:

1. We make an empirical study of the aspects of evolutionary linguistics applied to a large collection of historical data.
2. We develop a framework that incorporates temporal dependencies and, as a result, improve the quality of text classification.

2   Related Work

There is some work available regarding time aspects and empirical studies in text classification. Mourao et al. present an empirical analysis of temporal data on news classification [11]. The impact of empirical methods in information extraction is described by Cardie in [3]. Salles et al. analyse the impact of temporal effects in automatic document classification [14]. Dalli and Wilks [4] investigate the opposite research question. They predict the date of document using the distribution of word frequencies over time. Mihalcea and Nastase [10] identify time period by analysing changes in word usage over time. They make word disambiguation in order to predict time period. In our work we predict a category of a document and assume that a certain category is more common in a certain time period.

The main contribution of our work as compared to previous efforts can be summarise as follow: obtaining an insight of the role of EL in document classification, an empirical study of temporal characteristics and the improvement of classifier performance by integrating temporal information.

3 Data Description and General Approach

We use a dataset of historical notary acts provided by Brabant Historical Information Center[1]. The documents are available in Dutch. Archive notary acts correspond to a wide rage of topics such as sale agreements, inheritance acts, resolutions, etc. It was identified 88 different categories in total.

Volunteers manually digitised notary acts and to some of them assigned a single category that briefly describes document type. However a large number of documents still has to be classified. The overall collection consists of $234,325$ notary acts and $115,673$ of them contain an assigned category and also a specified date of the document.

Another important characteristic is that the dataset is not balanced regarding the number of documents per each category. The largest categories are *transport (property transfer), verkoop (sale agreement)* and *testament (inheritance act)* and they contain around 20 %, 15 % and 11 % of classified documents respectively. However there are a lot of very small categories that have a support value less than 1 %.

We start by pre-processing the documents and remove punctuation marks and non-alphabetical symbols. In the next step we transform the text to the lower case. AFter that, we create a special feature for each remaining token and apply *term frequency inverse document frequency* (TF-IDF) [8] to compute feature vector. The output of the feature extraction step is a feature set with numerical values. Initially we try a number of classifiers to predict a category of the documents and continue to use the one that has the highest performance on our dataset. We use classifiers from the scikit-learn python tool[2] [13].

4 Empirical Study

In this section, we describe time partitions and show an impact of the training set on the classifier accuracy. Afterwards, we present an analysis of different factors such as the sampling effect within the given time frame, category distribution over time, time-specific vocabulary and correlation between term frequency and time periods.

Identifying Time Frames. We split a large time period into smaller pieces. We define a set of time partitions as $\mathcal{T}$. Each $\mathcal{T}_i$ is described by two time points t_i and t_{i+1} that are the beginning and the ending of a time frame respectively. A document $\mathcal{D}_i$ belongs to the $\mathcal{T}_i$ when $t_i \leq date(\mathcal{D}_i) \leq t_{i+1}$. First, we consider major historical events and follow the time line proposed by the Rijksmuseum[3], later we present an approach to obtain optimal time periods in a data-driven way. We identify seven major periods in Dutch history presented in Table 1.

We do not consider periods after 1918 since they are relatively recent and notary acts are not publicly available yet.

---

[1] http://www.bhic.nl/.

[2] http://scikit-learn.org/.

[3] http://goo.gl/YZvP9q.

Table 1. The timeline of Dutch history obtained from Rijksmuseum.

1433 - 1565	Burgundian and Habsburg period	1
1566 - 1649	Revolt and the Rise of the Dutch Republic	2
1650 - 1715	Republic at war with its neighbours	3
1716 - 1779	Dutch republic	4
1780 - 1810	Patriots, Batavian Republic and the French	5
1811 - 1848	Kingdom of the Netherlands	6
1849 - 1918	Modernisation	7

Size of the Training Set. We also illustrate the impact of the size of the training set on the classifier accuracy. It allows to clarify the effect of the training size and to distinguish it from the time effect. The difference in a number of training examples in each time period affects the classifier accuracy. We use a number of classifiers, namely: *Support Vector Machines* classifier with a linear basis kernel function, *multinomial naive Bayes, nearest centroid, passive aggressive, perceptron, ridge regression* and *stochastic gradient descent* [15]. We divide data into fixed subsets (training and test) and vary a size of the training data from 1.000 to 20.000 examples. Figure 1(a) demonstrates a clear dependency between the overall classifier accuracy and the number of training examples. The more training examples we use, the better accuracy a classifier achieves. We compared a

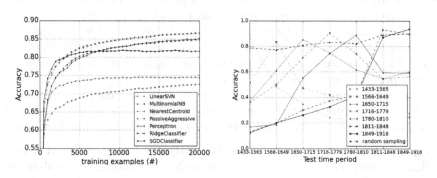

(a) Classification accuracy as a function of a training set size.

(b) Accuracy results as a function of different training and test time periods.

Fig. 1. Analysis of classification accuracy. In the case (a) each line on the graph represents applied classifiers such as: Support Vector Machines, multinomial naive Bayes, nearest centroid, passive aggressive, perceptron, ridge regression and stochastic gradient descent. In the case (b) the lines in the graph indicate the performance of SVM classifiers trained on one specific time period, applied on all the different time periods. It can be seen clearly that a classifier trained on period T_i when tested on period T_i (cross-validated performance figures) outperforms all other classifiers

number of classifiers and a SVM constantly showed the highest performance. Therefore we choose the SVM as a classifier for further experiments.

Sampling Effect within Given Time Frame. To demonstrate the effect of sampling within a particular time period on the text classification results we associate each document $d_i \in \mathcal{D}$ to the appropriate time period $\mathcal{T}_i$. The number of documents in each period varies significantly. The total number of unique categories in every $\mathcal{T}_i$ is also different. Table 2 shows the statistics about splitting the dataset into time periods.

Table 2. The number of documents and categories in each time period

	1433-1565	1566-1649	1650-1715	1716-1779	1780-1810	1811-1848	1849-1918
Categories	45	46	70	78	75	52	34
Documents	6166	3594	11550	25914	17301	26087	25538

The idea of this experiment is to construct training and test sets using documents from non-overlapping time frames. More specifically, we use the partition $\mathcal{T}_i$ to train a classifier and test it consequentially on all $\mathcal{T}_j$ with $i \neq j$. We evaluate a change in the classification results when the average time difference between training and test documents gradually increases.

We divide the data collection into partitions according to the identified timeline. Then we train a classifier on one partition and evaluate on all the others. When training and test belong to the same time period we apply 10-fold cross validation. Figure 1(b) demonstrates the overall performance accuracy. Each division on the X axis on the plot represents a fixed time period which was used for training a classifier and dots show results on test sets from different time periods. Clearly we see that all the peaks on the graph occur when training and test time partitions are equal $\mathcal{T}_{train} = \mathcal{T}_{test}$.

In order to compare our results we used random sampling to construct a training set. To avoid the fact that classifiers are sensitive to the number of training examples, we randomly select from every $\mathcal{T}_i$ equal number of documents.

Category Distribution over Time. We analyse category distribution over the time. Figure 2 represents a percentage of each type of documents in different time periods. We denote as *other* the categories that are not in the list of top-10 the most frequent categories. We see that the proportion of other categories gradually decreases over time leaving space for the top-10 categories.

Dealing with a large number of small categories requires additional efforts. They do not always have a sufficient number of training examples and can easily be confused with larger neighbours. In our previous work [6] we clustered categories according to their frequencies and identified small categories in two steps. In this work we analyse how time segmentation affects the classification results for both large and small categories.

Category distributions also confirm the existence of time dependencies in a dataset.

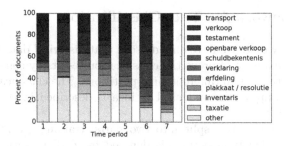

Fig. 2. The distribution of top 10 categories in each time period

Temporal Term Weighting. In this section we test if the occurrence of the term is independent of the time period. To do so, we use χ^2 statistic [12]. We compute χ^2 for each unique term in the corpus. We do not consider terms which occur less than 0.5 % in a collection.

Table 3 shows a number of terms and their corresponding p-value across the overall collection (not the balanced subset). The larger the p-value is, the more terms meet these requirements. The probability of 0.1 is the maximum bound when the observed deviation from the null hypothesis is significant.

Table 3. Time period analysis. Number of terms bounded by p-value

p-value	0.25	0.2	0.15	0.1	0.05	0.025	0.02	0.01	0.005	0.0025	0.001	0.0005
Number of terms	837	713	598	471	306	212	187	136	98	77	57	44

From that data it is possible to identify time-dependent named-entity groups of words such as: general terms, person names, professions. Table 4 shows p-values of time-dependent variations of the person name *Hendrik*, typical professions, their absolute frequencies and p-values.

Table 4. Example of time dependent names and professions and their p-values

Word	p-value	Freq	Word	Translation	p-value	Freq
Henrick	0.0002	4821	Arbeider	*Worker*	0.0000	2404
Hendricx	0.0003	1123	Bouwmans	*Builders*	0.0000	557
Henricks	0.0254	1023	Raaijmaker	*Wheelmaker*	0.0147	636
Hendricus	0.0488	3848	Biggelaar	-	0.0102	1071

We see that the official form of the name *Hendrik* has time-dependent variations such as: *Henrick, Hendricx, Henricks, Hendricus*, etc.

5   EL-Framework for Text Classification

General Framework. We have already seen that historical data contains time dependencies. We aim to improve the classification results by combining several models that are trained on different time partitions. The classification task can be done by any appropriate model. The idea is described in the pseudo-code from Algorithm 1.

The original data set is split into two parts: training set $\mathcal{D}$ and test set $\mathcal{R}$. A set of identified time periods is denoted by $\mathcal{T}$. For every $\mathcal{T}_i$ in $\mathcal{T}$ (line 1) the algorithm constructs corresponding subsets $\mathcal{D}^{'} \in \mathcal{D}$ such that $\{d_i \in \mathcal{D}^{'} | date(d_i) \in \mathcal{T}_i\}$ with the corresponding target categories $\mathcal{C}^{'} \in \mathcal{C}$ such that $\{d_i \in \mathcal{D}^{'}, c_i \in \mathcal{C}^{'} | category(d_i) = c_i$ and $\mathcal{R}^{'} \in \mathcal{R}$ such that $\{r_i \in \mathcal{R}^{'} | date(r_i) \in \mathcal{T}_i\}$ (lines 2-4). On the next step (line 5) we learn a prediction model $\mathcal{M}_i$ for the time partition $\mathcal{T}_i$ on the identified training subset: $(\mathcal{D}^{'}, \mathcal{C}^{'})$ that has only the documents from partition $\mathcal{T}_i$. We use a model M_i to predict a category only for the documents from the same time partition t_i (line 6). As a result we have a number of models $\{\mathcal{M}_1, \ldots \mathcal{M}_n\}$, one model for each time period. We choose a model depending on the date of a document that we need to classify.

Input: Training set $\mathcal{D} = \{d_1, ..., d_n\}$ with category-labels $\{c_1, ..., c_k\}$.
 Test set $\mathcal{R} = \{r_1, ..., r_h\}$.
 Set of categories $\mathcal{C} = \{c_1, ..., c_k\}$ and set of time periods $\mathcal{T}$.
Output: Predicted labels $\mathcal{N}$ for all test instances $\mathcal{R}$
1: **for** each time period $\mathcal{T}_i$ in $\mathcal{T}$ **do**
2:   $\mathcal{D}^{'} \in \mathcal{D}$: $\{d_i \in \mathcal{D}^{'} | date(d_i) \in \mathcal{T}_i\}$
3:   $\mathcal{C}^{'} \in \mathcal{C}$: $\{c_i \in \mathcal{C}^{'}, d_i \in \mathcal{D}^{'} | category(d_i) = c_i\}$
4:   $\mathcal{R}^{'} \in \mathcal{R}$: $\{r_i \in \mathcal{R}^{'} | date(r_i) \in \mathcal{T}_i\}$
5:   $\mathcal{M}_i \leftarrow TrainModel(\mathcal{D}^{'}, \mathcal{C}^{'})$ # Learn a model on a specific time period
6:   $\mathcal{N}_i \leftarrow Classify(\mathcal{R}^{'}, \mathcal{M}_i)$ # Classify data
7:   $\mathcal{N} \leftarrow \mathcal{N} \cup \mathcal{N}_i$
8: **end for**
9: **return** $\mathcal{N}$

Algorithm 1. EL-framework

Optimal Time Frame Identification. One of the benefits of the described approach is that it can be used as a framework of any text classification algorithm. It requires already predefined time periods. In Sect. 4 we identified time periods based on historical events. However, historical events give arbitrary time periods and may correspond to linguistic changes in the language. Another approach is to cluster years.

In the first step we merge all of the documents from the same year together. As a result we have one large single document per each year. Then we construct a feature vector using the TF-IDF vectorizer as described in Sect. 3. TF-IDF feature extraction is more appropriate for this task than term-frequency because

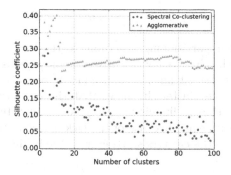

Fig. 3. Silhouette coefficient

it assigns a higher weight to infrequent words. After that we apply clustering. In this paper we compare two clustering techniques: *Spectral Co-clustering algorithm* [5] and *Agglomerative Hierarchical Cluster* [17]. Before apply clustering technique we remove from the original dataset all numbers, years, category names and non-alphabetical characters. This step is necessary in order to avoid biases in clustering.

6   Experiments and Results

We evaluate the designed EL-framework with Rijksmuseum time division and EL-framework with time frames according to year clustering and compared the results to two baselines. As the first baseline, we use the standard text classification method and as the second baseline, we use a sliding window (+/- decades). In the second case a classifier is trained on a decade before and a decade after a classifying year. We apply 10-fold cross-validation when training and testing examples belong to the same data partition. We evaluated the performance of the applied algorithms in standard metrics such as: overall accuracy and the macro-average indicators (precision, recall and f-score).

Cluster Evaluation. In order to evaluate the year clustering technique and find an appropriate number of clusters we compute the Silhouette coefficient [2] and vary the number of clusters from 2 to 100. The Silhouette coefficient is an unsupervised clustering evaluation measure when the ground truth labels are unknown. The higher the Silhouette coefficient is, the better clusters are found. Figure 3 shows the Silhouette coefficient of Spectral Co-clustering and Agglomerative Hierarchical Cluster for different number of clusters. We use *cosine similarity* to calculate the intra-cluster distance and nearest-cluster distance for each sample. We see that the Silhouette coefficient achieves the maximum value when the number of clusters $k = 5$ for Spectral Co-clustering and the number of clusters $k = 10$ for Agglomerative Hierarchical Clustering.

We consider the number of clusters $\{5, 7, 10\}$ for experiments. Number of clusters equals to 5 or 10 yields the maximum Silhouette coefficient, number of clusters equal to 7 corresponds to the number of identified main historical events.

Figure 4 shows year partitioning according to the two described clustering approaches. Most of the clusters have relatively homogeneous structure without forcing temporal cluster constrains. Years from early periods and recent periods never occur in the same cluster. It shows that the clustering is not random and confirms the existence of linguistic changes over time. In early periods the structure is less homogeneous, because of the lack of documentation standards. However, we clearly see the clusters starting from the beginning of 18th century and from 1811 onwards.

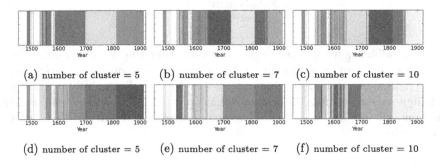

(a) number of cluster = 5   (b) number of cluster = 7   (c) number of cluster = 10

(d) number of cluster = 5   (e) number of cluster = 7   (f) number of cluster = 10

Fig. 4. Comparison of different year clusters: (a)-(c) after applying the Spectral Co-clustering algorithm [5], (d)-(f) after applying Agglomerative Hierarchical Cluster algorithm. The white space on the graph indicates that there are no documents in some years.

Cluster Analysis. We apply the χ^2 statistic to analyse the reasons of cluster homogeneity. All of the visualised clusters are similar, therefore we use in this experiment Spectral Co-clustering with the number of clusters equal to 7. Table 5 presents the number of terms and their corresponding p-value.

The number of terms that are cluster dependent is much higher than the number of terms that are dependent on arbitrary time partitioning, compare Table 5 with Table 3 respectively. It means that the current time partitioning is more optimal. There are different groups of terms with low p-values. Among of them occur general terms including verbs, professions, names, etc. For instance, words such as: *gulden (guilder), schepenen (alderman), beroep (profession), pastoor (pastor), burgemeester (mayor), goederen (goeds), verklaren (to declare)* have a large correlation with the clusters.

6.1 EL-Framework Evaluation

We compare the results of EL framework with two other approaches: standard text classification method and a sliding window decade based approach (see

Table 5. Cluster analysis. Number of terms bounded by p-value

p-value	0.25	0.2	0.15	0.1	0.05	0.025	0.02	0.01	0.005	0.0025	0.001	0.0005
Number of terms	9083	8619	8079	7476	6698	6144	6000	5581	5248	4961	4631	4423

Table 6). The EL-framework demonstrates an improvement in the main evaluation metrics. Overall, the classification accuracy increases almost 1 %, the three macro-average indicators (precision, recall and f-score) increase up to 2 %. The standard approach and sliding window that we take as a baseline already produces very high results, that is why it is very difficult to achieve contrasting difference. Improving the results from 90 % to 91 % means that we remove 10 % of the errors. It is easier to improve 10 % if the performance is only, for instance, around 40 % than when the performance is already 90 %. The EL-framework achieves the maximum improvements using Spectral Co-clustering for year partitioning with the number of clusters equal to 7. We exclude years and class labels to make clustering.

We see a large difference in the performance between overall accuracy and macro-average indicators in all experiments. The original dataset is not balanced: 20 % of the data belongs to the largest category and there are several very small categories that do not have enough examples for training the classifier.

Table 6. Overall accuracy and macro average indicators. TC stands for text classification

	Overall accuracy	Precision	Recall	f-score
Baseline 1: Standard TC	90.01 %	73.93 %	54.84 %	0.6297
Baseline 2: Sliding window	89.53 %	74.89 %	53.64 %	0.6238
EL + Spectral, $k = 5$	90.67 %	**75.87%**	55.90 %	0.6437
EL + Spectral, $k = 7$	**90.83%**	74.45 %	55.71 %	0.6373
EL + Spectral, $k = 10$	90.69 %	74.62 %	55.43 %	0.6361
EL + Aggl., $k = 5$	90.67 %	75.83 %	55.83 %	0.6431
EL + Aggl., $k = 7$	90.65 %	75.65 %	**56.03%**	**0.6438**
EL + Aggl., $k = 10$	90.59 %	75.80 %	55.81 %	0.6429

We also evaluate the performance of applied techniques per every as an average per century as it shown on Fig. 5. The standard text classification uses more training examples, however it never achieves the maximum performance compared to the EL-framework with an optimal time partitioning. The difference in performance between the EL-framework and the standard technique is positive in many centuries but the former depends on the selected time partitioning strategy.

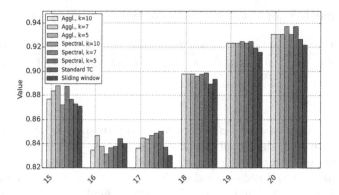

Fig. 5. Overall accuracy averaged per century that corresponds to EL-framework with different time partitioning and standard text classification.

The number of documents per year also varies a lot and the number of documents in some periods is less than in others. In many cases the available amount of training data is sufficient to make a high quality prediction within a time period. However we leave the identification of optimal size constrained time periods to future work.

7   Conclusions and Future Work

In this paper, we demonstrated temporal characteristics of the data applied to a collection of historical notary acts. We analysed dependency between time periods and correlated terms, class distributions and sampling effect. Then we designed a framework to incorporate temporal dependencies into the overall text classification process. We used main historical events to determine time periods. Moreover, we applied clustering techniques in order to obtain optimal time partitions automatically. This is a novel view of the text classification problem which demonstrated improvements in the results.

The presented empirical study of the temporal data aspects and the designed EL-framework make a significant contribution into the text classification area.

Acknowledgments. Mining Social Structures from Genealogical Data (project no. 640.005.003) project, part of the CATCH program funded by the Netherlands Organization for Scientific Research (NWO).

References

1. Almeida, T.A., Almeida, J., Yamakami, A.: Spam filtering: how the dimensionality reduction affects the accuracy of naive bayes classifiers. J. Internet Serv. Appl. **1**(3), 183–200 (2011)

2. Aranganayagi, S., Thangavel, K.: Clustering categorical data using silhouette coefficient as a relocating measure. In: Proceedings of the International Conference on Computational Intelligence and Multimedia Applications (ICCIMA 2007), vol. 2 pp. 13–17, IEEE Computer Society, Washington, DC (2007)

3. Cardie, C.: Empirical methods in information extraction. AI Mag. **18**, 65–79 (1997)

4. Dalli, A., Wilks, Y.: Automatic dating of documents and temporal text classification. In: Proceedings of the Workshop on Annotating and Reasoning About Time and Events. ARTE 2006, pp. 17–22. Association for Computational Linguistics (2006)

5. Dhillon, I.S.: Co-clustering documents and words using bipartite spectral graph partitioning. In: Proceedings of the Seventh ACM SIGKDD International Conference on Knowledge Discovery and Data Mining. KDD 2001, pp. 269–274. ACM (2001)

6. Efremova, J., Montes García, A., Calders, T.: Classification of historical notary acts with noisy labels. In: Hanbury, A., Kazai, G., Rauber, A., Fuhr, N. (eds.) ECIR 2015. LNCS, vol. 9022, pp. 49–54. Springer, Heidelberg (2015)

7. Efremova, J., Ranjbar-Sahraei, B., Calders, T.: A hybrid disambiguation measure for inaccurate cultural heritage data. In: The 8th Workshop on LaTeCH, pp. 47–55 (2014)

8. Ikonomakis, M., Kotsiantis, S., Tampakas, V.: Text classification using machine learning techniques (2005)

9. Leong, C.K., Lee, Y.H., Mak, W.K.: Mining sentiments in sms texts for teaching evaluation. Expert Syst. Appl. **39**(3), 2584–2589 (2012)

10. Mihalcea, R., Nastase, V.: Word epoch disambiguation: finding how words change over time. In: ACL (2), pp. 259–263. The Association for Computer Linguistics (2012)

11. Mourão, F., Rocha, L., Araújo, R., Couto, T., Gonçalves, M., Meira Jr., W.: Understanding temporal aspects in document classification. In: Proceedings of the 2008 International Conference on Web Search and Data Mining. WSDM 2008, pp. 159–170. ACM, USA (2008)

12. Pearson, K.: On the criterion that a given system of deviations from the probable in the case of a correlated system of variables is such that can be reasonably supposed to have arisen from random sampling. Phil. Mag. **50**, 157–175 (1900)

13. Pedregosa, F., et al.: Scikit-learn: machine learning in python. J. Mach. Learn. Res. **12**, 2825–2830 (2011)

14. Salles, T., da Rocha, L.C., Mourão, F., Pappa, G.L., Cunha, L., Gonçalves Jr, M.A., Wrigley Jr, W.: Automatic document classification temporally robust. JIDM **1**(2), 199–212 (2010)

15. Sammut, C., Webb, G.I.: Encyclopedia of Machine Learning. Springer, Berlin Heidelberg (2010)

16. Sebastiani, F.: Machine learning in automated text categorization. ACM Comput. Surv. **34**(1), 1–47 (2002)

17. Zhao, Y., Karypis, G., Fayyad, U.: Hierarchical clustering algorithms for document datasets. Data Min. Knowl. Discov. **10**, 141–168 (2005)

Evaluation of the Impact of Corpus Phonetic Alignment on the HMM-Based Speech Synthesis Quality

Marc Evrard[✉], Albert Rilliard, and Christophe d'Alessandro

LIMSI-CNRS, Audio & Acoustic Group, B.P. 133, 91403 Orsay Cedex, France
{marc.evrard,albert.rilliard,cda}@limsi.fr
http://www.limsi.fr

Abstract. This study investigates the impact of phonetization and phonetic segmentation of training corpora on the quality of HMM-based TTS synthesis. HMM-TTS requires phonetic symbols aligned to the speech corpus in order to train the models used for synthesis. Phonetic annotation is a complex task, since pronunciation usually differs from spelling, as well as differing among regional accents. In this paper, the infrastructure of a French TTS system is presented. A corpus whose phonetic label occurrences were systematically modified (number of schwas and liaisons) and label boundaries were displaced, was used to train several systems, one for each condition. A perceptual evaluation of the influence of labeling accuracy on synthetic speech quality was conducted. Despite the degree of annotation changes, the synthetic speech quality of the five best systems remained close to that of the reference system, built upon the corpus whose labels were manually corrected.

Keywords: HTS · HMM-based speech synthesis · TTS · Subjective evaluation · MOS · Phonetic labeling · Phonetic alignment · French speech synthesis

1 Introduction

HMM-based speech synthesis models are trained on speech corpora consisting generally of sentences read by a speaker, and annotated with phonetic labels, as well as other linguistic features. The process of annotating a corpus starts with the grapheme-to-phoneme (GP) conversion, which is a complex problem. State-of-the-art systems are still imperfect for most languages [5]. Also, the GP conversion is a deterministic process, while the speaker phoneme realization is not. This is particularly true for the schwa, which is not realized systematically the same way by different speakers, in different situations. In French, there are also liaisons between words, whose realizations particularly vary between speakers [12].

After each utterance of the corpus is converted to phonemes, the alignment process segments the speech utterances and assigns each phoneme label to its corresponding speech segment.

© Springer International Publishing Switzerland 2015
A.-H. Dediu et al. (Eds.): SLSP 2015, LNAI 9449, pp. 62–72, 2015.
DOI: 10.1007/978-3-319-25789-1_7

Various errors can thus arise from the corpus annotating process: a phonetic label — actually performed in the speech corpus — may be missing, a superfluous phonetic label may be added, or a phonetic label may be misplaced.

During the synthesis process in HMM-based speech synthesis systems, the correspondence between label units and speech units is not direct. The speech acoustic parameters are indeed produced by statistic models that are learned from the training corpora. Thus, it is hard to determine the extent that these systems are sensitive to the corpus annotation error.

In this paper, the sensitivity of the annotation errors is tested. Different text-to-speech (TTS) systems are built, using the same speech corpus, with various altered annotations. Some systems have the number of schwa and liaison realizations artificially varied, and in others, the label alignment is ruffled. A set of sentences will be synthesized with the different systems and a subjective evaluation will be taken by subjects to assess the quality differences between these systems.

Section 2 presents the platform and the corpus used for these experiments. In Sect. 3, the conducted experiment based on labeling variations will be presented, and Sect. 4 describes the subjective evaluation method and its results. In Sect. 5, the observations will be discussed before ending by a conclusion in Sect. 6.

2   Description of the Synthesis Platform LIPS[3]

The LIMSI Parametric Speech Synthesis System (LIPS[3]) is a TTS platform built around the HMM-based speech synthesis system (HTS) [10]. The language processing modules, used at training and synthesis stages, were developed *in situ* for the French language.

2.1   HTS

The standard HTS implementation with the Speech Signal Processing Toolkit (SPTK) [3] extraction tools and vocoder were used. The versions were hts 2.3alpha with two patches applied, SPTK 3.7 and hts_engine 1.08.

The vocoder used for the main systems was the HTS default from hts_engine 1.08. It is a basic excitation-filter model, with an impulse excitation and a Mel Generalized Cepstral analysis (MGC) [9] to model the filter. In this application the MGC analysis was configured to use a pure Mel Cepstrum Envelope (MCEP) for the analysis ($\gamma = 0$), and a Mel Log Spectrum Approximation (MLSA) [4] for synthesis.

The default configuration of the HTS demo training script (for 48 kHz audio files) was used, except for the fundamental frequency (f_0) range, which was set to 110 – 500 Hz for the acoustic feature extraction.

2.2   The Language Processing Module

The GP conversion was developed on a core set of rules previously created at LIMSI and evaluated in [13]. It was exclusively rule based and consisted of seven stages: sentence chunking, normalization, basic part of speech (POS) tagging, standard phonetization, peculiar rules application, liaisons management and syllabation. The system was developed in C++ and GNU Flex. The linguistic features were extracted using an object-oriented Python library developed *in situ*.

2.3   The Corpus

The text corpus was designed and recorded by an industrial partner in a collaborative project. It consists of:

- 1402 sentences
- 10 313 words
- 15 552 syllables
- 36 362 phonemes

These 1402 sentences were used to record the main corpus. The speaker was a professional actress, first-language speaker of standard Parisian French. The recording took place in a soundproof vocal booth, using a condenser microphone with an omnidirectional polar pattern, at a close distance. It was performed at 48 kHz 24-bit depth, then normalized with 6 dB headroom and converted to 16-bit depth PCM files. Sentences were read using a neutral declarative style.

The corpus was phonetically labeled using the Ergodic hidden Markov models (EHMM) tool [7] from the Festvox tool suite [1], and manual corrections of phonetic labels were applied.

3   Description of the Experiment: Labeling Variation

In order to explore the robustness of HMM-based speech synthesis to imperfect matching of corpus labeling to speaker performance, different TTS systems were built using the same speech corpus, and varying its phonetic alignment. The manually corrected labels were taken as reference and systematic modifications were applied.

Two types of variations were tested with the labeled corpus:

1. the addition and suppression of phonemes
2. the shifting of the label boundaries

3.1   Label Addition and Suppression

Two typical sources of phonetic variation in French were investigated: the liaison and the schwa realizations. The GP rules were modified to artificially increase and decrease the realization of liaisons and schwas among the corpus labels.

Liaison Variations. The reference is again the corpus whose labels were manually corrected. The numbers of liaison changes are 162 additions and 671 deletions, for the increased and the reduced liaison corpora, respectively (an

Table 1. Number of liaison changes for the reduced liaison corpus on the left column, and for the increased liaison corpus on the right column.

Liaison	Suppression	Addition
/z/	303	117
/t/	227	44
/n/	131	1
/p/	10	0
Total	671	162
Ratio	1.85 %	0.45 %

increase of 0.45 % in the total number of phonemes, and a decrease of about 1.85 %). The changes affected mainly the /z/ and the /t/ liaisons, which are the most common ones in French. The number of changes by liaison type are detailed on Table 1.

Example of liaison realization variations with the sentence: "*Puis il remit avec orgueil son mouchoir dans sa poche.*" (which could be translated in English by "Then he proudly put his handkerchief in his pocket."):

Reference   /pɥi zil rəmi avɛk ɔʁgœj sɔ̃ muʃwar dɑ̃ sa pɔʃ/
Liaisons added   /pɥi zil rəmi tavɛk ɔʁgœj sɔ̃ muʃwar dɑ̃ sa pɔʃ/
Liaisons removed   /pɥi il rəmi avɛk ɔʁgœj sɔ̃ muʃwar dɑ̃ sa pɔʃ/

Note that, in these phonetized sentences, the words were separated with white spaces for clarity. Also, the liaisons were placed at the beginning of the next word by convention.

Schwa Variations. The induced changes are mainly aimed toward the realization of schwas at the end of words. Compared to the reference (the manually labeled corpus), there were 2430 additions and 269 deletions in the two corpora, respectively (an increase of 6.68 % in the total number of phonemes, and a decrease of about 0.74 %). They concerned both the words labeled as "functional word" (513 additions and 42 deletions) and those labeled as "content word" (1917 additions and 227 deletions) by the POS tagger.

Example of ending schwa realization variations with the sentence: "*Vous êtes le peuple souverain.*" (which could be translated in English by "You are the sovereign population."):

Reference   /vu zɛt lə pœplə suvʁɛ̃/
Schwas added   /vu zɛtə lə pœplə suvʁɛ̃/
Schwas removed   /vu zɛt lə pœpl suvʁɛ̃/

3.2 Boundary Shifts

Figure 1 shows an example of a sentence extracted from the corpus with its segmentation on a Praat textgrid [2]. The sentence is "*Nul besoin*

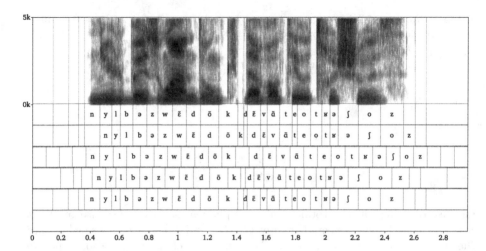

Fig. 1. Set of boundary changes applied to the manually labeled corpus, taken as reference (top and bottom tiers).

donc, d'inventer autre chose.", which is phonetically transcribed as: /nylbəzwɛ̃dɔ̃kdɛ̃vɑ̃teotʁəʃoz/ and could be translated in English by: "Hence, no need to invent something else." The upper part of the figure shows the sentence spectrogram (from 0 to 5 kHz), and below, the tiers show the different phonetic alignments produced. Three types of modifications were introduced. The first (**Tier 1**) and the last (**Tier 5**) hold the manually labeled corpus used as a reference. The other tiers show respectively:

Tier 2: shifting of the labels along the original boundaries to the right (**B2**, cf. Table 3)

Tier 3: all phoneme lengths are chosen identical to equally divide the sentence (Isochronous, **B1**)

Tier 4: iteratively shifting to the right the previous boundary by 50 % of the current interval (**B3**)

The shifting of phonemes to the right (**Tier 2**) is made possible without the need of swapping the first and last phonemes, because each sentence is starting and ending with at least one "silence" label. Since these labels are identical, this process simply adds one of them to the start and removes one from the end of the sentence. Also, as shown in Fig. 1, in this example, multiple "silence" labels appear around the edges of the sentences; these are inherent to the EHMM alignment process.

Figure 2 shows the time difference of the boundary position — expressed in milliseconds — relative to the manually labeled corpus. One can see that the highest value is reached near the end of the sentence for the isochronous segmentation (**B1**).

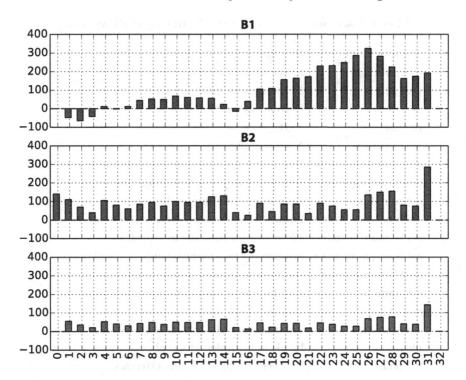

Fig. 2. Boundary position differences (in ms) when compared to the manually labeled corpus for each of the 32 phonemes of the sentence: /nylbəzwẽdɔ̃kdẽvãteotʁəʃoz/. **B1** corresponds to the isochronous segmentation, **B2**, the phonemes shifted to the right, and **B3**, the segmentation shifted 50 % to the right (cf. Table 3 for the complete list).

3.3   TTS System Training

Different TTS systems were built, using the same speech corpus. The altered phonetic alignment presented above were used to train each of the systems respectively.

The list of language features used for training the models were common across all systems. Out of the 53 features for the English Language — proposed by the HTS working group in [11], 19 were selected (cf. Table 2), based on the French language particularities, on the feature subset classification proposed in [6], and according to our own experiments.

A set of sentences (test set) were then synthesized with theses different systems. In order to have at disposal some natural examples of speech from the speaker for the perceptual evaluations, the test set was chosen from the main corpus. Also, to avoid synthesizing the exact same sentences that were used for training, these sentences were removed from the training set. Out of the 1402 of the corpus, 1333 sentences were kept from the training set, and 69 were allocated to the test set.

Table 2. List of used features for the HTS training of all systems.

Nom	Description
prev_prev_ph	Previous-previous phoneme
prev_ph	Previous phoneme
ph	Current phoneme
next_ph	Next phoneme
next_next_ph	Next-next phoneme
phone_from_syl_start	Position of the current phoneme in the syllable
phone_from_syl_end	(ditto counted from the syllable end)
syl_numphones	Number of phonemes in the syllable
syl_from_word_start	Position of the current syllable in the word
syl_from_word_end	(ditto counted from the word end)
syl_from_phrase_start	Position of the current syllable in the phrase
syl_from_phrase_end	(ditto counted from the phrase end)
syl_vowel	Vowel in the current syllable
word_numsyls	Number of syllable in the word
word_accent	Prominence of the current word
phrase_end	Finale punctuation of the phrase
utt_numsyls	Number of syllables in the utterance
utt_numwords	Number of words in the utterance
utt_numphrases	Number of phrases in the utterance

4 Analysis of the Subjective Evaluations

4.1 Conditions of the Test

Systems. Eight systems are tested: M, L−, L+, S−, S+, B1, B2 and B3 (cf. Table 3), along with the natural performance (O) for reference.

Stimuli. A set of ten sentences (nine syllables and six words by sentence on average) were synthesized with each system.

Subjects. A group of 13 subjects (mean age = 27) took the test. They had to rate the overall quality of each sentence on a MOS (mean opinion score) on a 1 to 5 scale (1: bad, 2: low, 3: average, 4: good, 5: excellent) to evaluate the global quality of the stimuli. The sentences were presented to listeners in a random order. One additional sentence, produced in each of the TTS systems and the natural voice, was presented at the beginning of the test, in order to allow the listeners to first assess the quality of the sounds they would have to judge.

Table 3. TTS systems used for the perceptual experiments.

Name	Systems
O	Natural
M	Manually corrected labels
L–	Less liaisons
L+	More liaisons
S–	Less schwas
S+	More schwas
B1	Isochronous segmentation
B2	Phone shifted right
B3	50 % shift

Table 4. Output of the ANOVA performed on MOS scores, for each factor (see text for labels) and their two-way interaction. The table reports the F-statistics for the sum of square total (SST), the factor's and the error's degrees of freedom (df), the associated p level and the effect size (η^2).

Factor	SST	df	df error	F	p	η^2
TTS	827	8	1080	140.86	0.000	0.51
SENT	129	9	1080	19.53	0.000	0.14
TTS*SENT	161	72	1080	3.05	0.000	0.17

4.2 Results

The MOS scores given by all subjects were pooled and analyzed using a two way analysis of variance (ANOVA) based on the fixed factors TTS system ("TTS", 10 levels) and the sentence ("SENT", 10 levels). Results of the ANOVA are presented in Table 4.

The two main factors induced significant changes in the subject's MOS ratings, although the interaction is not significant. The factor "TTS" explains most of the observed variance (cf. the η^2 column [8] in Table 4): the type of phonetization and segmentation is the main source of variation for the perceived quality of the TTS output. A post-hoc Tukey HSD test shows that the 10 systems (including natural speech) are grouped in 5 clusters (ordered in descending mean MOS scores in Table 5).

5 Discussion

Perceptual test results show that the five best TTS systems have a comparable quality. The two best ones (**M** and **L+**) are also perceived as significantly better than the three worst ones. The modifications that lead to the smallest quality loss are the changes in the phonetization output. Conversely, two out of three

Table 5. Tukey HSD test clusters.

Name	TTS system	Mean	Group
O	Natural	4.9692	A
M	Manually corrected labels	2.8615	B
L+	More liaisons	2.8076	B
S−	Less schwas	2.7692	BC
S+	More schwas	2.6538	BC
B1	Isochronous segmentation	2.6385	BC
L−	Less liaisons	2.4692	CD
B2	Phone shifted right	2.1538	D
B3	50 % shift	1.7615	E

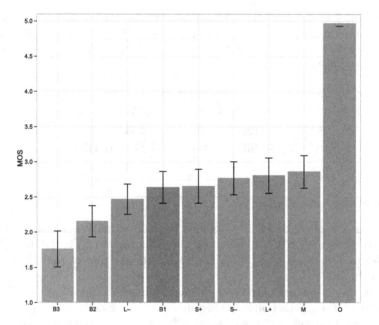

Fig. 3. MOS results obtained by the production systems presented to listeners, and their grouping into significantly different clusters according to a Tukey HSD test.

systems built upon the changes in placement of phoneme boundaries lead to a significant loss of perceived quality (Fig. 3).

Even though the number of changes is certainly an important factor to explain the quality loss, it is not the only one: adding more than 6 % of unperformed schwas in the labels and force-align with the speech signal of the corpus does not lead to a significant quality loss to the systems trained on this corpus. On the contrary, "forgetting" liaisons that are actually performed has a stronger effect on the quality output of the corresponding trained systems. One can assume that superfluous phonemes in the input phonetic string are given a minimal duration by the alignment tool, and thus leads to small quality decrease; on the contrary, performed but unlabeled phonemes will change the modeling of the adjacent phonemic entities, and brings more quality loss.

The perceptual test also concludes, to some degree, the resilience of the learning process, since only two of the three boundary shifts — whose magnitude is strong and whose occurrence is rather unlikely in practice — imposed on the corpus, lead to a perceived quality loss on the corresponding systems. One can note that the "50 % shift" has a more dramatic effect on the quality, than the "phone shift to the right". The fact that the "50 % shift" leads to an alignment solution that maximizes mixes among phoneme labels (all the units are located on phoneme transitions) may explains the higher quality of the "phone shift to the right", which — to some extent — respects the stable parts of phonemes, and leads to better models. The fully "isochronous" (equal duration) segmentation gives a system whose quality is comparable with the reference. This results support the resilience of the learning process, and gives confidence in a fully-automatic approach to corpus labeling, for statistical TTS systems.

6   Conclusion

The perceived quality of systems trained on artificially degraded corpora shows that the HTS system — and by extension HMM-based speech synthesis in general — is fairly robust to their training corpus labeling errors. According to these results, the phonetic alignment precision should not be seen as a priority for the training corpora of HMM-based speech synthesis. Indeed, even an annotation as trivial as using a segmentation in equal part of the sentence only slightly degrades the synthesized speech quality.

The observation of a significant quality degradation linked to phonemes deletion supports the hypothesis of a greater sensitivity of the learning process to missing labeling. This should push GP designers to favor the realization of the phonemes for ambiguous cases (e.g. for French non-mandatory liaisons or schwa). This strategy could be applied to develop TTS systems for sociocultural and regional variants. A GP system maximizing non-mandatory phonetic realization may be a good solution to deal with e.g. non standard schwa realization (typical of southern French), or liaisons deletions that are observed in some sociocultural variants [12].

A next step for the analysis of phonetic variation sensitivity for speech synthesis systems could be to use a fixed text corpus for assessing the synthesized speech quality, according to different phonetic realizations in the audio corpus. This reversed operation would be applicable to the typical condition of expressive speech synthesis using common text for the different expressive corpora.

Acknowledgments. This work is presented in the context of the "AND T-R" project (FUI-11 OSEO/DGCIS) granted by the *région Ile-de-France* and the *conseil général de la Seine-Saint-Denis* and the *ville de Paris*.

References

1. Anumanchipalli, G.K., Prahallad, K., Black, A.W.: Festvox: Tools for creation and analyses of large speech corpora. In: Workshop on Very Large Scale Phonetics Research, UPenn (2011)
2. Boersma, P., van Heuven, V.: Praat, a system for doing phonetics by computer. Glot Int. **5**(9/10), 341–345 (2001)
3. Imai, S., Kobayashi, T., Tokuda, K., Masuko, T., Koishida, K., Sako, S., Zen, H.: Speech Signal Processing Toolkit (SPTK) (2009)
4. Imai, S.: Cepstral analysis synthesis on the mel frequency scale. In: IEEE International Conference on Acoustics, Speech, and Signal Processing, ICASSP 1983, **8**, pp. 93–96. IEEE (1983)
5. Jouvet, D., Fohr, D., Illina, I.: Evaluating grapheme-to-phoneme converters in automatic speech recognition context. In: 2012 IEEE International Conference on Acoustics, Speech and Signal Processing (ICASSP), pp. 4821–4824. IEEE (2012)
6. Le Maguer, S., Barbot, N., Boeffard, O.: Evaluation of contextual descriptors for HMM-based speech synthesis in french. In: 8th International Speech Communication Association (ISCA) Speech Synthesis Workshop, pp. 153–158 (2013)
7. Prahallad, K., Black, A.W., Mosur, R.: Sub-phonetic modeling for capturing pronunciation variations for conversational speech synthesis. In: Proceedings of 2006 IEEE International Conference on Acoustics, Speech and Signal Processing (ICASSP), **1** (2006)
8. Rietveld, T., Van Hout, R.: Statistics in Language Research: Analysis of Variance. Walter de Gruyter, Berlin (2005)
9. Tokuda, K., Kobayashi, T., Masuko, T., Imai, S.: Mel-generalized cepstral analysis – a unified approach to speech spectral estimation. In: ICSLP (1994)
10. Tokuda, K., Nankaku, Y., Toda, T., Zen, H., Yamagishi, J., Oura, K.: Speech synthesis based on hidden Markov models. Proc. IEEE **101**(5), 1234–1252 (2013)
11. Tokuda, K., Zen, H., Black, A.W.: An HMM-based speech synthesis system applied to English. In: Proceedings of 2002 IEEE Workshop on Speech Synthesis, pp. 227–230 (2002)
12. Woehrling, C., de Mareuil, B.: Identification d'accents regionaux en francais: perception et analyse. Revue Parole **37**, 55 (2006)
13. Yvon, F., De Mareüil, P.B., Aubergé, V., Bagein, M., Bailly, G., Béchet, F., Foukia, S., Goldman, J.F., Keller, E., Pagel, V., et al.: Objective evaluation of grapheme to phoneme conversion for text-to-speech synthesis in french. Comput. Speech Lang. **12**(4), 393–410 (1998)

Decoding Distributed Tree Structures

Lorenzo Ferrone[1]([✉]), Fabio Massimo Zanzotto[1], and Xavier Carreras[2]

[1] Università degli studi di Roma Tor Vergata,
Via del Politecnico 1, 00133 Roma, Italy
lorenzo.ferrone@gmail.com, fabio.massimo.zanzotto@uniroma2.it
[2] Xerox Research Centre Europe, 6 chemin de Maupertuis,
38240 Meylan, France
xavier.carreras@xrce.xerox.com

Abstract. Encoding structural information in low-dimensional vectors is a recent trend in natural language processing that builds on distributed representations [14]. However, although the success in replacing structural information in final tasks, it is still unclear whether these distributed representations contain enough information on original structures. In this paper we want to take a specific example of a distributed representation, the distributed trees (DT) [17], and analyze the reverse problem: can the original structure be reconstructed given only its distributed representation? Our experiments show that this is indeed the case, DT can encode a great deal of information of the original tree, and this information is often enough to reconstruct the original object format.

Keywords: Parsing · Distributed representation · Tree kernel · Distributed tree kernel

1 Introduction

Typical Natural Language Processing methods are designed to produce a discrete, symbolic representation of the linguistic analysis. While this type of representation is natural to interpret, specially by humans, it has two main limitations. First, as a discrete structure, it is not immediately clear how to compute similarities between different structures. Second, it is not immediate to exploit such discrete linguistic representations in downstream applications. Thus, when dealing with discrete structures, one needs to define similarity measures and features that are appropriate for the application at hand.

Recently there has been a great deal of research on the so-called distributional representations [14]. These approaches represent linguistic annotations as vectors in d-dimensional spaces. The aim of these methods is that similarity across different annotations will be directly captured by the vector space: similar linguistic constructs should map to vectors that are close to each other. In other words, the Euclidean distance is meant to provide a similary measure. Because such d-dimensional vectors aim to capture the essence of the linguistic structure,

© Springer International Publishing Switzerland 2015
A.-H. Dediu et al. (Eds.): SLSP 2015, LNAI 9449, pp. 73–83, 2015.
DOI: 10.1007/978-3-319-25789-1_8

one can use them directly as inputs in downstream classifiers, thus avoiding the need of feature engineering to map a discrete structure to a vector of features. Word embeddings [12] are a prominent example of this trend, with many succesful applications. Noticeably, in the recent years there has been a large body of work about modeling semantic composition using distributional representations [2–4,9,13,15,19].

Stemming from distributed representations [14], real-valued vectors have been also used to represent complex structural information such as syntactic trees. Distributed Trees [17] are low-dimensional vectors that encode the feature space of syntactic tree fragments underlying tree kernels [5]. The main result behind this method is that the dot product in this low-dimensional vector space approximates tree kernels. In other words, the tree kernel between two syntactic trees (which computes tree similarity by looking at all subtrees) can be approximated by first mapping each tree into a low-dimensional distributed tree, and then computing the inner product. Thus, distributed trees can be seen as a compression of the space of all tree fragments down to a d-dimensional space, preserving tree similarity. Since the vector space representing distributed trees captures tree similarity (in the tree kernel sense), it must be that such vectors are good representations of syntactic structures. This hypothesis has been confirmed empirically in downstream applications for question classification and recozniging textual entailment [17].

However, a legitimate question is: what is exactly encoded in these vectors representing structural information? And, more importantly, is it possible to decode these vectors to recover the original tree? Auto-encoders from the neural network literature, such as those used to induce word embeddings, are designed to be able to encode words into vectors and decode them. Syntactic trees, however, are structures that are far more complex.

In this paper, we propose a technique to decode syntactic trees from a vector representation based on distributed trees. We pose the problem as a parsing problem: given a sentence, and a reference vector representing its syntactic tree, what is the syntactic tree that is most similar to the reference vector? Our solution is based on CKY, and the fact that we can compute the similarity between a partial tree and the reference vector representing the target tree.

In experiments we present a relation between the number of dimensions of the distributed tree representation and the ability to recover parse trees from vectors. We observe that, with sufficient dimensionality, our technique does in fact recover the correct trees with high accuracy. This confirms the idea that distributed trees are in fact a compression of the tree. Our methodology allows to direclty evaluate the compression rate at which trees can be encoded.

The rest of the paper is organized as follows: first we will give a brief background on Distributed Trees and then we will explore the process of going back from this representation to a symbolic one, we will introduce the famous CYK algorithm and then present our variant of it that deals with distributed trees. Finally, we will present an experimental investigation of this new algorithm, and we will then conclude with a few remarks and plans on possible direction of future works.

2 Background: Encoding Structures with Distributed Trees

Encoding Structures with Distributed Trees [17] (DT) is a technique to embed the structural information of a syntactic tree into a dense, low-dimensional vector of real numbers. DT were introduced in order to allow one to exploit the modelling capacity of tree kernels [5] but without their computational complexity. More specifically for each tree kernel TK [1,6,10,16] there is a corresponding distributed tree function [17] which maps from trees to vectors:

$$DT: T \to \mathbb{R}^d$$
$$t \mapsto DT(t) = \mathbf{t}$$

such that:

$$\langle DT(t_1), DT(t_2) \rangle \approx TK(t_1, t_2) \tag{1}$$

where $t \in T$ is a tree, $\langle \cdot, \cdot \rangle$ indicates the standard inner product in $\mathbb{R}^d$ and $TK(\cdot, \cdot)$ represents the original tree kernel. It has been shown that the quality of the approximation depends on the dimension d of the embedding space $\mathbb{R}^d$.

To approximate tree kernels, distributed trees use the following property and intuition. It is possible to represent subtrees $\tau \in S(t)$ of a given tree t in distributed tree fragments $DTF(\tau) \in \mathbb{R}^d$ such that:

$$\langle DTF(\tau_1), DTF(\tau_2) \rangle \approx \delta(\tau_1, \tau_2) \tag{2}$$

where δ is the Kronecker's delta function. Hence, distributed trees are sums of distributed tree fragments of trees, that is:

$$DT(t) = \sum_{\tau \in S(t)} \sqrt{\lambda}^{|\mathcal{N}(\tau)|} DTF(\tau)$$

where λ is the classical decaying factor in tree kernels and $\mathcal{N}(\tau)$ is the set of the nodes of the subtree τ. With this definition, the property in Eq. 1 holds.

Distributed tree fragments are defined as follows. To each node label n we associate a random vector $\mathbf{n}$ drawn randomly from the d-dimensional hypersphere. Random vectors of high dimensionality have the property of being quasi-orthonormal (that is, they obey a relationship similar to 2). The following functions are then defined:

$$DTF(\tau) = \bigodot_{n \in \mathcal{N}(\tau)} \mathbf{n}$$

where $\odot$ indicates the shuffled circular convolution operation[1], which has the property of preserving quasi-orthonormality between vectors.

---

[1] The circular convolution between $\mathbf{a}$ and $\mathbf{b}$ is defined as the vector $\mathbf{c}$ with component $c_i = \sum_j a_j b_{i-j \bmod d}$. The shuffled circular convolution is the circular convolution after the vectors have been randomly shuffled.

To compute distributed trees, there is a last function $SN(n)$ for each node n in a tree t that collects all the distributed tree fragments of t where n is the head. This is recursively defined as follows:

$$SN(n) = \begin{cases} \mathbf{0} \text{ if } n \text{ is terminal} \\ \mathbf{n} \odot \bigodot_i \sqrt{\lambda}\, [\mathbf{n_i} + SN(n_i)] \text{ otherwise} \end{cases}$$

where n_i are the direct children of n in the tree t. With a slight abuse of notation we can also think of the node n as the (sub-)tree headed by n itself, in this way we can aslo say that the function SN takes a tree as input. From the definition of SN it can be shown that distributed trees can also be computed with the more efficient equation:

$$DT(t) = \sum_n SN(n)$$

We now have all the equations to develop our idea of reconstructing trees from distributed trees. The fact that the approximation can get arbitrarily good with the increasing of the dimension of the vector space suggests that the vector $DT(t)$ encodes in fact all the information about the tree t. However, it is not immediately obvious how one could go about using this encoded information to recreate the symbolic version of the parse tree.

3 Going Back: Reconstructing Symbolic Trees from Vectors

Our aim here is to investigate if the information stored in the distributed tree correctly represents trees. The hypothesis is that this is a correct representation if we can reconstruct original trees from distributed trees.

We treated the problem of reconstructing trees as a parsing problem. Hence, given the distributed tree $DT(t)$ representing a syntactic tree of a given sentence s, we want to show that starting from s and $DT(t)$ it is possible to reconstruct t More formally, given a sentence s and its parse tree t, we use the following information in order to try to reconstruct t:

- the sentence s itself;
- the distributed vector of the tree: $\mathbf{t} = DT(t)$;
- a big context-free grammar G for English that generates also t.

To solve the problem we implemented a variant of the CYK algorithm that uses the information contained in the vector to guide the reconstruction of the original parse among all the possible parses of a given sentence. In the next section we will first give a brief recap on the CYK algorithm, and later we will present our variant.

3.1 Standard CYK Algorithm

The CYK algorithm is one of the most efficient algorithms for parsing context-free grammars. It takes as input a sentence $s = a_1, a_2, \ldots, a_n$ and a context-free grammar G in Chomsky-Normal Form, that is, each rule is of one of these two forms:

$$A \rightarrow B\ C$$
$$A \rightarrow w$$

where A, B, C are non-terminal symbol and w is a terminal symbol (a word). The grammar itself is composed of a list of symbols $R_1\ R_2, \ldots, R_r$ among which there is a subset R_S of starting symbols.

In its most common form the CYK algorithm works by constructing a 3-dimensional table P, where the cell $P[i, j, k]$ contains a list of all the ways that the span from word j to word $i + j - 1$ could be obtained from the grammar symbol R_k. Moreover, each rule is also linked to its children nodes: in this way it is possible to navigate the table P and reconstruct a list of possible parse trees for the original sentence.

$$P_{ijk} = [R_k \rightarrow A_1\ B_1, R_k \rightarrow A_2\ B_2, \ldots]$$

Collectively, in the cell $P[i, j]$ are stored all the possible derivations that could span the sentence from word j to word $i + j - 1$.

The CYK algorithm has subsequently been extended in various forms in order to deal with probabilistic grammars and be able thus to retrieve the k-best parse trees of a given sentence. Our proposal is similar in this intent, but instead of a k-best list of parse trees, it uses a beam search at each cell, guided by the distributed vector of the sentence that we want to reconstruct. In the following section we expose our algorithm.

3.2   CYK over Distributed Trees

In our variant of the algorithm we use the information encoded in the distributed vector ($\mathbf{t}$) of the parse tree of the sentence to guide the new parsing stage in the following manner: in each cell $P[i, j]$ we store not just a list of ways to obtain the rule active at the moment; instead, we store a list L of the m-best entire partial trees built up to that point, ordered by their similarity with $\mathbf{t}$.

More in detail, when we are in the cell $P[i, j]$ for all rules r in the grammar of the form $A \rightarrow B\ C$, we check for all trees that we already built in $P[k, j]$ and $P[i - k, j + k]$ that starts with $B \rightarrow \bullet\bullet$ and $C \rightarrow \bullet\ \bullet$, respectively. Let's call these two lists L_B and L_C, each composed of at most m trees. We create then the list L_{BC} composed of at most m^2 trees that have A as root, a subtree from L_B as the left child and one coming from L_C as the right child. For each one of these trees t_ℓ we compute:

$$\text{score} = \frac{\langle \text{SN}(t_\ell), \mathbf{t} \rangle}{\langle \text{DTF}(r), \mathbf{t} \rangle}$$

where $\text{DTF}(r)$ implies that we are thinking of the rule r as a tree (composed just of a root and two terminal children). We then sort the list L_{BC} in decreasing order according to score and store only the first m trees.

The pseudocode for our algorithm (also including the more advanced extension presented in the next section) is presented in (Algorithm 1).

Algorithm 1. CYK, DTK variant

Input: sentence $s = w_1, w_2, \ldots, w_n$, grammar G, distributed vector $\mathbf{t}$
1: **Initialization:** fill an $n \times n$ table P with zeroes
2: **for** $i = 1$ to n **do**
3:      **for all** symbol $A \in G$ **do:**
4:          rule $\leftarrow (A \to w_i)$
5:          ruleScore $\leftarrow \frac{\langle \mathrm{SN(rule)}, \mathbf{t} \rangle}{\| \mathrm{SN(rule)} \|}$
6:          **append** (rule, ruleScore) to $P[i, 0]$
7:      **end for**
8:      **sort** $P[i, 0]$ decreasingly by **score**
9:      take the first m element of $P[i, 0]$
10: **end for**
11: **for** $i = 2$ to n **do**
12:     **for** $j = 1$ to $n - i + 1$ **do**
13:         **for** $k = 1$ to $i - 1$ **do**
14:             **for all** pairs $(B \to \bullet\, \bullet, C \to \bullet\, \bullet) \in P[k, j] \times P[i - k, j + k]$ **do**
15:                 **for all** rule $(A \to B\ C) \in G$ **do**
16:                     ruleScore $\leftarrow \frac{\langle \mathrm{DTF(rule)}, \mathbf{t} \rangle}{\| \mathrm{DTF(rule)} \|}$
17:                     **if** ruleScore $\geq$ threshold **then**
18:                         tree $\leftarrow A \to [B \to \bullet\, \bullet, C \to \bullet\, \bullet]$
19:                         score $\leftarrow \langle \mathrm{DT(tree)}, \mathbf{t} \rangle$
20:                         **append** (tree, score) to $P[i, j]$
21:                     **end if**
22:                 **end for**
23:             **end for**
24:         **end for**
25:         **sort** $P[i, j]$ decreasingly by **score**
26:         take the first m element of $P[i, j]$
27:     **end for**
28: **end for**
29: **if** $P[n, 1]$ is **not** empty **then**
30:     **return** $P[n, 1]$
31: **else**
32:     **return** Not parsed
33: **end if**

As we do not store each possible parse tree it may happen that this algorithm will fail to recognize a sentence as belonging to the grammar, when this happens we can either flag the result as a fail to reconstruct the original parse tree, or try again increasing the value of m. In our experiment we decided to flag the result as an error, leaving the possibilities of increasing m to further experimentation.

3.3 Additional Rules

In order to increase both efficiency and accuracy, we introduced a few more rules in our algorithm:

- a filter: in this way we don't cycle through all the rules in the grammar but only on those with a score more then a given threshold. Such score is computed again as a similarity between the rule r (viewed as a tree) and $\mathbf{t}$:

$$\text{rulescore} = \frac{\langle \text{DTF}(r), \mathbf{t} \rangle}{\|\text{DTF}(r)\|}$$

  the threshold is defined as:

$$\frac{\lambda^{\frac{3}{2}}}{p}$$

  The significance of this threshold is the following: first recall that DTK is an approximation of TK, which in turn is a weighted count of common subtrees between two trees. The numerator $\frac{\lambda^{\frac{3}{2}}}{p}$ is then the exact score (not an approximation) that a rule would get if it appeared exactly once in the whole tree, while the parameter $p > 1$ relaxes the requirement by lowering this threshold, in order to take in account the fact that we are dealing with approximated scores. In other words, the rules that pass this filter are those that should appear at least one time in t, with some room to account for approximation error.
- a reintroduction of new rules to avoid the algorithm getting stuck on the wrong choice: that is, after the sorting and trimming of the list of the first m trees in each cell, we also add another m trees (again, sorted by score) but for which the root node is different than the node of the first tree in the cell
- a final reranking, according to the DTK between each element in our list of final candidate, and the original tree.

4 Experiment

The pipeline of the experiment is the following:

1. Create a (non probabilistic) context-free grammar G from the input file. This is just a collection of all the production rules (either terminal or not) that appear in the input file
2. For each tree t in the testing set:
   (a) compute the distributed vector $\text{DT}(t)$;
   (b) parse the sentence using the CYK algorithm as explained in Sects. 4 and 5;
   (c) check if the resulting parse tree is equal to the original one;
3. Compute average labeled precision, recall, and f-score for the entire dataset.

4.1 Setup

As testbed for our experiment we used the Wall Street Journal section of the PennTree Bank. Sections 00 to 22 have been used to generate the grammar. Note that the generated grammar is not a probabilistic one, nor there is any learning involved. Instead, it is just a collection of all the rules that appear in parse trees in those sections, more precisely the resulting grammar contains 95 (non-terminal) symbols and 1706 (non-terminal) production rules. The test has been performed on Sect. 23, consisting of a total of 2389 sentences. As a preprocessing step we also first binarise each tree in the PTB so that the resulting grammar is in Chomsky Normal form.

There are four parameters in our model that can be changed: a first set of parameters pertains to the DTK encoding, while another set of parameters pertains to the CYK algorithm. The parameter relative to the DTK are the dimension d of the distributed trees, for which we tried the values of $1024, 2048, 4096, 8192$ and 16384, and the parameter λ for which we tried the values $0.2, 0.4, 0.6, 0.8$.

For the second set of parameter instead we have m, which is the size of the list of best trees that are kept at each stage and p which is the threshold of the filter. For the parameter m we tried the values up to 10 but found out that increasing this value doesn't increase the performance of the algorithm, while at the same time increasing significantly the computational complexity. For this reason we fixed the value of this parameter to 2, and only report results relative to this value. Finally, for the filter threshold we tried values of $p = 1.5, 2$ and 2.5. In the following section we report the results for the given parameters on the test set.

4.2 Results

In this section we report our results on the dataset. In Table 1 we report the percentage of *exactly* reconstructed sentences. The parameter λ is kept fixed at 0.6, while the dimension d and filter p vary. The same results are also presented in Fig. 1.

Table 1. Percentage of correctly reconstructed sentences. $\lambda = 0.6$

p	1.5	2	2.5
$d = 1024$	22.26 %	23.5 %	23.25 %
$d = 2048$	48.8 %	60.46 %	52.32 %
$d = 4096$	77.9 %	81.39 %	75.58 %
$d = 8192$	91.86 %	88.28 %	87.5 %
$d = 16384$	92.59 %	92.54 %	**92.79 %**

In Table 2 we report the average precision and recall on the entire dataset for different values of p, fixed $\lambda = 0.6$, and varying the dimension d. In Fig. (2) we graph the f-measure relative to those same parameter.

Table 2. Average precision and recall. $\lambda = 0.6$

p	1.5	2	2.5
$d = 1024$	0.89	0.78	0.71
$d = 2048$	0.964	0.912	0.85
$d = 4096$	0.984	0.967	0.951
$d = 8192$	0.994	0.994	0.99
$d = 16384$	**0.995**	**0.995**	0.994

(a) precision

p	1.5	2	2.5
$d = 1024$	0.285	0.43	0.477
$d = 2048$	0.58	0.754	0.78
$d = 4096$	0.846	0.923	0.929
$d = 8192$	0.959	0.959	0.967
$d = 16384$	0.965	0.965	**0.976**

(b) recall

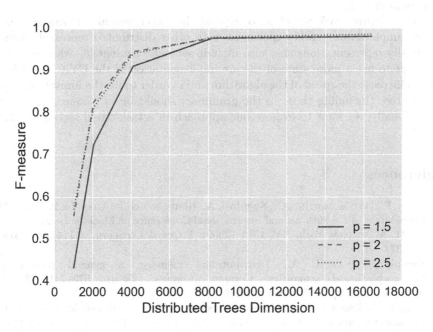

Fig. 1. Average F-measure on the entire dataset as the dimension increase. $\lambda = 0.6$

As we can see the number of correctly reconstructed sentences grows significantly with the increasing of the dimension (as expected) topping at 92.79 % for $d = 16384, p = 2.5$ and $\lambda = 0.6$. On the other hand lower values of d while yielding a low percentage of reconstructed sentences, still can provide a high precision with value as low as 1024 resulting in a precision of 0.89.

In conclusion it seems that the main parameter to influence the algorithm is the dimension d, which was what we expected, because the quality of the approximation depends on d, and thus the amount of information that can be stored in DT(t) without too much distortion. The parameter p on the other hand does not seem to influence the final results nearly as much. As we can see in the tables the results, especially for high dimension d are nearly the same for all the values of p that we tried.

5   Conclusions and Future Work

We showed under our setting that it is possible to reconstruct the original parse tree from the information included in its distributed representation. Together with the work on distributed representation parsing [18] we envision that it would be possible to create a symbolic parse tree of a sentence from *any* distributed representation, not necessarily derived directly from the correct parse tree, but which may be learned in some other way, for example as output of a neural network approach.

As for future work we plan to expand the experimental investigation on a more ample dataset, and experiment with other distributed representations, specifically representations that also include semantic content [8]. Moreover we also want to use a more state-of-the art implementation of the CYK algorithm both to increase the speed of the algorithm and in order to lift the limitation that all the trees (including those in the grammar) should be in Chomsky Normal Form. Finally, we want to explore our approach in a task-based setting as [11] and [7].

References

1. Aiolli, F., Da San Martino, G., Sperduti, A.: Route kernels for trees. In: ICML 2009 Proceedings of the 26th Annual International Conference on Machine Learning, pp. 17–24. ACM, New York, NY, USA (2009). http://doi.acm.org/10.1145/1553374.1553377
2. Baroni, M., Lenci, A.: Distributional memory: a general framework for corpus-based semantics. Comput. Linguist. **36**(4), 673–721 (2010). http://dx.doi.org/10.1162/coli_a_00016
3. Baroni, M., Zamparelli, R.: Nouns are vectors, adjectives are matrices: representing adjective-noun constructions in semantic space. In: Proceedings of the 2010 Conference on Empirical Methods in Natural Language Processing, pp. 1183–1193. Association for Computational Linguistics, Cambridge, MA, October 2010. http://www.aclweb.org/anthology/D10-1115
4. Clark, S., Coecke, B., Sadrzadeh, M.: A compositional distributional model of meaning. In: Proceedings of the Second Symposium on Quantum Interaction (QI-2008), pp. 133–140 (2008)
5. Collins, M., Duffy, N.: Convolution kernels for natural language. In: Dietterich, T.G., Becker, S., Ghahramani, Z. (eds.) Advances in Neural Information Processing Systems 14, pp. 625–632. MIT Press, Cambridge (2001)

6. Collins, M., Duffy, N.: New ranking algorithms for parsing and tagging: kernels over discrete structures, and the voted perceptron. In: Proceedings of ACL 2002 (2002)

7. Dagan, I., Glickman, O., Magnini, B.: The PASCAL RTE challenge. In: PASCAL Challenges Workshop, Southampton, U.K. (2005)

8. Ferrone, L., Zanzotto, F.M.: Towards syntax-aware compositional distributional semantic models. In: Proceedings of COLING 2014, The 25th International Conference on Computational Linguistics: Technical Papers, pp. 721–730. Dublin City University and Association for Computational Linguistics, Dublin, Ireland, August 2014. http://www.aclweb.org/anthology/C14-1068

9. Grefenstette, E., Sadrzadeh, M.: Experimental support for a categorical compositional distributional model of meaning. In: Proceedings of the Conference on Empirical Methods in Natural Language Processing, EMNLP 2011, pp. 1394–1404. Association for Computational Linguistics, Stroudsburg, PA, USA (2011). http://dl.acm.org/citation.cfm?id=2145432.2145580

10. Kimura, D., Kuboyama, T., Shibuya, T., Kashima, H.: A subpath kernel for rooted unordered trees. In: Huang, J.Z., Cao, L., Srivastava, J. (eds.) PAKDD 2011, Part I. LNCS, vol. 6634, pp. 62–74. Springer, Heidelberg (2011)

11. Li, X., Roth, D.: Learning question classifiers. In: Proceedings of the 19th International Conference on Computational Linguistics, COLING 2002, vol. 1, pp. 1–7. Association for Computational Linguistics, Stroudsburg, PA, USA (2002). http://dx.doi.org/10.3115/1072228.1072378

12. Mikolov, T., Chen, K., Corrado, G., Dean, J.: Efficient estimation of word representations in vector space. CoRR abs/1301.3781 (2013). http://arxiv.org/abs/1301.3781

13. Mitchell, J., Lapata, M.: Vector-based models of semantic composition. In: Proceedings of ACL 2008: HLT, pp. 236–244. Association for Computational Linguistics, Columbus, Ohio, June 2008. http://www.aclweb.org/anthology/P/P08/P08-1028

14. Plate, T.A.: Distributed Representations and Nested Compositional Structure. Ph.D. thesis (1994). http://citeseerx.ist.psu.edu/viewdoc/summary?doi=10.1.1.48.5527

15. Turney, P.D., Pantel, P.: From frequency to meaning: vector space models of semantics. J. Artif. Intell. Res. (JAIR) **37**, 141–188 (2010)

16. Vishwanathan, S.V.N., Smola, A.J.: Fast kernels for string and tree matching. In: Becker, S., Thrun, S., Obermayer, K. (eds.) NIPS, pp. 569–576. MIT Press, Cambridge (2002)

17. Zanzotto, F.M., Dell'Arciprete, L.: Distributed tree kernels. In: Proceedings of International Conference on Machine Learning, 26 June–1 July 2012

18. Zanzotto, F.M., Dell'Arciprete, L.: Transducing sentences to syntactic feature vectors: an alternative way to "parse"? In: Proceedings of the Workshop on Continuous Vector Space Models and Their Compositionality, pp. 40–49. 8 August 2013. http://www.aclweb.org/anthology/W13-3205

19. Zanzotto, F.M., Korkontzelos, I., Fallucchi, F., Manandhar, S.: Estimating linear models for compositional distributional semantics. In: Proceedings of the 23rd International Conference on Computational Linguistics (COLING), August 2010

Combining Continuous Word Representation and Prosodic Features for ASR Error Prediction

Sahar Ghannay[1]([⊠]), Yannick Estève[1], Nathalie Camelin[1], Camille Dutrey[2], Fabian Santiago[2], and Martine Adda-Decker[2]

[1] LIUM - University of Le Mans, Le Mans, France
{sahar.ghannay,yannick.esteve,nathalie.camelin}@univ-lemans.fr
[2] LPP - Université Sorbonne Nouvelle, Paris, France
martine.adda-decker@univ-paris3.fr

Abstract. Recent advances in continuous word representation have been successfully used in several natural language processing tasks. This paper focuses on error prediction in Automatic Speech Recognition (ASR) outputs and proposes to investigate the use of continuous word representation (word embeddings) within a neural network architecture.

The main contribution of this paper is about word embeddings combination: several combination approaches are proposed in order to take advantage of their complementarity. The use of prosodic features, in addition to classical syntactic ones, is evaluated.

Experiments are made on automatic transcriptions generated by the LIUM ASR system applied on the ETAPE corpus. They show that the proposed neural architecture, using an effective continuous word representation combination and prosodic features as additional features, outperforms significantly state-of-the-art approach based on the use of Conditional Random Fields. Last, the proposed system produces a well calibrated confidence measure, evaluated in terms of Normalized Cross Entropy.

Keywords: ASR error detection · Neural networks · Continuous word representations · Prosodic features

1 Introduction

Recent advances in the field of speech processing have led to significant improvements in speech recognition performances. However, errors recognition is still unavoidable, whatever the quality of the ASR systems. This reflects their sensitivity to the variability, *e.g.* to acoustic conditions, speaker, language style, *etc.* These errors may have a considerable impact on applications based on the use of automatic transcriptions, like information retrieval, speech to speech translation, spoken language understanding, *etc.*

Error prediction aims to improve the exploitation of ASR outputs by downstream applications, but it is a difficult task because there are several types of errors, which can range from a simple mistake on the number agreement to

© Springer International Publishing Switzerland 2015
A.-H. Dediu et al. (Eds.): SLSP 2015, LNAI 9449, pp. 84–95, 2015.
DOI: 10.1007/978-3-319-25789-1_9

the insertion of an irrelevant word for the overall understanding of the words sequence.

In this paper, we propose a neural architecture to predict ASR errors. We focus on the use of continuous word representations which have proven their efficiency for many Natural Language Processing (NLP) tasks: part-of-speech tagging, chunking, named entity recognition, and semantic role labeling [2,6,23]. They were introduced through the construction of neural language models [4,21].

In the framework of the ASR error prediction task, we investigate the use of several kinds of continuous word representations. We use different available implementations: a variant of the Collobert and Wetson word embeddings [23], word2vec [15] and GloVe [20] and propose different approaches to combine them in order to take advantage of their complementarity.

We focus on the combination of ASR posterior probabilities, lexical, syntactic and prosodic features as well as the effective continuous word representation combination within a neural network classifier for efficient misrecognized word prediction. While prosodic features have previously been used to identify ASR error at the utterance level, their use in word error prediction have been less studied.

Last, we are interested in the confidence measure produced by the neural system, which have been proven to be a well calibrated measure in [26].

The paper is organized along the following lines: Sect. 2 presents related work on errors prediction task, on continuous words representations and on prosodic features. In Sect. 3, different types of word embeddings are reported as well as the approaches to combine them. Section 4 describes the proposed error prediction system. The experimental setup and results are described in Sect. 5, just before the conclusion (Sect. 6).

2 Related Work

For two decades, many studies have focused on the ASR error prediction task. Recently, several approaches are based on the use of Conditional Random Field (CRF). In [19], authors focused on detecting error regions generated by Out Of Vocabulary (OOV) words. They propose an approach based on a CRF tagger, which takes into account contextual information from neighboring regions instead of considering only the local region of OOV words. A similar approach for other kinds of ASR errors is presented in [3]; the authors propose an error prediction system based on a CRF tagger using various ASR, lexical and syntactic features.

Recently, a neural network trained to locate errors in an utterance using a variety of features is also presented in [25]. Some of these features are captured from forward and backward recurrent neural network language model in order to capture long distance word context within and across previous utterances. The other features are extracted from two complementary ASR systems.

Continuous word representations are successfully used in several NLP tasks as additional features. Turian *et al.* [23] evaluate different types of word representations: Brown clusterings, the Collobert and Wetson embeddings [6], the hierarchical log-bilinear (HLBL) [16] embeddings of words, and their combination by a simple concatenation. Evaluations are conducted on chunking and named entity recognition task.

Our use of prosodic features is motivated by previous researches [14, 22] and [11]. Stoyanchev et al. [22] have proven that the combination of prosodic and syntactic features are helpful to localize misrecognized words in a speaker turn. They propose two approaches for misrecognized word detection. The first one is a simple classification of words in a single stage. The second approach based on two stages: first they predict utterance misrecognition for each utterance in ASR hypothesis; second they predict whether each word in the ASR hypothesis is misrecognized or not. Hirschberg et al. [14] find that prosodic features are useful for misrecognized utterance detection. Sharon et al. [11] discover that misrecognized words have extreme prosodic values.

In this paper, we propose to integrate an effective continuous word representation combination with additional features in a neural network architecture designed for ASR error prediction.

3   Continuous Word Representations

Different approaches have been introduced to calculate word embeddings through neural networks.

Considering the framework of ASR error prediction, we need to capture syntactic information in order to use them to analyze sequences of recognized words, but we also need to capture semantic information to measure the relevance of co-occurrences of words in the same ASR hypothesis. Thus, we use and evaluate three kinds of word embeddings coming from different available implementations, as previously presented in [10].

3.1   Description of Embeddings

Three 100-dimensional word embeddings were computed from a large textual corpus, composed of about 2 billions of words. This corpus was built from articles of the French newspaper "Le Monde", from the French Gigaword corpus, from articles provided by Google News, and from manual transcriptions of about 400 hours of French broadcast news.

The embeddings presented in [23] revisited by Turian, are based on the existence/absence of n-grams, in the training data. In our experiments, we use 5-grams for these word embeddings. These word embeddings are called *tur* further in the paper. We also use embeddings computed by the word2vec tool, estimated with the continuous bag-of-words (CBOW) approach. Like for the *tur* word embeddings, a window size of 5 was chosen. These word embeddings are called *w2v*. The embeddings vectors presented in [20] are based on the analysis of co-occurrences of words in a window. We choose a window size of 15, in order to capture more semantic information than in the *tur* and *w2v* embeddings. We call these word embeddings *glove*.

3.2   Continuous Word Representation Combination

In order to take advantage of the complementarity of the word embeddings described above, we propose to combine them, using several approaches.

Simple Concatenation. For the first approach, we simply use the concatenation of the three word embeddings types. As a consequence, each word is represented by a 300-dimensional vector which is the concatenation of the three word representations in this order: *glove*, *tur* and *w2v*. This representation is called *GTW-300* further in the paper.

Principal Component Analysis. For the second approach, the Principal Component Analysis (PCA) technique, a statistical procedure commonly used to compress data, is applied. It consists in an orthogonal linear transformation that transposes the data into a new coordinate system such as the new axis are ordered according to the decreasing variance of the projected data. Once this new coordinate system is obtained, the data can be represented with less components. Last components are discarded because they are considered to convey less information. The PCA is applied on the *GTW-300* representation. According to this representation, the matrix composed of all words is first mean centering using Z-scores. The new coordinate system is then obtained computing PCA using the correlation method. The data is then projected onto the new basis considering only the first 100 or 200 components, respectively further denoted *GTW-PCA100* and *GTW-PCA200*.

Auto-Encoders. Last, we investigate the use of ordinary (O) and denoising (D) auto-encoders [24] to combine the three word embeddings. These auto-encoders are composed of one hidden layer with 100 (GTW-100) or 200 (GTW-200) hidden units each. They take as input the concatenation of the three different embedding vectors and as output a vector of 300 nodes. For each word, the vector of numerical values produced by the hidden layer will be used as the combined *GTW-D/GTW-O* word embedding. The difference between ordinary and denoising auto-encoders comes from the random corruption of the inputs during the training process in the case of a denoising auto-encoder. This corruption was proposed with the idea of making the auto-encoder more generalist by discovering more robust features than a classical auto-encoder. We use auto-encoders with a non-linear hidden layer: it has been shown that such auto-encoders may get the ability to capture multi-modal aspect of the training features, and could be more efficient than PCA to capture main factors of variation in the training data [13]. More, to our knowledge auto-encoders are not used for combination, but just used to learn a compressed, distributed representation for a set of data, typically for the purpose of dimensionality reduction.

4   Error Prediction System

The error prediction system has to attribute the label *correct* or *error* to each word based on a set of features. This attribution is made by analyzing each recognized word within its context.

4.1   Set of Features

In this section, we describe the features collected for each word and how they are extracted. Some of these features are nearly the same as the ones presented in [3]. The word feature vector is the concatenation of different features: **ASR confidence scores** are the posterior probabilities generated from the ASR system (PAP). The word posterior probability is computed over confusion network, which is approximated by the sum of the posterior probabilities of all transitions through this word and that compete with it; **lexical features** are derived from the word hypothesis output from the ASR system. They include the *word length* (the number of letters in the word), and three binary features indicating if the three 3-grams containing the current word have been seen in the training corpus of the ASR language model; **syntactic features** are provided by the MACAON NLP tool chain[1] [17] applied to the ASR outputs. They consist on *part-of-speech tags* (POS), *dependency labels* and *word governors*. The *orthographic representation* of a **word** is used in CRF approaches, as presented in [3]. With our neural approach, we will use different *continuous word representations* which permit us to take advantage of generalizations extracted during the construction of the continuous vectors.

All these features were the ones used in our previous work presented in [10]. In this paper, prosodic features are also integrated and evaluated.

Prosodic Features. Automatic forced alignment is carried out using the LIMSI ASR system [8]. As well as extra-lexical events such as silences, breath and hesitations, words and corresponding phone segments are located (time-stamped) in the speech signal. Forced alignments are carried out twice: (i) using the manual / reference transcription and (ii) using the hypothesis of the LIUM ASR system [7]. As a result, we get among others word and phone segment durations which are relevant to prosody, rhythm and speech rate. The system's pronunciation dictionary includes major pronunciation variants (*e.g.* optional schwas in word-internal and word-final positions, optional liaison consonants, *etc.*) and the best matching pronunciation variant is selected during alignments.

The audio signal is processed using `Praat` [5] (standard analysis parameters) to measure fundamental frequency (f_0) every 5 ms, depending on the method used in [1]. f_0 is the acoustic parameter corresponding to the perceived voice frequency. Speech f_0 variations are highly relevant to prosody, in particular if they can be linked to phone segments and their embedding words and phrases [18]. Figure 1 shows an ASR error where we can see the same hypothesis (hyp.) for two

---

[1] http://macaon.lif.univ-mrs.fr.

distinct word sequences in the reference (ref.). We can see how prosody patterns can contribute to identify the right words among different possible homophone word sequences. The phone time-codes delivered during forced alignment are used to retrieve the corresponding f_0 values both for reference and hypothesis transcriptions, on three points for each vocalic segment. Since the properties of central segments show more acoustic stability, we use measures taken at the middle of vocalic segments in order to calculate f_0 values at the word level: (i) f_0 mean of the word (in Hertz) and (ii) delta between f_0 of the first vocalic segment of the word and f_0 of the last vocalic segment of the word (in both Hertz and semitone).

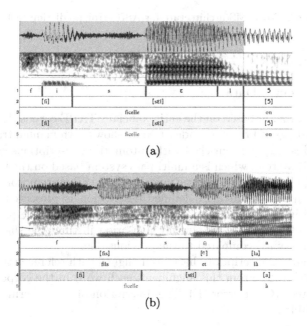

Fig. 1. Illustration of prosodic parameters to speech recognition. Two samples of [fisɛl]: (a) corresponds to the French words "ficelle (on)" ("twine (we)"); (b) corresponds to "fils et l(à)" ("son and th(en)") which was wrongly transcribed "ficelle" by the ASR system, the phone sequences of the pronunced and recognized words being identical. Lines under the spectrogram: 1 = phones; 2 = ref. syllables; 3 = ref. words; 4 = hyp. syllables; 5 = hyp. words.

4.2   Architecture

We have presented in a previous study a neural architecture based on a multi-stream strategy to train the network, named multilayer perceptron multi stream (MLP-MS). The MLP-MS architecture depicted in Fig. 2 is used in order to better integrate the contextual information from neighboring words. A detailed description of this architecture is presented in [10].

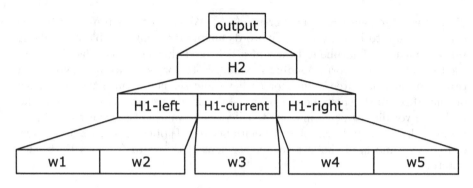

Fig. 2. MLP-MS architecture for ASR error prediction task.

5   Experiments

5.1   Experimental Data

Experimental data are based on the entire official ETAPE corpus [12], composed by audio recordings of French Broadcast News shows with manual transcriptions (reference). This corpus is enriched with automatic transcriptions generated by the LIUM ASR system, which is a multi-pass system based on the CMU Sphinx decoder, using GMM/HMM acoustic models. This ASR system won the ETAPE evaluation campaign in 2012. A detailed description is presented in [7]. This corpus is used in the previous study [10], but for this study we propose a new normalization of ETAPE. This new normalisation justifies the slight difference of the results obtained compared to the previous one.

The automatic transcriptions have been aligned with reference transcriptions using the *sclite*[2] tool. From this alignment, each word in the corpora has been labeled as correct (C) or error (E). The description of the experimental data is reported in Table 1.

5.2   Experimental Results

This section reports the experimental results made on the data set using the ASR error detection system *MLP-MS*. The performance of the proposed approach is compared with a state-of-the-art system based on the CRF tagger provided by *Wapiti*[3] and applied to the set of features presented in Sect. 4.1.

The performance is evaluated by using recall (R), precision (P) and F-measure (F) for the erroneous word prediction and global Classification Error Rate (CER). CER is defined as the ratio of the number of misclassifications over the number of recognized words. The calibrated confidence measure produced by the error prediction systems is evaluated by the Normalized Cross Entropy (NCE). The significance of our results is measured using the 95 % confidence interval.

---

[2] http://www.icsi.berkeley.edu/Speech/docs/sctk-1.2/sclite.htm.
[3] http://wapiti.limsi.fr.

Table 1. Description of the experimental corpus.

ASR	Name	#words REF	#words HYP	WER
Sphinx GMM	Train	349K	316K	25.9
	Dev	54K	50K	25.2
	Test	58K	53K	22.5

Comparison of Word Representation. A set of experiments is performed in order to evaluate the impact of the different types of word embeddings as well as their combination. The ASR error prediction system (MLP-MS) is trained on the train corpus and is applied on the development set (Dev).

Experimental results, presented in Table 2, show that our proposition to combine word embeddings is helpful and yields significant improvements in terms of *CER*, compared to the use of the single embeddings. The best CER reductions comprised between 5 % and 5.3 % are achieved using *GTW-D200* and *GTW-O200* compared to the CRF baseline system. These two systems are used for the remaining experiments.

Table 2. Comparison of different types of word embeddings used as additionnal features in MLP-MS error detection system on Dev corpus.

		Label error			Global
Approach	Representation	P	R	F	CER
Neural	glove	67.80	53.23	59.64	10.60
	tur	70.63	48.61	57.58	10.54
	w2v	72.25	46.65	56.69	10.49
	GTW 300	69.68	52.23	59.71	10.38
	GTW-PCA100	70.32	50.88	59.04	10.39
	GTW-PCA200	71.82	47.37	57.09	10.48
	GTW-O100	74.93	45.26	56.43	10.28
	GTW-O200	71.78	54.35	61.86	**9.86**
	GTW-D100	69.69	55.25	61.63	10.12
	GTW-D200	69.61	58.24	63.42	**9.89**
CRF	discrete	69.67	51.89	59.48	10.41

Table 3 shows the performance of the best embeddings vectors (*GTW-D200* and *GTW-O200*) applied to the Test corpus, compared to the baseline system. Based on the confidence intervals, we observe that our systems yield significant improvement compared to the state-of-the-art CRF approach. The best gains achieved in terms of CER reduction are respectively 6.14 % and 7.84 %.

Table 3. Performance of the best embeddings vectors applied to the Test corpus.

| | | Label error | | | Global | |
Corpus	Approach	P	R	F	CER	95 % confidence interval
Test	CRF	68.34	49.66	57.52	8.79	[8.55 ; 9.04]
	GTW-O200	**71.00**	54.76	61.83	**8.10**	[7.87 ; 8.33]
	GTW-D200	68.23	**58.35**	**62.90**	8.25	[8.01 ; 8.49]

Performance of Prosodic Features. This section reports the evaluation of the effect of combining prosodic features with the effective word embeddings combination (*GTW-D200* and *GTW-O200*), the ASR confidence scores, the lexical and the syntactic features. Experimental results reported in Table 4 show the usefulness of the prosodic features, which yield an improvement in terms of CER reduction compared to the results in Tables 2 and 3. Comparing to the baseline CRF system, our systems *GTW-D200+PROS* and *GTW-O200+PROS* achieve statistically significant improvements, by respectively 8.65 % and 9.45 % in terms of CER reduction. Even if we cannot prove that injecting prosodic features imply a statistically significant reduction of the CER because of lack of observations, it is noticeable that prosodic features imply always an improvement.

Based on the analysis we did in previous study [9], we found that our system has difficulty to detect isolated errors that do not trigger a significant linguistic rupture. By adding prosodic features into the *GTW-D200* system, we observe an improvement of 9.8 % of precision to detect such isolated errors.

Table 4. Performance of the prosodic features on Dev and Test corpus.

| | | Label error | | | Global | |
Corpus	Approach	P	R	F	CER	95 % confidence interval
Dev	CRF+PROS	69.89	52.86	60.20	10.29	[10.03 ; 10.56]
	GTW-O200+PROS	69.76	60.50	64.80	9.67	[9.40 ; 9.93]
	GTW-D200+PROS	**70.40**	**60.57**	**65.11**	**9.55**	[9.30 ; 9.81]
Test	CRF+PROS	68.95	51.82	59.17	8.57	[8.33 ; 8.81]
	GTW-O200+PROS	**69.01**	**60.95**	**64.73**	**7.96**	[7.73 ; 8.20]
	GTW-D200+PROS	68.68	60.65	64.42	8.03	[7.80 ; 8.27]

Calibrated Confidence Measure. Authors in [26] have tackled the problem of confidence measure calibration, in order to improve their quality. This post-processing step is viewed as a special adaptation technique applied to confidence measure to make optimal decisions. The use of artificial neural network is one of the proposed methods for this post-processing step. Based on the NCE scores presented in Table 5, we prove the validity of our approach to produce a well calibrated confidence measure, while the posterior probability provided by the LIUM ASR system is not calibrated at all. The CRF system also produces well calibrated confidence measure. Furthermore, the use of prosodic features improves the NCE score for all systems.

Table 5. NCE for the PAP and the probabilities resulting from MLP-MS and CRF.

Name	PAP	softmax proba GTW-D200	softmax proba GTW-O200	CRF
Without prosodic features				
Dev	-0.064	0.425	0.443	**0.445**
Tes	-0.044	0.448	**0.461**	0.457
With prosodic features				
Dev	-0.064	0.461	**0.463**	0.449
Test	-0.044	0.471	**0.477**	0.463

As shown in Fig. 3, the probabilities derived from our neural systems and CRF match with the probability of correct words. More, the curves are well aligned with the diagonal, especially for our neural systems with prosodic features.

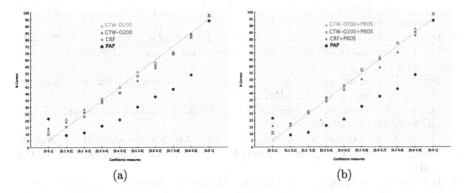

(a)   (b)

Fig. 3. Percentage of correct words based on PAP and confidence measures derived from MLP-MS and CRF: without (a) and with prosodic features (b)

6   Conclusion

In this paper, we have investigated the evaluation of different continuous word representations on an ASR error prediction task. We have also proposed several approaches to combine three types of word embeddings built by different methods. As well, we evaluated the use of prosodic features, in addition to classical syntactic ones.

Experiments performed on the automatic transcriptions of the ETAPE corpus generated by the LIUM ASR system. Results show that the proposed neural architecture, using an effective continuous word representation combination and the prosodic features as additional features, outperforms state-of-the-art approach based on the use of Conditional Random Fields (CRF).

More, our systems produce a well calibrated confidence measure evaluated in terms of NCE comparing to the posterior probability.

Acknowledgments. This work was partially funded by the European Commission through the EUMSSI project, under the contract number 611057, in the framework of the FP7-ICT-2013-10 call, by the French National Research Agency (ANR) through the VERA project, under the contract number ANR-12-BS02-006-01, and by the Région Pays de la Loire.

References

1. Adda-Decker, M., Gendrot, C., Nguyen, N.: Contributions du traitement automatique de la parole l'étude des voyelles orales du franais. Traitement Automatique des Langues **49**(3), 13–46 (2008)
2. Bansal, M., Gimpel, K., Livescu, K.: Tailoring continuous word representations for dependency parsing. In: Proceedings of the 52nd Annual Meeting of the Association for Computational Linguistics, vol. 2, Short Papers, pp. 809–815. Association for Computational Linguistics (2014)
3. Béchet, F., Favre, B.: ASR error segment localisation for spoken recovery strategy. In: IEEE International Conference on Acoustics, Speech and Signal Processing (ICASSP) (2013)
4. Bengio, Y., Ducharme, R., Vincent, P., Janvin, C.: A neural probabilistic language model, vol. 3, pp. 1137–1155. JMLR (2003)
5. Boersma, P., Weenink, D.: Praat, a system for doing phonetics by computer. Glot Int. **5**(9), 341–345 (2001)
6. Collobert, R., Weston, J., Bottou, L., Karlen, M., Kavukcuoglu, K., Kuksa, P.: Natural Language Processing (Almost) from Scratch. vol. 12, pp. 2493–2537. JMLR (2011)
7. Deléglise, P., Estève, Y., Meignier, S., Merlin, T.: Improvements to the LIUM French ASR system based on CMU Sphinx: what helps to significantly reduce the word error rate? In: Interspeech, Brighton, UK, September 2009
8. Gauvain, J.-L., Adda, G., Lamel, L., Lefvre, F., Schwenk, H.: Transcription de la parole conversationnelle. Traitement Automatique des Langues **45**(3), 35–47 (2005)
9. Ghannay, S., Camelin, N., Estève, Y.: Which ASR errors are hard to detect? In: Errors by Humans and Machines in Multimedia, Multimodal and Multilingual Data Processing (ERRARE 2015) Workshop, Sinaia, Romania, pp. 11–13 (2015)
10. Ghannay, S., Estève, Y., Camelin, N.: Word embeddings combination and neural networks for robustness in asr error detection. In: European Signal Processing Conference (EUSIPCO 2015), Nice, France, 31 August–4 September (2015)
11. Goldwater, S., Jurafsky, D., Manning, C.D.: Which words are hard to recognize? prosodic, lexical, and disfluency factors that increase speech recognition error rates. In: Speech Communication, pp. 181–200 (2010)

12. Gravier, G., Adda, G., Paulsson, N., Carr, M., Giraudel, A., Galibert, O.: The ETAPE corpus for the evaluation of speech-based TV content processing in the French language. In: Proceedings of the Eight International Conference on Language Resources and Evaluation (LREC 2012) (2012)

13. Hinton, G.E., Salakhutdinov, R.R.: Reducing the dimensionality of data with neural networks. Science **313**(5786), 504–507 (2006)

14. Hirschberg, J., Litman, D., Swerts, M.: Prosodic and other cues to speech recognition failures. Speech Commun. **43**(1), 155–175 (2004)

15. Mikolov, T., Chen, K., Corrado, G., Dean, J.: Efficient estimation of word representations in vector space. In: Proceedings of Workshop at ICLR (2013)

16. Mnih, A., Hinton, G.E.: A scalable hierarchical distributed language model. In: Koller, D., Schuurmans, D., Bengio, Y., Bottou, L. (eds.) Advances in Neural Information Processing Systems, pp. 1081–1088. Curran Associates Inc. (2009)

17. Nasr, A., Béchet, F., Rey, J.-F., Favre, B., Le Roux, J.: Macaon: An nlp tool suite for processing word lattices. In: Proceedings of the 49th Annual Meeting of the Association for Computational Linguistics: Human Language Technologies: Systems Demonstrations, pp. 86–91. Association for Computational Linguistics (2011)

18. Nemoto, R., Adda-Decker, M., Durand, J.: Investigation of lexical f0 and duration patterns in french using large broadcast news speech corpora. In: Proceedings of Speech Prosody (2010)

19. Parada, C., Dredze, M., Filimonov, D., Jelinek, F.: Contextual information improves OOV detection in speech. In: Human Language Technologies: Procedings of the North American Chapter of the Association for Computational Linguistics (NAACL 2010) (2010)

20. Pennington, J., Socher, R., Manning, C.D.: Glove: Global vectors for word representation. In: Proceedings of the Empiricial Methods in Natural Language Processing (EMNLP 2014), vol. 12 (2014)

21. Schwenk, H., Dchelotte, D., Gauvain, J.-L.: Continuous space language models for statistical machine translation. In: Proceedings of COLING/ACL, COLING-ACL 2006, pp. 723–730, Stroudsburg, PA, USA. Association for Computational Linguistics (2006)

22. Stoyanchev, S., Salletmayr, P., Yang, J., Hirschberg, J.: Localized detection of speech recognition errors. In: 2012 IEEE Spoken Language Technology Workshop (SLT), pp. 25–30, December 2012

23. Turian, J., Ratinov, L., Bengio, Y.: Word representations: A simple and general method for semisupervised learning, pp. 384–394 (2010)

24. Vincent, P., Larochelle, H., Bengio, Y., Manzagol, P.A.: Extracting and composing robust features with denoising autoencoders. In: Proceedings of the 25th International Conference on Machine Learning (2008)

25. Yik-Cheung, T., Lei, Y., Zheng, J., Wang, W.: ASR error detection using recurrent neural network language model and complementary ASR. In: Procedings of Acoustics, Speech and Signal Processing (ICASSP 2014), pp. 2312–2316 (2014)

26. Dong, Y., Li, J., Deng, L.: Calibration of confidence measures in speech recognition. IEEE Trans. Audio, Speech, Lang. Proces. **19**, 2461–2473 (2011)

Semi-extractive Multi-document Summarization via Submodular Functions

Fatemeh Ghiyafeh Davoodi$^{(\boxtimes)}$ and Yllias Chali

Department of Mathematics and Computer Science, University of Lethbridge,
4401 University Dr W, Lethbridge, Canada
{f.ghiyafehdavoodi,yllias.challi}@uleth.ca

Abstract. In this paper, we propose an Extractive Maximum Coverage KnaPsack (MCKP) based model for query-based multi document summarization which integrates three monotone and submodular measures to detect importance of a sentence including Coverage, Relevance, and Compression. We apply an efficient scalable greedy algorithm to generate a summary which has a near optimal solution when its scoring functions are monotone nondecreasing and submodular. We use DUC 2007 dataset to evaluate our proposed method and the result shows improvement over two closely related works.

Keywords: Text summarisation · Extractive summarization · Coverage · Relevance · Compression · Submodularity · Monotonicity

1 Introduction

Considering the volume of relevant information on the web, document summarization has become a must. Automatic document summarization aims at filtering out less informative pieces of documents and only presents the most relevant parts of document(s). Document summarization is "the process of distilling the most important information from the source (or sources) to produce an abridged version for a particular user (or users) and task (or tasks)" [21]. Summarization techniques are categorized into different classes based on different criteria as: (1) Single-document vs. Multi-documents summarization: In single-document summarization, a summary is generated from a single document, while in multi-document summarization, a summary is generated from multiple relevant documents. (2) Extractive vs. Abstractive summarization: Extractive methods select important sentences and paragraphs from the original document and concatenate them into shorter form, while abstractive summarization methods understand the original text and rewrite it in fewer words. (3) Query-based vs. Generic summarization: In query-based summarization, a summary is generated with regards to a specific query, while in generic summarization, a summary is generated for general purposes. Three essential criteria are typically considered in selecting a sentence in the query-based extractive multi-document summarization based on [22]: (1) relevance, (2) redundancy and (3) length. The relevance of each sentence

© Springer International Publishing Switzerland 2015
A.-H. Dediu et al. (Eds.): SLSP 2015, LNAI 9449, pp. 96–110, 2015.
DOI: 10.1007/978-3-319-25789-1_10

shows its relevance to the given query. Sentence redundancy depicts the degree of overlap between the candidate sentence and the generated summary. Finally, length is a constraint on the number of words in the final summary. Coverage is another measure which is considered in some other researches that considers coverage level of a sentence. Sentence compression also has been considered in the process of document summarization since 2000 [14].

We use MCKP to model the summarization problem as it is a good fit for the summarization problem and it is proved to have a great performance [9]. Although, there has been some research on modeling summarization using MCKP, integration of extraction and compression, or employ submodulairty in document summarization, our research differs from those in the following aspects:

- Introducing compression into MCKP modeling in the process of generating a summary. This is the first attempt to investigate the potential of applying sentence compression in a MCKP model for the task of summarization to the best of our knowledge.
- Integrating approximation techniques and compression to improve the quality of summarization. The works in [16–18] take advantage of approximation when their functions have some specific properties (submodularity and monotonicity). However their functions lack the compression. We integrate compression in their approximation algorithm and it as another measure to select important sentences to generate more accurate summary.
- Considering semantic similarity measure to calculate the relevance of a sentence and a query to better detect correlation of words better. The majority of researches use word-matching based measures, which lack the consideration of semantic relations between words. So, we employ a WordNet based measure to calculate the semantic similarity between a sentence and a query.

The remaining of this paper is organized as follows: Sect. 2 will briefly present an overview of previous works on automatic document summarization. In Sect. 3, We introduce our semi-extractive document summarization model which we call Comp-Rel-MCKP Summarizer and explain its preprocessing phase and problem formulation as well as how we solve the problem and generate a summary using different algorithms. In Sect. 4, we introduce our Dataset and evaluation measure. In addition, results of our proposed model besides comparison to some of previous researches are shown. Section 5 will conclude our discussion.

2   Background on Submodularity

Submodularity is widely used in many research areas including game theory, economic, combinatorial optimization, and operations research. Recently it is also considered in NLP research [16–18,23] since submodular functions can helps improving scalability.

2.1   Definition

Submodularity is considered as a property of a set of function [23]. Let $V = \{v_1, v_2, ..., v_n\}$ be a set of objects. A set function $\mathcal{F} : 2^V \to \mathbb{R}$ maps subsets of the ground set, $S \subseteq V$, into real values. There are many equivalent definitions of submodularity and one of them is as follows:

Definition 1. *For any $R \subseteq S \subseteq V$, function $\mathcal{F} : 2^V \to \mathbb{R}$, is Submodular if:*

$$\mathcal{F}(S \cup \{v\}) - \mathcal{F}(S) \leq \mathcal{F}(R \cup \{v\}) - \mathcal{F}(R) \tag{1}$$

Definition 1 is equivalent to the property of diminishing returns and means that a set function $\mathcal{F}$ is submodular if the incremental value of the function for the superset S, is not greater than the incremental value for the subset R by adding a new element v to both sets. Submodular functions can be *Monotone* which is defined as follows:

Definition 2. *For any $R \subseteq S \subseteq V$, function $\mathcal{F} : 2^V \to \mathbb{R}$, is Monotone Submodular if:*

$$\mathcal{F}(R) \leq \mathcal{F}(S) \tag{2}$$

3   Related Work on Automatic Document Summarization

Automatic document summarization were introduced in late 1950s [1,20]. Since then, various models have been introduced for automatic document summarization in both *Extractive* and *Abstractive* areas, however as our focus is on *Extractive*, we confine our literature review to solely extractive document summarization approaches. Most of the extractive methods use greedy-like or global selection approaches. One of the widely used greedy approaches is *Maximum Marginal Relevance (MMR)* [3]. It gives a penalty to sentences that are similar to the already-chosen sentences in the summary and selects sentences having highest value of relevance. More complicated summarization methods which also use MMR are introduced by Erkan et al. [8], Goldstein et al. [13], Radev et al. [27] and Dang [5]. The summarization method of Schiffman [28] is another example of a greedy-like approach. They rank sentences based on some corpus-based features and location-based features Then, their method produces a summary by sequentially selecting top-ranked sentences until reaching the desired length. Filatova and Hatzivassiloglou [9] also used a greedy algorithm and their work was the first attempt in which document summarization is formulated as a Maximum Coverage with KnaPsack constraint (MCKP) problem. To generate a summary, the algorithm selects sentences with greatest total *Coverage* of words, while implicitly minimizing information overlap within the summary. They believe that the coverage measure simultaneously minimize the redundancy and there is no need to have a seperate measure of redundancy. Yih et al. [31] also use MCKP, but they consider location-based information of a sentence in addition to the *Coverage* measure for sentence scoring and apply stack decoding to solve it.

Takamura and Okumura [30] also present a MCKP based model for generic summarization and generate summary using both greedy-like and global selection methods. Our approach to model the document summarization problem is similar to their approach. However, it differs from their approach since ours is for query-based document summarization and we consider both query-based and important-based features to calculate relevance, while they consider just important-based features. In addition, we augment our model with a compression measure which is missed in their model.

Recently, submodularity has been used in document summarization [16–18,23] which results in utilize greedy algorithms with performance guarantees for the summarization process. Lin et al. [16] formulate summarization as a submodular function consisting of two measures of Redundancy and Coverage under a budget constraint. They propose a greedy algorithm with a $(1 - \frac{1}{\sqrt{e}})$ performance guarantee to generate a summary. This greedy algorithm needs objective function be monotone and submodular. They improve their previous works in [17] which uses two measures of *Relevance* and *Diversity* to rank sentences and apply their modified greedy algorithm proposed in [16] to generate a summary. They reward *Diversity* instead of penalizing *Redundancy* since *Redundancy* violates the monotonicity of the objective function. Employing submodular functions in our proposed model is inspired by Lin and Bilmes's work [17]. However, our scoring functions are different. They use Diversity as a replacement measure for Redundancy and a different Relevance measure, while our model is based on MCKP and we use three measures of coverage which implicitly contains redundancy, relevance and compression to score sentences.

Global selection is another strategy to generate a summary. One of the first global selection approaches is introduced by McDonald [22] which is an improvement to Goldstein's method [13] by considering the MMR as a knapsack problem. They employ a dynamic programming algorithm and Integer Linear Program (ILP) to maximize the optimality of the generated summary. Gillick et al. [11,12] also introduced a global selection approach using a concept-based ILP approach to generate a summary in which three measures of *Relevance*, *Coverage*, and *Compression* are considered. Berg-Kirkpatrick et al. [2] proposed a supervised tree-based approach to compress and extract sentences simultaneously. They model their joint approach as an ILP in which objective function is based on *Coverage* and *Compression* which is based on subtree deletion model (in terms of number of cut choices in parse tree of a sentence). They used an approximate solver for their compressive-extractive summarizer to generate a summary. Their work is another closely related work to ours, however, they do not employ submodular functions and greedy algorithm, but instead use Integer Linear Programming. Their proposed method is for generic summarization in which they use supervised learning, while our method is for query-based summarization and we use unsupervised learning. Chali and Hasan's document summarization method [4] is also a global selection approach. They also used ILP to formulate query-based multi-document summarization, but they consider three measure of *Relevance*, *Redundancy*, and *Compression* in their work.

4   Semi-extractive Document Summarization

We can divide our semi-extractive document summarization into the following steps/phases: *Preprocessing, Problem formulation,* and *Solving the problem.* Each step is further discussed in the next sections.

4.1   Preprocessing

Each document D is decomposed into its sentences and shown as $D = \{s_1, ..., s_{|D|}\}$. Next, each sentence s_i is decomposed to its words e_{ij} (i.e. $s_i = \{e_{i1}, ..., e_{i|s_i|}\}$). Each word is then represented by its stem using Porter Stemmer [26]. Next step is detecting stop words. However, unlike most of researches, the detected stop words are not removed from the documents in our approach because removing stop words from a sentence affect its grammatical correctness. We, though, detect the stop words in our work in order to bypass them in sentence scoring process.

4.2   Problem Formulation

Our proposed document summarization method is based on MCKP. The goal of document summarization as a MCKP is to cover as many conceptual units as possible using only a small number of sentences. However, in query-based summarization methods, relevance of the generated summary to a given query and compression ratio of the summary are also important. So, in our summarization technique which we call **Comp-Rel-MCKP** document summarization, three measures of **Coverage**, **Relevance**, and **Compression** are considered. Goal of *Comp-Rel-MCKP document summarization* is to generate a summary while maximizing the value of all three measures. In the next sections, each measure will be discussed in more details.

Coverage: Coverage measure represents coverage level of conceptual units by any given set of textual units [9]. Two different coverage functions are defined for measuring the coverage level of a summary and a sentence. *Coverage* function for summary, $Cov(S)$, shows how the generated summary S covers D and is defined as follows:

$$Cov(S) : \sum_j z_j \forall j, e_j \in \{S\} \tag{3}$$

where z_j is 1 when word e_j is covered in the summary S, and 0 otherwise. $Cov(S)$, considers the number of unique words in the summary as the coverage score.

Coverage function for sentence s_i, $Cov(s_i)$ is similar to $Cov(S)$, but it considers summary S in its measurement. That is, $Cov(s_i)$ measures the number of unique words in the sentence s_i which are not covered by the already selected sentences in the summary S. $Cov(s_i)$ is defined below:

$$Cov(s_i) : \sum_j z_j \forall j, e_j \in \{s_i\} \text{and} e_j \notin \{S\} \tag{4}$$

The aforementioned definition of *Coverage* function has the advantage of implicitly encompassing the notion of redundancy because redundant sentences cover fewer words.

Relevance: *Relevance* measure represents the importance of a given set of textual units as well as its correlation with a given query. *Relevance* function is considered as a combination of a set of query-oriented and importance-oriented measures. The query-oriented measures consider the similarity between a sentence and the given query while the importance-oriented measures calculate the importance of a sentence in a given document [4, 7, 29] regardless of the query. Relevance function for a sentence or a summary is calculated in the same way and is defined as follows:

$$Rel(S) : \sum_i sim(s_i, q) + imp(s_i) \forall i, s_i \in \{S\} \quad (5)$$

where $sim(s_i, q)$ and $imp(s_i)$ consider the query-oriented and importance-oriented features respectively and reveal similarity of sentence s_i to the given query q and the importance of the sentence s_i regardless of considering the query q respectively.

Many works use vocabulary matching between the query q and the sentence s_i to calculate $sim(s_i, q)$. They consider the number of words that the sentence s_i overlaps ith query q as their similarity score [17]. Vocabulary matching similarity measure is easy to calculate, however, it fails to detect any semantic similarity between words, *i.e.* "teaching" and "education". One remedy to this problem is to exploit WordNet-based measures which consider the semantic relations between words. To calculate semantic similarity between sentence s_i and query q, $sim(s_i, q)$, both sentence s_i and query q are represented as a vector of words (bag of words) after tokenization and stop word removal process. Then, the semantic similarity is calculated as follows[1].

Suppose Fig. 1 illustrates the vectors of words representing sentence s_i and query q. For each word e_{sj}, $(j = 1, ..., m)$ in the vector of words s_i, we find the semantic similarity of e_{sj} to all words e_{qk}, $(k = 1, ..., n)$ in the vector of words q using FaITH similarity measure of WordNet proposed by Pirro [25]. We then assign e_{sj} to e_{qk} which has the highest similarity. As word types of e_{sj} and e_{qk} are unknown, we have to look them up in all four parts of WordNet (noun, verb, adjective, and adverb parts), and assign the highest similarity among four similarity that we come up with as the similarity of e_{sj} and e_{qk}. After assigning all the words in the vector of words s_i to a word in the vector of words q and calculating the pair similarities, the total semantic similarity of s_i and q is the result of summing up all the pair similarities divided by the total number of words in the vectors of words s_i and q, $(m + n)$.

---

[1] The way we find the semantic similarity between two vectors of words, representing the sentence and query is inspired by the maximum weighted matching problem in a bipartite graph.

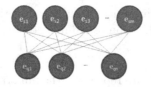

Fig. 1. Vectors of words representing s_i and q

To calculate $imp(s_i)$ which represents the importance a sentence, we combine TF-IDF measure and inverse position of the sentence. In addition, position of a sentence is also used as a good indicator of importance in document summarization (state of arts) as early sentences tend to be more important [4,10]. Thus, the importance of sentence s_i, $imp(s_i)$, is defined as:

$$imp(s_i) = \alpha \sum_{j \in s_i} TF - IDF(e_{kj}) + \beta \frac{1}{Pos(s_i)} \forall j, e_j \in s_i \text{and} s_i \in \{S\} \quad (6)$$

where $TF - IDF(e_{kj})$ is the *Term Frequency* and *Inverse Document Frequency* for word e_j within its original document d_k and $Pos(s_i)$ indicates the position of the sentence s_i within its original document d_k.

Compression: Sentence compression plays a key role in summary generation as it reduces space wasting and enhances the chance of including more relevant information. It detects insignificant part of a sentence as deletable parts while keeping important part such that the readability and correctness of the sentence are preserved. Consider the following sentence[2] as a candidate to be added to a summary.

"Thousands of jobless demonstrated across France on Monday, to press the Socialist-led government for a bigger increase in unemployment benefits and a Christmas bonus, according to the official way of accounting unemployment."

In this example, deletable parts are shown in gray color and as it can be seen, removing deletable parts of the sentence preserves significant information, readability, and grammatical correctness of the sentence. In the process of sentence compression which is viewed as a word deletion process, we remove deletable parts of a sentence using Berg's compression method [2]. We define the compression function as:

$$Comp(S) : \sum_i d_{ij} \forall i, s_i \in \{S\}, \forall d_{ij} \in \{s_i\} \quad (7)$$

where $DS(D)$ contains deleted choices in the document sets and where d_{ij} denotes a constant which is 1 if word e_j is deleted from sentence s_i. $DS(D)$ is defined by using Berkeley parser [24] to generate a constituency parse tree for each sentence and then apply thirteen subtree deletion features trained in [2] to find words that can be deleted from each sentence.

---

[2] This sentence is from DUC 2007, topic D0701A.

To formulate the problem of document summarization as a **Comp-Rel-MCKP**, let K, summary length, is measured by the number of words and x_i be a variable which is 1 if sentence s_i is selected in generating a summary, and 0 otherwise. In addition, let a_{ij} be a constant which is 1 if sentence s_i contains word e_j, and 0 otherwise. Word e_j is considered as covered when at least one sentence containing e_j is selected in the summary.

Considering all three described measures, the described goal and summary length constraints, our objective function is defined as:

$$MaximizeF(S) = \alpha Cov(S) + \beta Rel(S) + \gamma Comp(S)$$

$$= \alpha \sum_j z_j + \beta \sum_i (sim(s_i, q) + imp(s_i))x_i + \gamma \sum_i (d_{ij})x_i$$

$$\text{subject to } \sum_i c_i x_i \leq K, \sum_i a_{ij} x_i \geq z_j \tag{8}$$

$$\forall i, x_i \in \{0, 1\}, \forall j, z_j \in \{0, 1\}$$

$$\forall d_{ij} \in DS(D)$$

where α, β, and γ are damping factors for *Coverage*, *Relevance*, and *Compression* respectively. x_i is a variable which is 1 if sentence s_i is selected, and 0 otherwise. Let K, the summary length, be measured by the number of words. In addition, let a_{ij} be a constant which is 1 if sentence s_i contains word e_j, and 0 otherwise. Word e_j is considered as covered when at least one sentence containing e_j is selected in the summary. Let c_i be the cost of selecting s_i, e.g. the number of words or bytes in s_i. $DS(D)$ contains deleted choices in the document sets. In other words, for each sentence, it contains some parts that we can remove while keeping its grammatical correctness and informative parts. As we discussed before, d_{ij} denotes a constant which is 1 if word e_j is deleted from sentence s_i. So, our goal is to find a binary assignment on x_i with the best value for the measures such that the summary length is at most K.

To calculate $DS(D)$, first step is to generate a constituency parse tree, $p(s)$, for each sentence using Berkeley parser [24]. Second step is to detect subtrees in the parse tree of $p(s_i)$ as a set of $T = \{t_{1i}, t_{2i}, ..., t_{mi}\}$, where m is number of possible subtree in $p(s_i)$. Then, apply a method introduced by Berg-Kirkpatrick et al. [2] on each subtree in T, to detect deletable parts. Berg-Kirkpatrick et al. [2] introduced thirteen features trained on TAC dataset using human annotated data set of extracted and compresses sentences. Finding the features on the generated parse tree of a sentence will result in determining deletable subtrees or sometimes root of the parse tree.

4.3 Solving the Problem

To solve the proposed **Comp-Rel-MCKP** document summarizer, we apply the modified greedy algorithm for monotone and submodular which was introduced in [16]. Lin and Bilmes proved theoretically and empirically that their modified greedy algorithm solves the budgeted submodular maximization problem near-optimally and has a constant factor approximation of $(1 - \frac{1}{\sqrt{e}})$ when using scaled cost in the problem. It shows the worst case bound, however, in most cases the quality of the generated summary will be much better than this bound. The algorithm is based on sequentially selection. In each step, it selects sentence s_i with greatest ratio of score gained based on our objective function to the scaled cost. To get near optimal solution, the scoring function should be monotone and submodular [17], otherwise, this greedy algorithm cannot guarantee a near optimal summary. In next section, we show that how the proposed objective function which is monotone and submodular.

Coverage Function. Since penalizing redundancy violates the monotonicity property [16], we reward coverage instead, which implicitly have redundancy in its definition. The function $Cov(S)$ penalizes redundancy implicitly as redundant sentences cover fewer words and reward coverage by selecting sentences with greatest number of uncovered words. As soon as a sentence s_i is chosen to be in the summary S, all of words forming the sentence s_i, will be ignored in calculating coverage level of other sentences if they include the same word. The function $Cov(S)$ has the monotonicity property as coverage is improved by adding some sentences. It also has the submodularity property. Let's consider two summary sets $S(A)$ and $S(B)$, where $S(B) \subseteq S(A)$. Adding a new sentence s_i to $S(B)$ increases the value of the function $Cov(S)$ more than the increment resulting of adding s_i to $S(A)$ because the words forming the new sentence might have already been covered by those sentences that are in the larger summary $S(A)$ but not in the smaller summary $S(B)$.

Relevance Function. $Rel(S)$ combines a query-based and an importance-base functions. They are both monotone as similarity of summary S to the given query q is not improved by adding a sentence to it. This is because the selected sentence s_i is totally dissimilar to q and hence there is no added value for the query-related part in the worst case. In addition, the value of $imp(s_i)$, even though for last sentences in a document would result an increment in importance-based value of a summary. It also has the submodularity property. Let's consider two summary sets $S(A)$ and $S(B)$, where $S(B) \subseteq S(A)$. Adding a new sentence s_i to $S(B)$ increases the value of the both functions equals to the increment resulting of adding s_i to $S(A)$ because the same sentence is added to both summary $S(A)$ and $S(B)$ which results in a same increase in both $sim(s_i, q)$ and $imp(s_i)$.

Theorem 3. *Given functions* $F : 2^V \rightarrow \mathbb{R}$ *and* $f : \mathbb{R} \rightarrow \mathbb{R}$, *the composition* $F' = f \circ F : 2^V \rightarrow \mathbb{R}$ *(i.e.,* $F'(S) = f(F(S))$*) is nondecreasing submodular, if* f *is non-decreasing concave and* F *is nondecreasing submodular.*

Submodular functions have some similar properties to convex and concave functions [19] such as their closure under some operations including mixtures, truncation. So, using Theorem 3 and since summation preserves submodularity property, it is easy to see that $R(S)$ is submodular.

Compression Function. The $Comp(S)$ function is monotone as the compression level of the summary is not worsen by adding a sentence because some words might be deleted in the new sentence. It also has the submodularity property because the same sentence is added to both summary sets $S(A)$ and $S(B)$, where $S(B) \subseteq S(A)$. So, incremental value of $Comp(S)$ by adding the same new sentence is the same for both summary.

5   Experimental Results

We use the data form document understanding conference (DUC[3]) and focus on DUC 2007 dataset which is the latest dataset for query-based summarization and contains 45 different topics each with 25 relevant documents. We use ROUGE (Recall-Oriented Understudy for Gisting Evaluation) [15] package[4] to evaluate the results automatically which is a well-know package for comparing the system generated summaries to a set of reference summaries written by humans. ROUGE-N is widely used in multi-document summarization researches and also Lin and Bilmes [17] show that ROUGE-N is monotone and submodular, thus, we use ROUGE-N measure for our evaluation since our proposed method is also submodular. We focus on ROUGE-1 (unigram) and ROUGE-2 (bigram) scores and report precision, recall and F-measure for evaluation since these metrics are found to correlate well with human judgments and widely used to evaluate automatic summarizer [6,17,22].

5.1   Comparison with the State-Of-The-Art

Baseline: We adopt the baseline from DUC 2007. It concatenates leading sentences of all relevant documents up to the length limit.

Rel-MCKP: In this method, a summary is generated using the summarization method proposed by Takamura and Okumura [30]. They consider MCKP to model summarization and consider *Relevance* and *Coverage* measures in sentence selection process. They use two different greedy algorithms (one with proved performance gurantee) to generate a summary. These two systems are shown by *Rel-MCKP-Greedy* and *Rel-MCKP-Greedy-Per* respectively in the comparisons.

5.2   Experiments

In this section, we present the experimental results for our proposed method and compare it with the methods discussed in the previous section.

Experiment 1: In this experiment, we investigate the effect of cost scaling factor, r, which is used in the greedy algorithm introduced in [16] to adjust the scale of the cost. The result of our experiment for different cost scaling factors are shown in Fig. 2. The scaling factor ranges from 0.8 to 2. As the diagram shows scaling factor to 1.2, $r = 1.2$ results in better performance in form of recall, precision, and F-measure.

---

[3] http://duc.nist.gov/.

[4] ROUGE package is available at http://www.berouge.com.

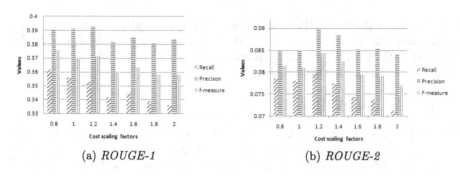

(a) *ROUGE-1*   (b) *ROUGE-2*

Fig. 2. Values of Recall, Precision, and F-measure for different scaling factors

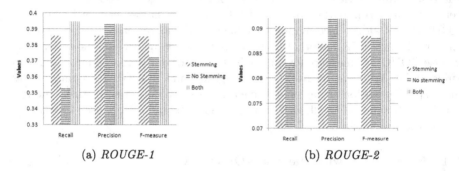

(a) *ROUGE-1*   (b) *ROUGE-2*

Fig. 3. Values of Recall, Precision, and F-measure for different stemming strategies

Experiment 2: In this experiment, we investigate the effect of stemming algorithm which is used to calculate the relevance of a sentence and a query. Stemming which is a process to reduces all words with the same root to a common form is widely used in NLP and document summarization field. Stemming was helpful in our method specially for finding the similarity of a pair of words containing plural nouns. Stemming plural nouns let us find the words easily in the WordNet, while non-stemmed words in plural form cannot be found in Word-Net. For example, words such as "schools" or "programs" cannot be found in WordNet. So, the similarity measures consider no correlation between them as they are not in the WordNet. On the other side, stemming algorithms such as Porter [26] do not find the stems of many words correctly. As an illustration, these algorithms eliminate "e" which exists at the end of most words such as "article" or "revoke", and result in a word which does not have any corresponding concept in WordNet, while looking up most non-stemmed words in WordNet such as "articles" leads to a matching concept in WordNet. Therefore, to have advantageous of both stemming and not stemming, we run an experiment in which we consider each word in both stemmed and original form in process of calculating similarity between pairs of words of a sentence and a query. Figure 3 illustrates the result of this experiment in form of ROUGE-1 and ROUGE-2.

In this experiment, the value of the scaling factor is 1.2 for all three cases and we consider three cases including (1) *Stemming* in which we apply Porter stemmer [26] and consider the stemmed word to measure the similarity between a

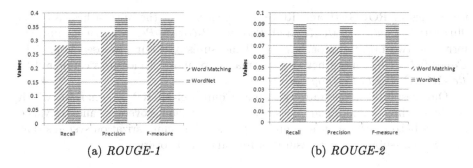

(a) ROUGE-1   (b) ROUGE-2

Fig. 4. Values of Recall, Precision, and F-measure for WordNet and Word Matching based similarity measures to calculate Relevance

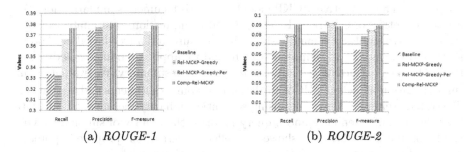

(a) ROUGE-1   (b) ROUGE-2

Fig. 5. Values of Recall, Precision, and F-measure for WordNet and Word Matching based similarity measures to calculate Relevance

sentence and a query, (2) *No Stemming* in which we consider words in the original form, and (3) *Both* in which we consider both stemmed and not-stemmed words in similarity measurement and consider the higher one as the similarity between them. As it was clear, considering both stemmed and not-stemmed words results in having better performance in form of all three measures of Recall, Precision, and F-measure. In spite of the above mentioned mistakes in stemming words, the performance of our summarizer outperforms in form of Recall, and F-measure in case of using stemmed words compared to the case using the original form of the words. Considering both stemmed and not-stemmed words increases the complexity of our calculation and makes our system slow, so, we consider just stemmed words in our similarity measurement.

Experiment 3: In this experiment, we investigate the effect of using semantic similarity measure in measuring the relevance of a sentence to the query. The value of Precision, Recall, and F-measure for ROUGE-1 and ROUGE-2 are illustrated in Fig. 4a and b. As we predicted, using WordNet based measures improve the value of all three measures of Recall, Precision, and F-measure compared to the word-matching based measure.

Experiment 4: In this experiment, we investigate the performance of different summarization approaches. The result is shown in Fig. 5 which compares the values of ROUGE-1 and ROUGE-2 measures of the approaches. As it is illustrated, *Comp-Rel-MCKP* and *Rel-MCKP-Greedy-Per* approaches outperform the two other approaches for all measures and our proposed approach, *Comp-Rel-MCKP*, has a better performance for most of measures compared to *Rel-MCKP-Greedy-Per* approach.

Our results demonstrate that our Comp-Rel-MCKP summarizer which combines three submodular measures of compression, coverage, and relevance achieves better performance compared to the other summarization systems that use two non-submodular measures of relevance and coverage.

6   Conclusion

We proposed an extractive MCKP model for multi-document summarization. We use three metrics of Coverage, Relevance, and Compression to estimate scores of sentences. We employ a semantic similarity measure and a tree-based method to identify Relevance and Compression score and we apply a modified greedy algorithm $(1 - \frac{1}{\sqrt{e}})$ performance guarantee to generate a summary. We evaluated our summarization model on DUC 2007 dataset and our experiments and evaluations illustrated stemming words and using semantic similarity measures to calculate the relevance of a sentence and a query increase the quality of summaries. Our results on the DUC data sets showed the effectiveness of our proposed approach compared to two recent MCKP based summarization models.

Acknowledgments. Authors would like to thank Taylor Berg-Kirkpatrick [2] for kindly providing us valuable information and details of their work on compression features and also Giuseppe Pirro [25] for kindly providing us with their API for accessing WordNet.

References

1. Baxendale, P.B.: Machine-made index for technical literature: an experiment. IBM J. Res. Dev. **2**(4), 354–361 (1958)
2. Berg-Kirkpatrick, T., Gillick, D., Klein, D.: Jointly learning to extract and compress. In: Proceedings of the 49th Annual Meeting of the Association for Computational Linguistics: Human Language Technologies, vol. 1, pp. 481–490. Association for Computational Linguistics (2011)
3. Carbonell, J., Goldstein, J.: The use of mmr, diversity-based reranking for reordering documents and producing summaries. In: Proceedings of the 21st Annual International ACM SIGIR Conference on Research and Development in Information Retrieval, pp. 335–336. ACM (1998)
4. Chali, Y.L., Hasan, S.A.: On the effectiveness of using sentence compression models for query-focused multi-document summarization. In: Proceedings of COLING 2012, pp. 457–474. Citeseer (2012)
5. Dang, H.T.: Overview of duc 2005. In: Proceedings of the Document Understanding Conference (2005)

6. Dasgupta, A., Kumar, R., Ravi, S.: Summarization through submodularity and dispersion. In: ACL, vol. 1, pp. 1014–1022 (2013)
7. Edmundson, H.P.: New methods in automatic extracting. J. ACM (JACM) **16**(2), 264–285 (1969)
8. Erkan, G., Radev, D.R.: Lexrank: graph-based lexical centrality as salience in text summarization. J. Artif. Intell. Res. 457–479 (2004)
9. Filatova, E., Hatzivassiloglou, V.: A formal model for information selection in multi-sentence text extraction. In: Proceedings of the 20th international conference on Computational Linguistics, p. 397. Association for Computational Linguistics (2004)
10. Gillick, D., Favre, B.: A scalable global model for summarization. In: Proceedings of the Workshop on Integer Linear Programming for Natural Langauge Processing, pp. 10–18. Association for Computational Linguistics (2009)
11. Gillick, D., Favre, B., Hakkani-Tur, D.: The icsi summarization system at tac 2008. In: Proceedings of the Text Understanding Conference (2008)
12. Gillick, D., Favre, B., Hakkani-Tur, D., Bohnet, B., Liu, Y., Xie, S.: The icsi/utd summarization system at tac 2009. In: Proceedings of the Text Analysis Conference Workshop, Gaithersburg, MD, USA (2009)
13. Goldstein, J., Mittal, V., Carbonell, J., Kantrowitz, M.: Multi-document summarization by sentence extraction. In: Proceedings of the 2000 NAACL-ANLP Workshop on Automatic summarization, vol. 4, pp. 40–48. Association for Computational Linguistics (2000)
14. Jing, H.: Sentence reduction for automatic text summarization. In: Proceedings of the Sixth Conference on Applied Natural Language Processing, pp. 310–315. Association for Computational Linguistics (2000)
15. Lin, C.Y.: Rouge: A package for automatic evaluation of summaries. In: Text Summarization Branches Out: Proceedings of the ACL-04 Workshop, pp. 74–81 (2004)
16. Lin, H., Bilmes, J.: Multi-document summarization via budgeted maximization of submodular functions. In: Human Language Technologies: The 2010 Annual Conference of the North American Chapter of the Association for Computational Linguistics, pp. 912–920. Association for Computational Linguistics (2010)
17. Lin, H., Bilmes, J.: A class of submodular functions for document summarization. In: Proceedings of the 49th Annual Meeting of the Association for Computational Linguistics: Human Language Technologies, vol. 1, pp. 510–520. Association for Computational Linguistics (2011)
18. Lin, H., Bilmes, J., Xie, S.: Graph-based submodular selection for extractive summarization. In: IEEE Workshop on Automatic Speech Recognition & Understanding, ASRU 2009, pp. 381–386. IEEE (2009)
19. Lovász, L.: Submodular functions and convexity. Mathematical Programming The State of the Art. Springer, Heidelberg (1983)
20. Luhn, H.P.: The automatic creation of literature abstracts. IBM J. Res. Dev. **2**(2), 159–165 (1958)
21. Mani, I., Maybury, M.T.: Advances in automatic text summarization, vol. 293. MIT Press, Cambridge (1999)
22. McDonald, R.: A study of global inference algorithms in multi-document summarization. In: Amati, G., Carpineto, C., Romano, G. (eds.) ECiR 2007. LNCS, vol. 4425, pp. 557–564. Springer, Heidelberg (2007)
23. Morita, H., Sasano, R., Takamura, H., Okumura, M.: Subtree extractive summarization via submodular maximization. In: ACL, vol. 1, pp. 1023–1032 (2013)

24. Petrov, S., Klein, D.: Learning and inference for hierarchically split pcfgs. In: Proceedings of the National Conference on Artificial Intelligence, vol. 22, p. 1663. AAAI Press, MIT Press; Menlo Park, Cambridge, London (1999, 2007)

25. Pirró, G., Euzenat, J.: A feature and information theoretic framework for semantic similarity and relatedness. In: Patel-Schneider, P.F., Pan, Y., Hitzler, P., Mika, P., Zhang, L., Pan, J.Z., Horrocks, I., Glimm, B. (eds.) ISWC 2010, Part I. LNCS, vol. 6496, pp. 615–630. Springer, Heidelberg (2010)

26. Porter, M.F.: An algorithm for suffix stripping. Program **14**(3), 130–137 (1980)

27. Radev, D.R., Jing, H., Styś, M., Tam, D.: Centroid-based summarization of multiple documents. Inf. Process. Manage. **40**(6), 919–938 (2004)

28. Schiffman, B., Nenkova, A., McKeown, K.: Experiments in multidocument summarization. In: Proceedings of the Second International Conference on Human Language Technology Research, pp. 52–58. Morgan Kaufmann Publishers Inc. (2002)

29. Sekine, S., Nobata, C.: Sentence extraction with information extraction technique. In: Proceedings of the Document Understanding Conference (2001)

30. Takamura, H., Okumura, M.: Text summarization model based on maximum coverage problem and its variant. In: Proceedings of the 12th Conference of the European Chapter of the Association for Computational Linguistics, pp. 781–789. Association for Computational Linguistics (2009)

31. Yih, W.-T., Goodman, J., Vanderwende, L., Suzuki, H.: Multi-document summarization by maximizing informative content-words. In: IJCAI (2007)

A Comparison of Human and Machine Estimation of Speaker Age

Mark Huckvale$^{(\boxtimes)}$ and Aimee Webb

Speech, Hearing and Phonetic Sciences,
University College London, London, UK
{m.huckvale,aimee.webb.11}@ucl.ac.uk

Abstract. The estimation of the age of a speaker from his or her voice has both forensic and commercial applications. Previous studies have shown that human listeners are able to estimate the age of a speaker to within 10 years on average, while recent machine age estimation systems seem to show superior performance with average errors as low as 6 years. However the machine studies have used highly non-uniform test sets, for which knowledge of the age distribution offers considerable advantage to the system. In this study we compare human and machine performance on the same test data chosen to be uniformly distributed in age. We show that in this case human and machine accuracy is more similar with average errors of 9.8 and 8.6 years respectively, although if panels of listeners are consulted, human accuracy can be improved to a value closer to 7.5 years. Both human and machines have difficulty in accurately predicting the ages of older speakers.

Keywords: Speaker profiling · Speaker age prediction · Computational paralinguistics

1 Introduction

The estimation of the age of a speaker from an analysis of his or her voice has forensic applications – for example the profiling of perpetrators of crimes [1], commercial applications – for example targeted advertising, and technological applications – for example adaptation of a speech recognition system to a speaker [2].

Many previous studies have looked at the performance of both human listeners and machine-learning systems for the estimation of age from speech. Unfortunately, variations in the data set, task and performance metric make these studies hard to compare. In our work we take the view that the natural task should be numerical estimation of the age of the speaker, and the natural performance metric should be the mean absolute error (MAE) of estimation. The MAE answers the question "how close is the average estimate to the actual age?"

A recent review of previous studies on human listener judgments of speaker age may be found in [3]. Of the studies reported which used numerical age estimation and MAE, most seem to suggest human performance has an MAE of about 10 years. Table 1 provides a summary.

© Springer International Publishing Switzerland 2015
A.-H. Dediu et al. (Eds.): SLSP 2015, LNAI 9449, pp. 111–122, 2015.
DOI: 10.1007/978-3-319-25789-1_11

Table 1. Previous studies on human listener age estimation

Study	MAE (yr)	Notes
Braun et al. [4]	10.5	German speakers and listeners
Braun et al. [4]	8.5	Italian speakers and listeners
Krauss et al. [5]	7.1	Limited age-range
Amilon et al. [6]	9.7	
Moyse et al. [7]	10.8	

There have also been many studies in the machine prediction of speaker age from speech, see [8] for a review of the state of the art. The studies also vary greatly in terms of the data set, audio quality, audio duration, audio feature set, recognition task and machine learning approach taken. Machine learning methods have included support vector machines, Gaussian mixture models (GMM), GMM supervectors, i-vectors and phoneme recognisers. Reference [9] provides system design and performance figures for a range of contemporary approaches together with a fusion of systems for age estimation of very short speech excerpts (<2 s). The different approaches only varied by a few percentage points (43.1–47.5 % age categories correctly identified) suggesting that the choice of machine learning algorithm is not a critical factor.

The results of some machine studies that addressed the problem of numerical age estimation evaluated with MAE are shown in Table 2.

The best performing system on adult speech described in [8] used the i-vector approach followed by support vector regression and demonstrated an MAE of 6.1 years. While at first glance this looks considerably better than the MAE figures quoted for human performance, it is important to note that the test data used in this study had a non-uniform age distribution, with significantly more speakers in the 20–29 age band than in other bands. This uneven distribution means that even a null model which always predicted the mean age of the training speakers would show an MAE of 10.6 years for female speakers and 10.1 years for male speakers. The superiority of the machine system might therefore have arisen from the unfair knowledge it had of the prior age distribution. Since all machine systems in Table 2 may have exploited a prior on the test speaker age, this makes it impossible to compare any of them fairly with the human listeners, who were not given that information.

The goals of this study are to make a fair comparison between human and machine speaker age estimation. This will be done by: (i) comparing human and machines on same test data, (ii) comparing them on the on same task – numerical age estimation, (iii) evaluating both using the same performance metric – MAE, and (iv) removing any advantage of knowing a prior on the test set by using a uniform test age distribution.

We describe the data set used for the task, results of a human listening task and results of a machine age estimation system constructed to be similar to the best performing systems in Table 2.

Table 2. Previous studies on machine age estimation

Study	MAE (yr)	Notes
Bocklet et al. [10]	0.8	Children 7–10 yrs
Feld et al. [11]	7.2–12.8	Same and cross-language
Dobry et al. [12]	9.29–10.00	Depending on gender
Bahari et al. [13]	7.48	Null model = 8.88
Bahari and van Hamme [14]	7.9	
Bahari et al. [8]	6.08	Null model = 10.3

2 Speech Corpus

The work described here uses the Accents of the British Isles corpus (version 2) available from The Speech Ark [15]. The ABI-2 corpus consists of recordings of 262 speakers covering 13 accent areas of the British Isles. Each speaker is recorded reading a range of English language materials; although for this work we used only the first part of the "accent diagnostic" passage which has a median duration of 39.2 s. The recordings are supplied as wide bandwidth audio of good quality, recorded using a close-talking microphone at 22050 samples/s.

The corpus was divided into a test set containing 52 speakers, and a training set of the remaining 210 speakers. The test set was chosen to have equal representation of men and women for all 5-year age bands between 15 and 80. Figure 1 shows the age distribution by gender for the test and training sets.

The mean age of the training set was 42.6 years. Used as a null prediction for the test set this value would score a mean absolute error of 16.7 years.

3 Human Prediction Performance

To obtain human age prediction performance a web-based data collection protocol was used. Listeners were able to listen to each test recording then make an age estimation using a sliding scale between 15 and 80. Estimates were recorded as whole numbers of years. Recordings were presented in a random order different for each listener. Listeners could make their age estimate at any time while the recording was playing, or could listen to the audio multiple times. Listeners conducted the test in their own homes, but were asked to listen over headphones. The web interface may be seen in Fig. 2.

An attempt was made to recruit listeners over a range of ages and genders, although the balance was not perfect. In all, 36 native English listeners completed the test; Table 3 shows their distribution by age and gender.

The raw age predictions are plotted against the true speaker ages in Fig. 3. The line of best fit has a slope of 0.68 and an intercept of 12.7 years. The correlation coefficient is 0.759 and the mean absolute error (MAE) of prediction is 9.79 years (male speakers only 10.1, female speakers only 9.51).

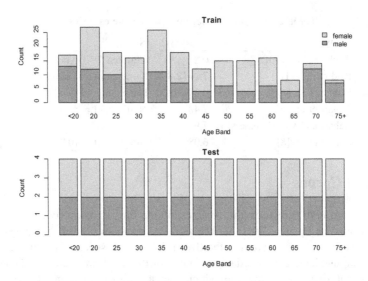

Fig. 1. Age and gender distribution for the train and test sets.

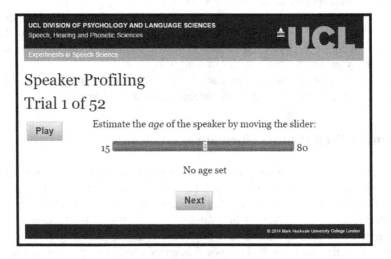

Fig. 2. Web experiment interface.

Table 3. Distribution of listeners by age and gender

Number	20–29	30–39	40–49	50–59	60–69
Male	4	4	3	3	2
Female	5	4	3	6	2

The MAE as a function of the age and sex of the speaker is shown in Table 4, and MAE as a function of the age and sex of the listener is shown in Table 5.

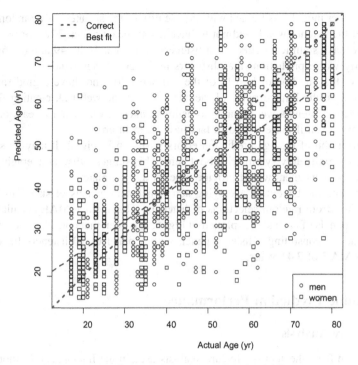

Fig. 3. Age predictions of 52 test speakers by 36 listeners.

Table 4. Mean Absolute error of prediction as a function of age and sex of the speaker

MAE(yr)	20–29	30–39	40–49	50–59	60–69	70–79
Male	7.42	9.32	10.52	7.99	13.85	11.14
Female	5.63	8.00	8.71	12.07	12.10	11.30

Table 5. Mean absolute error of prediction as a function of age and sex of the listener

MAE (yr)	20–29	30–39	40–49	50–59	60–69
Male	8.34	8.22	11.36	10.01	13.24
Female	9.95	9.39	10.57	9.38	10.10

Generalised linear mixed-effects models of the predictions were estimated using Markov chain Monte Carlo techniques with the MCMCglmm package [16]. The models were used to determine if the absolute error in age prediction was affected by the sex or age of the speaker, or the sex or age of the listener.

The speaker model was trained with the identity of the listener as a random factor. The sex of the speaker was found not to have significant effect. The age-band of the speaker did have significant effect, with the ages of speakers in the 20–29 age band being significantly better estimated than the other bands.

The listener model was trained with the identity of the speakers as a random factor. The sex of the listener was found not to have a significant effect. The age-band of the listener did have significant effect with listeners in age-bands 40–49 and 60–69 giving significantly worse predictions than listeners in the 20–29 age band.

The fact that the line of best fit of the estimates does not have a gradient of one might be due to the limited range of the age slider in the web task creating floor and ceiling effects. Listeners were unable to estimate ages lower than 15 years or greater than 80 years even if these would in fact have been in error.

A distribution of the age prediction errors is shown in Fig. 4. It may be seen that errors are approximately symmetric about zero. This suggests that an averaging of age predictions over listeners would provide a better age estimate.

To estimate the benefit of averaging across listeners, panels of size 2 to 12 were built post hoc from random selections of listeners. The average MAE calculated over 50 random panels of each size is plotted in Fig. 5. It is seen that considerable advantage may be had by consulting a listener panel, with a panel of 10 listeners for example having an MAE of 7.41 years.

4 Machine Prediction Performance

4.1 Feature Analysis

Following on from the acoustic feature analysis used in the Interspeech Computational Paralinguistics challenges, we have used the OpenSMILE toolkit [17] to generate a large feature vector for each audio recording. The specific set of parameters was those used for the 2014 challenge [18]. This feature set comprises 65 low-level descriptors which are extracted from short-term windows on the signal. These describe speech signal properties such as energy, spectral envelope, pitch and voice quality. The descriptors are then summarized over each file using a large number of statistical measures such as means, medians, quantiles, differences, and so on. The output is a vector of 6373 features for each file.

4.2 Machine Learning

The method chosen for learning the prediction model was Support Vector Regression (SVR) [19] as used by previous authors [10–13]. The "e1071" package for the "R" statistics library was the chosen implementation [20]. In support vector regression a subset of the training vectors are chosen to represent the optimal regression hyperplane.

To reduce the training complexity, a feature selection process was implemented. Only features which had an absolute value of correlation greater than an arbitrary threshold of 0.1 with the age of the speaker were passed to SVR. This selection was made on the training set only and left 2538 features. Performance was not strongly affected by the choice of this threshold providing enough features were included.

At the front end of the SVR, a radial-basis function kernel is applied – this provides an additional tunable non-linearity applied to the feature values. Also the SVR

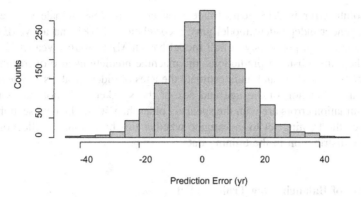

Fig. 4. Distribution of age prediction errors by human listeners.

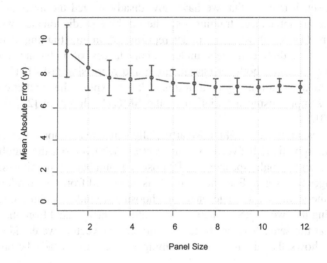

Fig. 5. Distribution of mean absolute error of age prediction by listener panel size. Average over 50 random panels. Bars show 1 s.d.

algorithm applies a feature normalization step to ensure all features have a similar dynamic range.

Optimal control parameters were found using a cross-validation procedure on subsets of the training data only. The optimal parameters were: C = 8, gamma = 0.25/number-of-features, epsilon = 0.1.

Separate SVR systems were trained for male and female speakers as in [8].

4.3   Raw Prediction Performance

The raw prediction performance of SVR is shown in Fig. 6. The line of best fit has a slope of 0.53 and an intercept of 18.9 years. The correlation coefficient is 0.82, and the mean absolute error is 9.13 years (male speakers only 7.98, female speakers only 10.29). A gender independent model gave a correlation of 0.81 and an MAE of 9.18 years. As mentioned previously, a null model has an MAE of 16.7 years.

Like the human listener predictions, the machine predictions also overestimate the ages of younger speakers and underestimate the ages of older speakers. Table 6 shows the MAE as a function of the age and sex of the speaker. It is noticeable that the greatest estimation errors are with the speakers older than 50. In the next section we try to rebalance the training set to investigate whether this bias is just a reflection of the uneven age distribution in the training data.

4.4   Effect of Balancing the Training Set

Since our original motivation was to make a fair comparison with human listeners on a balanced test set, it may be that we have now disadvantaged the machine system by only providing an unbalanced training set. The machine predictions are worse for the older speakers (Table 6) who are under-represented in the training data (Fig. 1). The training of predictive models under circumstances of imbalanced data is an ongoing area of research both for classification and regression tasks [21]. To explore the effect of imbalanced training data in this task, we explore the synthetic creation of training data samples using a variation of the SMOTE algorithm [22] designed for regression [23].

Here we present results in which we artificially generate additional training samples from linear interpolations between existing vectors. We even out the number of samples for male and female speakers and boost the number of training samples for speakers of ages >50 years. Each new sample is generated from two randomly-chosen instances of the same sex and age band by choosing a random point along the interpolation joining the two vectors. The new age value is interpolated from the ages of the two samples at the same fraction. In total a further 271 vectors were added.

Figure 7 shows the distribution of training samples by decade before and after balancing.

A new SVR model was trained on the re-balanced data. 3378 features were selected using the same correlation threshold. Cross validation on the training data suggested the best control parameters were now C = 32, gamma = 0.125/number of features, epsilon = 0.001.

Figure 8 shows the age predictions for the test set after training with the re-balanced training data. The line of best fit had a slope of 0.554 and an intercept of 17.8 years. The correlation was 0.852 and the MAE 8.64 years (male speakers only 7.87, female speakers only 9.42). A gender independent model gave a correlation of 0.81 and an MAE of 9.49 years. Table 7 shows the mean absolute error of prediction as function of the age and sex of the speaker.

While some performance improvements are seen in comparison to results with the original training set, overall the improvement is small. It may be that in this task, the SVR model does not gain any useful information from the synthetic samples.

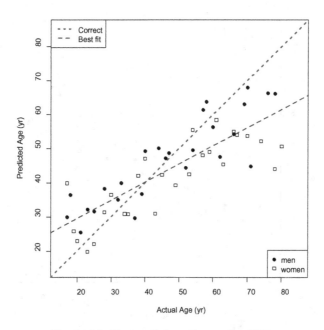

Fig. 6. Machine prediction of age using SVR.

Table 6. Mean absolute error of prediction as a function of age and sex of the speaker

MAE(yr)	20–29	30–39	40–49	50–59	60–69	70–79
Male	7.64	4.89	4.68	5.54	8.87	12.40
Female	3.12	4.46	7.86	7.72	11.00	23.97

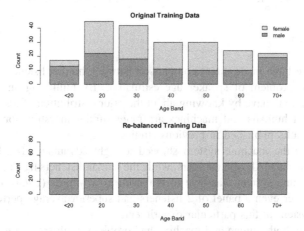

Fig. 7. Results of boosting the frequency of the older speakers in the training data.

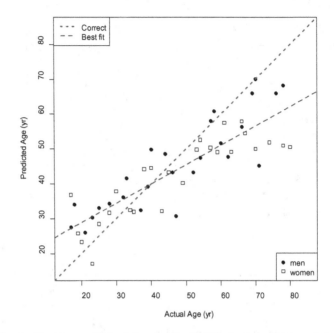

Fig. 8. Machine prediction of age using SVR trained on rebalanced data.

Table 7. Mean absolute error of prediction as a function of age and sex of the speaker using SVR trained on rebalanced data

MAE(yr)	20–29	30–39	40–49	50–59	60–69	70–79
Male	6.81	4.41	8.32	4.79	8.74	11.38
Female	4.10	4.68	6.45	5.27	9.45	23.04

5   Discussion

In this study we have made direct comparison between human listeners and machine learning on the problem of speaker age estimation. By nullifying any advantage a machine system may have by knowing about the prior distribution of test speakers, we have shown that humans and machines are more similar in estimation performance compared to results published in previous studies.

Nevertheless the machine system showed a slight advantage. The best machine performance had an MAE of 8.64 years, while the human listeners had an MAE of 9.79 years. The machine system was able to outperform two-thirds (25/36) of the human listeners. However even a panel of 2 listeners had superior average performance than the machine system in this particular experiment.

Interestingly, both human and machine had problems with the extremes of the age range, both showing lines of best fit with slopes significantly less than unity. We showed that boosting the number of older speakers in the training set had very little effect, perhaps because the SVR model did not extract any more information from the interpolated samples than it could extract from the original samples. The difficulty of predicting the ages of older speakers may be due to some inherent characteristics of the data – perhaps the voice characteristics of older speakers are more variable for a given age. This would fit with other research [24] that has shown how cognitive abilities become increasingly heterogeneous with advancing age. Further research into this issue, and improved machine performance, is likely to come from data sets with a larger number of speakers and a larger range of ages.

References

1. Tanner, D.C., Tanner, M.E.: Forensic Aspects of Speech Patterns: Voice Prints, Speaker Profiling, Lie and Intoxication Detection. Lawyers & Judges Publishing, Tucson (2004)
2. Pellegrini, T., Hedayati, V., Trancoso, I., Hämäläinen, A., Dias, M.: Speaker age estimation for elderly speech recognition in European Portuguese. In: Proceedings of InterSpeech 2014, Singapore, pp. 2962–2966 (2014)
3. Moyse, E.: Age estimation from faces and voices: a review. Psychologica Belgica **54**, 255–265 (2014)
4. Braun, A., Cerrato, L.: Estimating speaker age across languages. In: Proceedings of ICPhS 1999, San Francisco, pp. 1369–1372 (1999)
5. Krauss, R., Freyberg, R., Morsella, E.: Inferring speakers' physical attributes from their voices. J. Exp. Soc. Psychol. **38**, 618–625 (2002)
6. Amilon, K., van de Weijer, J., Schötz, S.: The impact of visual and auditory cues in age estimation. In: Müller, C. (ed.) Speaker Classification II. Lecture Notes in Computer Science LNCS(LNAI), vol. 4441, pp. 10–21. Springer, Heidelberg (2007)
7. Moyse, E., Beaufort, A., Brédart, S.: Evidence for an own-age bias in age estimation from voices in older persons. Eur. J. Aging **11**, 241–247 (2014)
8. Bahari, M., McLaren, M., van Hamme, H., van Leeuwen, D.: Speaker age estimation using i-vectors. Eng. Appl. Artif. Intell. **34**, 99–108 (2014)
9. Li, M., Han, K., Narayanan, S.: Automatic speaker age and gender recognition using acoustic and prosodic level information. Comput. Speech Lang. **27**, 151–167 (2013)
10. Bocklet, T., Maier, A., Nöth, E.: Age determination of children in preschool and primary school age with GMM-based supervectors and support vector machines/regression. In: Sojka, P., Horák, A., Kopeček, I., Pala, K. (eds.) TSD 2008. LNCS (LNAI), vol. 5246, pp. 253–260. Springer, Heidelberg (2008)
11. Feld, M., Barnard, E., van Heerden, C., Müller, C.: Multilingual spear age recognition: regression analyses on the Lwazi corpus. In: IEEE Workshop on Automatic Speech Recognition and Understanding, pp. 534–539 (2009)
12. Dobry, G., Hecht, R., Avigal, M., Zigel, Y.: Supervector dimension reduction for efficient speaker age estimation based on the acoustic speech signal. IEEE Trans. Audio Speech Lang. Process. **19**, 1975–1985 (2011)
13. Bahari, M., van Hamme, H.: Speaker age estimation and gender detection based on supervised non-negative matrix factorization. In: Proceedings of IEEE Workshop Biometric Measurements and Systems for Security and Medical Applications, pp. 1–6 (2011)
14. Bahari, M., van Hamme, H.: Speaker age estimation using hidden Markov model weight supervectors. In: IEEE International Conference on Information Science, Signal Processing and their Applications, pp. 517–521 (2012)

15. Speech Ark, Second Accents of the British Isles Corpus. www.thespeechark.com/abi-2-page.html
16. Hadfield, J.: MCMC methods for multi-response generalized linear mixed models: The MCMCglmm R package. J. Stat. Softw. **33**, 1–22 (2010)
17. Eyben, F., Weninger, F., Groß, F., Schuller, B.: Recent developments in opensmile, the Munich open-source multimedia feature extractor. In: Proceedings of the 21st ACM International Conference on Multimedia, Barcelna, Spain, pp. 835–838 (2013)
18. Schuller, B., Steidl, S., Batliner, A., Epps, J., Eyben, F., Ringeval, F., Marchi, E., Zhang, Y.: The INTERSPEECH 2014 Computational Paralinguistics Challenge: Cognitive and Physical Load. Interspeech 2014, Singapore (2014)
19. Smola, A., Schölkopf, B.: A tutorial on support vector regression. J. Stat. Comput. **14**, 199–222 (2004)
20. CRAN Project, E1071 package of functions from Dept. Statistics, TU Wein. cran.r-project.org/web/packages/e1071/index.html
21. Branco, P., Torgo, L., Ribeiro, R.: A survey of predictive modelling under imbalanced distributions. CoRR abs/1505.01658 (2015)
22. Chawla, N.V., Bowyer, K.W., Hall, L.O., Kegelmeyer, W.P.: SMOTE: synthetic minority over-sampling technique. J. Artif. Intell. Res. **16**, 321–357 (2002)
23. Torgo, L., Ribeiro, R.P., Pfahringer, B., Branco, P.: SMOTE for regression. In: Correia, L., Reis, L.P., Cascalho, J. (eds.) EPIA 2013. LNCS, vol. 8154, pp. 378–389. Springer, Heidelberg (2013)
24. Ardila, A.: Normal aging increases cognitive heterogeneity: analysis of dispersion in WAIS-III scores across age. Arch. Clin. Neuropsychol. **22**, 1003–1011 (2007)

Acoustical Frame Rate and Pronunciation Variant Statistics

Denis Jouvet[1,2,3](✉) and Katarina Bartkova[4]

[1] Speech Group, Inria, LORIA, Villers-lès-Nancy 54600, France
denis.jouvet@inria.fr
[2] Université de Lorraine, LORIA, UMR 7503, Villers-lès-Nancy 54600, France
[3] CNRS, LORIA, UMR 7503, Villers-lès-Nancy 54600, France
[4] ATILF - Analyse et Traitement Informatique de la Langue Française,
44 Av De La Libération, BP 30687, 54063 Nancy Cedex, France
katarina.bartkova@atilf.fr

Abstract. Speech technology enables computing statistics on word pronunciation variants as well as investigating various phonetic phenomena. This is achieved through a forced alignment of large amounts of speech signals with their possible pronunciations variants. Such alignments are usually performed using a 10 ms frame shift acoustical analysis. Therefore, the three emitting state structure of conventional acoustic hidden Markov models introduces a minimum duration constraint of 30 ms for each phone segment. This constraint is not critical at low speaking rates, but may introduce artefacts at high speaking rates. Thus, this paper investigates the impact of the acoustical frame rate on corpus-based phonetic statistics. Statistics on pronunciation variants obtained with a shorter frame shift (5 ms) are compared to the statistics resulting from the standard 10 ms frame shift. Statistics are computed on a large speech corpus of more than 3 million running words, and are analyzed with respect to the estimated local speaking rate. Results exhibit some discrepancies between the two sets of statistics, in particular for high speaking rates where the usual acoustic analysis frame shift of 10 ms leads to an under-estimation of the frequency of the longest pronunciation variants.

Keywords: Speech modeling · Speech-text alignment · Acoustical frame rate · Corpus-based phonetic statistics

1 Introduction

Many phonetic studies rely on a segmentation of the speech signal into words and phones. Manual segmentation, especially at the phone level, is a lengthy and tedious task. Moreover the agreement between annotators is not perfect even with respect to the existence of some phone segments [26] in spontaneous speech. Another approach consists in relying on automatic segmentation at the phone and at the word levels of large amounts of speech data. In such an approach the

© Springer International Publishing Switzerland 2015
A.-H. Dediu et al. (Eds.): SLSP 2015, LNAI 9449, pp. 123–134, 2015.
DOI: 10.1007/978-3-319-25789-1_12

manual transcription of the speech signal into words is still required, but this is a much easier task than the phonetic segmentation itself. Knowing the sequence of words corresponding to a speech segment, all the possible pronunciations into sequences of phones are derived, and automatic alignment tools find the sequence of phones (among all the possible phone sequences corresponding to the different pronunciation variants) that best matches with the speech signal.

Speech-text alignments are typically performed on speech segments that are the size of a sentence (e.g., [4,13,27]) using hidden Markov models (HMM) with feature vectors computed every 10 ms. Although context dependent phone models provide the best performance in speech recognition because of a better modeling of the co-articulation between adjacent sounds, more accurate boundaries are usually obtained with context independent phone models [20]. Boundary statistical corrections were proposed for context-dependent based modeling [29], and segmentation constrained training [15] was investigated, as well as the impact of the model topology [25].

Many phone segmentation procedures were improved and evaluated for corpus-based speech synthesis (e.g. [19]). In this context, higher acoustic analysis frame rates corresponding to 3 ms [29], 4 ms [3] or 5 ms [23] frame shifts are used for improving the phone boundary precision. Moreover, boundary refinement post-processing was also proposed, for example, using other features or techniques targeted towards the detection of transitions [29]. Other proposed approaches consist in using multiple features [23], multiple models [21] or multiple systems [17]. It should be noted that all these approaches have been developed for corpus-based speech synthesis, so they are dealing with good quality speech signal, well-articulated speech, and the sequence of phones corresponding to each sentence is assumed to be known (because of the controlled recording that is carried out for such corpora).

Another research direction relates to the use of automatic speech-text alignment for conducting phonetic and linguistic studies on large speech corpora [1]. This includes the study of the schwa and of liaisons [5,8,9], as well as the study of pronunciation variants [2] and the analysis of other phenomena [22,24]. In these approaches speaker-independent models are required, and the acoustic models typically rely on 10 ms frame shift between adjacent feature vectors. Consequently, the three emitting states of the acoustic models lead to a minimal duration of three frames (i.e., 30 ms) for each phone segment. As such minimum phone duration is a constraint for the phone segmentation process, this paper focuses on the analysis of the impact of the acoustic analysis frame rate on corpus-based phonetic statistics. Several pronunciation phenomena are studied and analyzed with respect to the local speaking rate. The mute "e" is particularly studied as the phonetic realization or the elision of the corresponding sound (/ə/) is one main adjustment variable of the speaking rate in French (further comments are given in Sect. 4.1). Another aspect studied is the pronunciation of some consonantal clusters in word final position (detailed in Sect. 4.2).

The paper is organized as follows. Section 2 presents the speech data and the modeling used while Sect. 3 details the speech-text alignment process. Then,

Sect. 4 analyzes the frequency of some pronunciation variants with respect to the local speaking rate, using speech-text alignments obtained with 5 ms and 10 ms frame shifts. A conclusion ends the paper.

2   Speech Data and Modeling

The speech corpora used in the experiments come from the ESTER2 [12] and the ETAPE [14] evaluation campaigns, and from the EPAC [10,11] project. The ESTER2 and EPAC data are French broadcast news collected from various radio channels. They contain mainly prepared speech (speech from the journalists). A large part of the data is of studio quality, though some parts are of telephone quality. The ETAPE data corresponds to debates collected from various radio and TV channels. Thus this corresponds mainly to spontaneous speech. Only the training subsets of these corpora are used in the experiments reported in this paper. This amounts to almost 300 hours of speech signal for which a manual orthographic transcription, at the word level, is available.

The speech material was analyzed by computing 13 Mel frequency cepstral coefficients (MFCC) per frame. The whole data set was analyzed two times, once with the standard 10 ms frame shift, and once with the reduced 5 ms frame shift. First and second order temporal derivatives were then added to the static coefficients to produce a 39 coefficient vector. For the 5 ms frame shift, the indexes of the frames involved in the computation of the temporal derivatives were modified in order to correspond to the same temporal window as in the 10 ms frame shift case.

Using the conventional modeling approach, each phone was modeled by a three emitting state hidden Markov model. The training process involved several steps. First, using the standard pronunciation of each word, a first model was trained, and then used to realign the training data in order to associate with each speech segment the sequence of estimated pronunciation variants of the words. A second model was then trained from these alignments, and the training data were re-aligned a second time using this second model. Finally a third, more detailed, model was trained (7500 shared densities – senones), and was then used for determining the speech-text alignments that are later analyzed in the paper. The trained acoustic models have 64 Gaussian components per density.

This training procedure was applied for each frame shift (5 and 10 ms frame shifts), and for each pronunciation lexicon (see details in Sect. 3.1), using the Sphinx toolkit [28].

3   Speech-Text Alignment

All the training data were aligned with the trained acoustic models, thus providing the phone and word segmentations.

3.1 Lexicon and Pronunciation Variants

The lexicon contains more than 60,000 words. Whenever possible the pronunciation variants of the words were extracted from available lexicons (BDLEX [7] and in-house lexicons). For the words not present in these lexicons, the pronunciation variants were obtained automatically using joint multigram models (JMM) and conditional random field (CRF) based grapheme-to-phoneme converters [16]. On average, there are 2.25 pronunciations variants per word in the training lexicon. Most of the pronunciation variants come from the mute "*e*" (schwa /ə/ which can be pronounced or not at the end of many words, or in internal position in some French words), and from the liaisons (i.e. introduction of a liaison conso-nant which may be pronounced when the following word starts by a vowel). This corresponds to the standard lexicon which is used in the experiments reported later on in Sects. 3.2 and 4.1.

An extended lexicon was created for the last set of experiments (in Sect. 4.2). It corresponds to the standard lexicon to which extra pronunciation variants were added for words that end by a cluster /plosive liquid/ such as /t ʁ/ in final position in the word "*ministre*" (minister). In fact, in order to analyze the pronunciation of such final clusters, all the possible pronunciation variants were generated, considering the phonetic realization or the elision of any of the corresponding phonemes, as well as the pronunciation of a final schwa. Hence for the word "*ministre*" we get eight pronunciations variants, as shown in Table 1.

Table 1. The eight pronunciation variants for the word "*ministre*" ("minister") in the extended pronunciation lexicon.

/m i n i s t ʁ ə/	[+t][+ʁ][+ə]	/m i n i s ʁ ə/	[-t][+ʁ][+ə]
/m i n i s t ʁ/	[+t][+ʁ][-ə]	/m i n i s ʁ/	[-t][+ʁ][-ə]
/m i n i s t ə/	[+t][-ʁ][+ə]	/m i n i s ə/	[-t][-ʁ][+ə]
/m i n i s t/	[+t][-ʁ][-ə]	/m i n i s/	[-t][-ʁ][-ə]

3.2 Example of Phone Segmentation

Figure 1 displays an example of speech alignment. The panel "man" displays the manual segmentation, and the panels ".f05ms" and ".f10ms" display the automatic segmentations achieved with the standard pronunciation lexicon and, respectively, the 5 and 10 ms frame shifts. The French sentence of this example is "... *Madame la Ministre merci* ..." ("... Madame Minister thanks ...") pronounced in a rather rapid speaking rate. Our expert phonetician has not observed any presence of a /t/ at the end of the word "*Ministre*", but just a short /ʁ/ and a short schwa /ə/. As the pronunciation variant without /t/ is not present in the standard pronunciation lexicon used, the automatic alignments found, in both cases, that the pronunciation variant providing the best match is /m i n i s t ʁ ə/. However, with the 5 ms frame shift, the part /t ʁ ə/ corresponds to three short segments (and the /t ʁ/ segment almost correspond to the /ʁ/ segment of the manual annotation), whereas for the 10 ms frame shift, the 30 ms

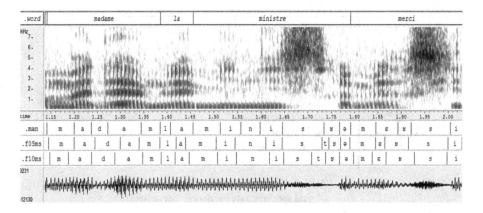

Fig. 1. Example of manual and automatic phone segmentations (".man" indicates the manual segmentation, ".f05ms" the automatic segmentation using 5 ms frame shift, and ".f10ms" the automatic segmentation using 10 ms frame shift). The speech segment corresponds to "... *Madame la Ministre merci* ..." ("... Madame Minister thanks ...") pronounced in a rather rapid speaking rate.

phone minimum constraint force the /t/ to a wrong temporal position (where it overlaps with the actual /s/ sound of the manual segmentation).

This example shows that having a shorter phone minimum duration constraint helps when dealing with rapid speaking rate, although it is sometime difficult to decide in fast speaking rate if a sound is reduced (in duration) or is discarded by the speaker.

4 Impact of Frame Rate on Statistics

This section analyzes, with respect to the local speaking rate, statistics on pronunciation variants estimated using 5 and 10 ms frame shifts. The local speaking rate is computed for each word using a local window of at most seven words (three words before and three words after the current word), similar to what was done in [18]. A smaller number of words may be considered if a long pause (more than 100 ms) is present in this window. In the reported statistics, the local speaking rate is expressed in phones per second.

4.1 Mute "*e*"

Figure 2 shows the frequency of the variants corresponding to the pronunciation of the mute "*e*" in function words followed by a word starting with a consonant. The frequency of these pronunciation variants are displayed for the function words "*que*" (meaning "that", 150,000 occurrences) and "*de*" (meaning "of", 38,000 occurrences). The last curves correspond to the frequency of pronunciation of the mute "*e*" estimated over all similar function words (340,000 occurrences).

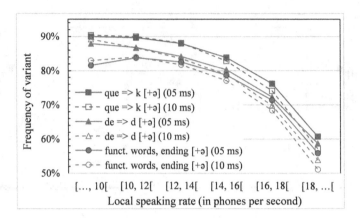

Fig. 2. Frequency of pronunciation of the final schwa, estimated using 5 or 10 ms frame shifts, in short words such as *"que"* (that) and *"de"* (of), and average over all similar monosyllabic function words.

For the word *"que"*, at low speaking rate, in 90 % of the cases the final schwa is pronounced. The frequency of pronunciation of the schwa gets lower as the speaking rate increases. The figure shows that at low speaking rate, the 5 and 10 ms frame shifts lead to rather similar frequency values. However, as the speaking rate increases, the difference between the statistics estimated with the 5 and 10 ms frame shifts gets larger, the 5 ms frame shift leading to somewhat higher values.

Variation of the speaking rate does not affect all the phonetic units in the same way. The variation of the speaking rate is achieved either by a faster (or slower) movement of the articulators or by omission (or insertion) of some phonetic units. In French, one main adjustment variable of the speaking rate is the pronunciation of the schwa vowel (/ə/). When the articulation rate increases, the length of the schwa can, not only be substantially shortened, but this vowel can completely disappear even in monosyllabic French function words, where the schwa is the only vowel. When the articulation rate is slowing down, there is a tendency to utter all the schwa vowels of a word, and also to add epenthetic schwas after each word final consonant (especially before a pause or a consonantal syllable attack). The schwa vowel that never occurs in a stressed position is generally more affected by duration reduction than the other French vowels; therefore shorter acoustic models are better adapted to their detection.

Figure 3 displays similar statistics for the pronunciation of the final schwa, before a word starting by a consonant, computed on more than 213,000 occurrences of one syllable words such as *"quatre"* (four), *"entre"* (between), and on more than 167,000 occurrences of two or more syllable words such as *"Europe"*, *"histoire"* (history), ... Monosyllabic function words, such as *"le"*, *"de"*, *"que"*, ..., are not considered in those statistics. Again, statistics computed using 5 ms frame shifts lead to larger frequency values for the pronunciation of the ending schwa.

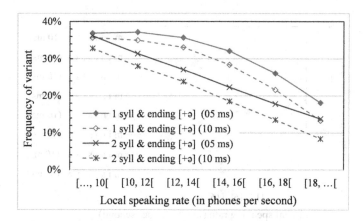

Fig. 3. Frequency of pronunciation of the final schwa, estimated using 5 or 10 ms frame shifts, in one syllable words and in polysyllabic words.

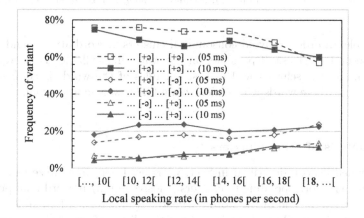

Fig. 4. Frequency of pronunciation of the first and/or the second schwa in polysyllabic words, estimated using 5 or 10 ms frame shifts.

Figure 4 concerns the pronunciation of the mute "*e*" in internal positions of words. The statistics displayed here are extracted from 7,500 occurrences of words including two mute "*e*" in internal positions, as for example "*revenir*" (come back) which can be pronounced as:

$$/ʁ\ ə\ v\ ə\ n\ i\ ʁ/\ [+ə]\ \ldots [+ə]\ \ldots$$
$$\text{or}\ /ʁ\ ə\ v\ n\ i\ ʁ/\quad [+ə]\ \ldots [-ə]\ \ldots$$
$$\text{or}\ /ʁ\ v\ ə\ n\ i\ ʁ/\quad [-ə]\ \ldots [+ə]\ \ldots$$

According to the pronunciation rules of standard French, when several adjacent syllables contain neutral schwa like vowels, every second schwa can be elided from the pronunciation. This rule is applied especially when the speed of pronunciation is increasing. Also, there is a preference to keep the schwa in the

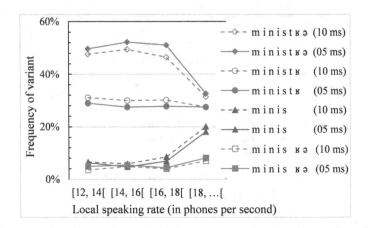

Fig. 5. Frequency of some pronunciation variants for the cluster "*tre*" /t ʁ ə/ at the end of the word "*ministre*" (minister), estimated using 5 or 10 ms frame shifts.

first syllable, in order to avoid consonantal clusters at word attacks and also to place a secondary stress, if possible, on the word first syllable. This preference of maintaining the schwa vowel in first syllables of the words is confirmed by our statistics, where words with elided schwa like vowels in first syllable are very seldom.

4.2   Analysis of Final Clusters

The extended lexicon described in Sect. 3.1 is used here to analyze the speech-text alignments corresponding to clusters /plosive liquid/ in word final position, such as /t ʁ/ at the end of the word "*ministre*" (minister). All possible variants corresponding to the phonetic realization or the elision of any of the phonemes of the cluster are set possible in the extended lexicon, as well as a possible insertion of a final schwa (cf. example for the word "*ministre*" in Table 1). Figure 5 displays the frequency, with respect to the local speaking rate, of some variants observed for the word "*ministre*", which occurs 3,200 times in the data. However, because of a too small number of occurrences associated to the lowest speaking rates ([...-10[and [10-12[phones per second), results are reported only for speaking rates higher than 12 phones per second, for which there are more than 400 occurrences in each speaking rate bin.

Results shows that the frequency of the full pronunciation /m i n i s t ʁ ə/ is higher when estimated using the 5 ms frame shift. Two interesting facts are related to the omission of /t/ and of the whole cluster /t ʁ/ in final position at high speaking rates.

The simplification of consonantal clusters especially at word final positions is very frequent in French (but also in other languages). In fact, the final syllabic codas ending with consonants of increasing degree of sonority (example occlusive followed by a liquid) are not favored by the French language and the last liquid can disappear from the pronunciation (example /k a t ʁ ə/ ⇒ /k a t ə/) [6]. The last simplified consonant pronunciation can be strengthened by an additional final schwa like vowel. In more complex word final consonant clusters such as /s t ʁ/ several cluster simplifications can be carried out. As the two adjacent consonants /s/ and /t/ share the same place of articulation, the occlusive can disappear, with or without the liquid, generating this way several pronunciation variants.

The analysis is then generalized to all the words ending with a cluster of type /t liquid/. To contrast, the analysis statistics are reported separately for the 14,000 occurrences of such words where the cluster is preceded by a vowel sound and for the 2,700 occurrences of words where the cluster is preceded by the consonant /s/ which has the same place of articulation as /t/ (i.e., part of a larger final cluster /s t liquid/).

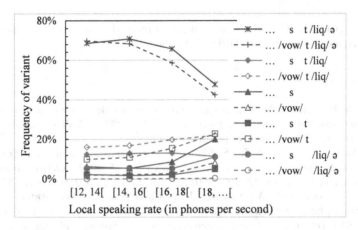

Fig. 6. Frequency of some pronunciation variants for the cluster /t liquid/ in final position of words when this cluster is preceded by a vowel or by the consonant /s/ which has the same place of articulation as /t/.

Figure 6 shows the frequency of omission of the consonant /t/ in the cluster /t liquid/, and the frequency of omission of the whole cluster /t liquid/, are much higher when the preceding sound is the consonant /s/, which has the same place of articulation as /t/, rather than when the preceding sound is a vowel.

5   Conclusion

This paper has investigated the impact of the frame shift used in acoustic analysis on computing corpus-based phonetic statistics through forced speech-text alignment of large speech corpora. The three emitting states of the conventional acoustic hidden Markov models introduce a minimum duration constraint of three frames for each phone segment. With the usual 10 ms frame shift, this corresponds to a 30 ms minimum duration for each phone, which is problematic at high speaking rates. In this paper we compared corpus-based phonetic statistics achieved with a shorter frame shift (5 ms) to those obtained from the usual 10 ms frame shift. Statistics computed on a large speech corpus of more than 3 million running words are analyzed with respect to the various speaking rates. Results exhibit some discrepancies between the two sets of statistics, in particular for high speaking rates, where the usual acoustic analysis frame shift leads to an under-estimation of the frequency of the longest pronunciation variants.

A complementary analysis of pronunciation variants for some word ending clusters was also conducted. The results show that at high speaking rates, final /plosive liquid/ clusters may be omitted, especially when the cluster is preceded by a consonant which shares the same place of articulation as the first consonant (plosive) of the cluster. Further studies will refine such pronunciation statistics, and consider their handling in speech recognition systems.

Acknowledgments. This work has been partly realized thanks to the support of the Région Lorraine and the CPER MISN TALC project.

References

1. Adda-Decker, M.: De la reconnaissance automatique de la parole à l' analyse linguistique de corpus oraux. In: Proceedings of JEP' 2006, XXVIes Journées dEtude sur la Parole, Dinard, France, pp. 389–400 (2006)
2. Adda-Decker, M., Lamel, L.: Systèmes d' alignement automatique et études de variantes de prononciation. In: Proceedings of JEP' 2000, XXIIIes Journées d' Etudes sur la Parole 19–23 juin 2000, Aussois, France, pp. 189–192 (2000)
3. Adell, J., Bonafonte, A., Gómez, J.A., Castro, M.J.: Comparative study of automatic phone segmentation methods for TTS. In: Proceedings of ICASSP' 2005, IEEE International Conference on Acoustics, Speech and Signal Processing, Philadelphia, USA, pp. 309–312 (2005)
4. Bigi, B., Hirst, D.: SPeech Phonetization Alignment and Syllabification (SPPAS): a tool for the automatic analysis of speech prosody. In: Proceedings of Speech Prosody, Shanghai, China, pp. 1–4 (2012)
5. Bürki, A., Gendrot, C., Gravier, G., Linarès, G., Fougeron, C.: Alignement automatique et analyse phonétique: comparaison de différents systèmes pour l' analyse du schwa. Traitement Automatique des Langues **49**(3), 165–197 (2008)
6. Côté, M.-H.: Phonetic salience and consonant cluster simplification. In: Bruening, B., Kang, Y., McGinnis, M. (eds.) PF: Papers at the Interface. MIT Working Papers in Linguistics, vol. 29, pp. 229–262 (1997)
7. de Calmès, M., Pérennou, G.: BDLEX: a Lexicon for spoken and written French. In: Proceedings of LREC' 1998, 1st International Conference on Language Resources and Evaluation, Grenada, Spain, pp. 1129–1136, 28–30 May 1998
8. De Mareüil, P.B., Adda-Decker, M., Gendner, V.: Liaisons in French: a corpus-based study using morpho-syntactic information. In: Proceedings of ICPhS' 2003, 15th International Congress of Phonetic Sciences, Barcelona, Spain (2003)
9. Demuynck, K., Laureys, T.: A comparison of different approaches to automatic speech segmentation. In: Sojka, P., Kopeček, I., Pala, K. (eds.) TSD 2002. LNCS (LNAI), vol. 2448, pp. 277–284. Springer, Heidelberg (2002)

10. EPAC Corpus: Orthographic transcriptions. ELRA catalogue (http://catalog.elra. info), ref. ELRA-S0305

11. Estève, Y., Bazillon, T., Antoine, J.-Y., Béchet, F., Farinas, J.: The EPAC corpus: manual and automatic annotations of conversational speech in French broadcast news. In: Proceedings of LREC' 2010, Seventh International Conference on Language Resources and Evaluation, Valetta, Malta, 19–21 May 2010

12. Galliano, S., Gravier, G., Chaubard, L.: The ESTER 2 evaluation campaign for rich transcription of French broadcasts. In: Proceedings of INTERSPEECH' 2009, 10th Annual Conference of the International Speech Communication Association, Brighton, UK, pp. 2583–2586, 6–10 September (2009)

13. Goldman, J.P.: EasyAlign: an automatic phonetic alignment tool under Praat (2011)

14. Gravier, G., Adda, G., Paulsson, N., Carré, M., Giraudel, A., Galibert, O.: The ETAPE corpus for the evaluation of speech-based TV content processing in the French language. In: Proceedings of LREC' 2012, 8th International Conference on Language Resources and Evaluation, Istanbul, Turkey, 23–25 May 2012

15. Huggins-Daines, D., Rudnicky, A.I.: A Constrained Baum-Welch Algorithm for Improved Phoneme Segmentation and Efcient Training. CMU report (2006)

16. Illina, I., Fohr, D., Jouvet, D.: Grapheme-to-phoneme conversion using conditional random fields. In: Proceedings of INTERSPEECH' 2011, 12th Annual Conference of the International Speech Communication Association, Florence, Italy, 27–31 August 2011

17. Jarifi, S., Pastor, D., Rosec, O.: A fusion approach for automatic speech segmentation of large corpora with application to speech synthesis. Speech Commun. 50(1), 67–80 (2008)

18. Jouvet, D., Fohr, D., Illina, I.: Detailed pronunciation variant modeling for speech transcription. In: Proceedings of INTERSPEECH' 2010, 11th Annual Conference of the International Speech Communication Association, Makuhari, Japan, 26–30 September 2010

19. Kawai, H., Toda, T.: An evaluation of automatic phone segmentation for concatenative speech synthesis. In: Proceedings of ICASSP' 2004, IEEE International Conference on Acoustics, Speech and Signal Processing, Montreal, CA, vol. I, pp. 677–680 (2004)

20. Kessens, J.M., Strik, H.: On automatic phonetic transcription quality: lower word error rates do not guarantee better transcriptions. Comput. Speech Lang. 18(2), 123–141 (2004)

21. Kominek, J., and Black, A.W.: A family-of-models approach to HMM-based segmentation for unit selection speech synthesis. In: Proceedings of INTERSPEECH' 2004, 8th International Conference on Spoken Language Processing, Jeju Island, Korea, 4–8 October 2004

22. Kuperman, V., Pluymaekers, M., Ernestus, M., Baayen, H.: Morphological predictability and acoustic duration of interfixes in Dutch compounds. J. Acous. Soc. Am. 121(4), 2261–2271 (2007)

23. Mporas, I., Ganchev, T., Fakotakis, N.: Speech segmentation using regression fusion of boundary predictions. Comput. Speech Lang. 24(2), 273–288 (2010)

24. Nakamura, M., Iwano, K., Furui, S.: Differences between acoustic characteristics of spontaneous and read speech and their effects on speech recognition performance. Comput. Speech Lang. 22(2), 171–184 (2008)

25. Ogbureke, K.U., Carson-Berndsen, J.: Improving initial boundary estimation for HMM-based automatic phonetic segmentation. In: Proceedings of INTER-SPEECH' 2009, 10th Annual Conference of the International Speech Communication Association, Brighton, UK, pp. 884–887, 6–10 September 2009
26. Raymond, W.D., Pitt, M.A., Johnson, K., Hume, E., Makashay, M.J., Dautricourt, R., Hilts, C.: An analysis of transcription consistency in spontaneous speech from the buckeye corpus. In: Proceedings of INTERSPEECH' 2002, 7th International Conference on Spoken Language Processing, Denver, Colorado, USA, 16–20 September 2002
27. Sjlander, K.: An HMM-based system for automatic segmentation and alignment of speech. In: Proceedings of Fonetik, Lvnger, Sweden, pp. 93–96 (2003)
28. Sphinx (2011). http://cmusphinx.sourceforge.net/
29. Toledano, D.T., Gómez, L.A.H., Grande, L.V.: Automatic phonetic segmentation. IEEE Trans. Speech Audio Process. **11**(6), 617–625 (2003)

The Influence of Boundary Depth on Phrase-Final Lengthening in Russian

Tatiana Kachkovskaia[(⊠)]

Department of Phonetics, Saint Petersburg State University,
Saint Petersburg, Russia
kachkovskaia@phonetics.pu.ru

Abstract. As shown for many languages, words are lengthened at ends of major prosodic boundaries, and the lengthening effect is most prominent at ends of intonational phrases and utterances. However, it is not clear whether the deeper boundary—end of utterance—shows more lengthening than the other one—end of intonational phrase. The present paper is aimed at answering this question by analysing the duration of Russian stressed and post-stressed vowels in open and closed syllables in words immediately preceding prosodic boundaries; the study is based on corpus data. The results show that for most speakers boundary depth only affects the lengthening of absolute-final vowels, either stressed or post-stressed; vowels in closed or penultimate syllables show similar lengthening effect for both types of boundaries. Additionally, our data show that in Russian stressed vowels are the main carriers of final lengthening, compared to post-stressed vowels.

Keywords: Phrase-final lengthening · Segmental duration · Boundary depth

1 Introduction

Phrase-final lengthening—segmental lengthening at ends of prosodic units—is a complex phenomenon influenced by a number of factors, such as pitch movement type (sentence type) [8], segmental context [2], pausation [12], boundary depth [1] etc. The latter is of particular interest since the data found in literature is scarce and to some extent contradictory.

If we distinguish between two types of prosodic boundaries—signalling the end of the utterance (utterance-final) and signalling the end of the intonational phrase but not the end of the utterance (IP-final)—we might suppose that the deeper the boundary, the greater the degree of phrase-final lengthening; in other words, phrase-final lengthening as a prosodic boundary marker is supposed to have more effect at larger prosodic units (in our case—utterances). However, published experimental data either point to the opposite [12] or to the fact that boundary depth is marked by different speakers differently [1]. Our recent pilot study for Russian [6], where the duration of the word "Ludmila" was measured in

© Springer International Publishing Switzerland 2015
A.-H. Dediu et al. (Eds.): SLSP 2015, LNAI 9449, pp. 135–142, 2015.
DOI: 10.1007/978-3-319-25789-1_13

different positions within the phrase, enabled us to suppose that utterance-final lengthening might be weaker than intonational-phrase-final lengthening. Since the data was obtained for only one word, answering this question requires a more accurate analysis based on a large speech corpus.

Thus, the main question asked in the present paper is as follows: *How does boundary depth influence the degree of phrase-final lengthening?*

Additionally, the current analysis will be able provide information on how the lengthening is distributed between the segments of the utterance-final word. In our previous study [4] we have shown that at the end of non-utterance-final intonational phrase the lengthening is observed primarily on stressed vowels, while post-stressed vowels show little lengthening when in open syllables, and no lengthening at all when in closed syllables. Therefore, a question arises whether this holds true for utterance-final words.

2   Material

For the present study the Corpus of Professionally Read Speech (CORPRES) [10] was chosen. The corpus contains texts of different speaking styles recorded from 4 male and 4 female speakers—approx. 30 h of manually segmented speech with phonetic and prosodic annotation. Prosodic information includes the type of the intonation contour, pause type, and prominence.

From this corpus 4 of the 8 speakers were taken, 2 male (A, M) and 2 female (C, K), since they recorded more material than others (approx. 5 h for each speaker).

3   Method

3.1   Obtaining Data from the Corpus

In order to get all the necessary information about the segments (vowels) in question, a Python script was written to process the annotation files of the corpus. For each segment we obtained information about its duration and context.

It is also worth noting here that for the purposes of this study only the voiced parts of vowels were included in the analysis.

3.2   Factors Included in the Analysis

Based on our previous data for Russian as well as papers for other languages we assume that the following factors influence the degree of phrase-final lengthening: segmental context, pitch movement type and prominence, boundary depth, pausation, stress pattern, length of the prosodic unit, speech tempo.

Thus, contextual parameters for each analysed segment of the corpus included the length of prosodic units (clitic group and intonational phrase) where the segment occurs and the position of each prosodic unit within the higher one; the CV-pattern of the clitic group where the segment occurs (e. g. "cVccv" for /'maska/, "mask"; "c" stands for the consonant, uppercase "V" stands for the stressed vowel, lowercase "V" stands for the unstressed vowel); the presence of a pause after the intonational phrase; the type of intonation contour and its location. (The CV-pattern of each clitic group was calculated based on its acoustic transcription. This means, in particular, that in cases where phoneme /j/ was vocalized, it was treated as a vowel.)

Each of the factors listed above were either varied, taken as constant or normalized. Besides boundary depth, which is the object of the present study, we decided to consider stress pattern as the second parameter since we were interested in how the lengthening is distributed among the segments of the final word. In order to obtain statistically reliable results, we decided to process only words ending in -cV, -cVc, -cVcv, -cVcvc, and -cVccv, which are the most frequent CV-patterns occurring word-finally in Russian.

Based on the frequency data, we have selected only intonational phrases containing from 2 to 6 clitic groups and at least 5 syllables. We also limited our choice to intonational phrases with no internal pauses[1], no prominent words, and *no* nuclear stress on the IP-final or utterance-final clitic group (e. g. intonational phrase /u'darʲil jiˈvo nʲiˈtrʲi ˈraza [pause]/ ("kicked him not three times [pause] [, but ... times]") with nuclear stress on /ˈtrʲi/ (three)).

To eliminate pausation as a factor influencing the degree of lengthening, we observed only those intonational phrases and utterances which were followed by a physical pause.

3.3 Duration Normalization

In order to be able to compare duration values for different types of segments (e.g., closed and open vowels, which differ in inherent duration) it is reasonable to calculate *normalized* duration values. Here the formula given in [12, formula (4)] was used, which allowed us to compensate for the average duration of the segment, its standard deviation, and tempo:

$$\tilde{d}(i) = \frac{d(i) - \alpha\mu_p}{\alpha\sigma_p}$$

where $\tilde{d}(i)$ is the normalized duration of segment i, $d(i)$ is its absolute duration, α is the tempo coefficient, and μ_p and σ_p are the mean and standard deviation of the duration of the corresponding phone p.

The tempo coefficient (α) was calculated using formula provided in [12, formula (6)]:

$$\alpha = \frac{1}{N} \sum_{i=1}^{N} \frac{d_i}{\mu_{p_i}}$$

---

[1] Despite the fact that pauses are often used to mark prosodic boundaries, it was not uncommon in our corpus to onserve short IP-internal pauses. Since the role of these short pauses is no quite clear so far, we decided to refrain from analysing IPs with internal pauses.

where d_i is the duration of segment i, and μ_{p_i} is the mean duration of the corresponding phone.

3.4 Statistical Analysis

To estimate the influence of different factors on segment duration, statistical analysis was carried out using R. For normally distributed data ANOVA and pairwise t-tests were used with Welch's correction for unequal variance if necessary; for non-normally distributed data Kruskal-Wallis test was used instead.

4 Results

Our analysis included a comparison of clitic groups occurring in three positions:

- position 1: IP-medial, the clitic group *not* followed by a pause;
- position 2: IP-final, but not utterance-final, with a following pause;
- position 3: utterance-final, with a following pause.

The mean normalized duration values for stressed and post-stressed vowels are provided in Tables 1 and 2, respectively. The values are grouped according to the CV-pattern of the end of the clitic group where the vowel occurs.

4.1 Boundary Depth

As shown in Table 1, there is a clear tendency that **stressed** vowel duration values are higher in IP-final position (position 2) compared with utterance-final position (position 3). This difference is statistically significant ($p<0.05$)

- for **3 out of 4** speakers for clitic groups ending in -cV;
- for **1 out of 4** speakers for clitic groups ending in -cVc;
- for **1 out of 4** speakers for clitic groups ending in -cVcv;
- for **0 out of 4** speakers for clitic groups ending in -cVcvc;
- for **1 out of 4** speakers for clitic groups ending in -cVccv.

Therefore, this tendency is rather weak in all cases but the first, where the stressed vowel occurs in absolute-final position (ultimate open syllable).

The same general tendency is observed for the **post-stressed** vowels (see Table 2). The difference between these two positions is statistically significant ($p<0.05$):

- for **2 out of 4** speakers for clitic groups ending in -cVcv;
- for **1 out of 4** speakers for clitic groups ending in -cVcvc;
- for **3 out of 4** speakers for clitic groups ending in -cVccv.

Similarly to stressed vowels, for post-stressed vowels this effect is more pronounced for ultimate open syllables.

We might conclude that in Russian both stressed and post stressed vowels show more phrase-final lengthening at ends of non-utterance-final intonational

Table 1. Mean normalized duration values for *stressed* vowels in clitic groups ending in differenc CV-patterns, for 3 positions: (1) IP-medial, (2) IP-final, (3) utterance-final (4 speakers). Sample sizes are given in brackets.

		speaker			
		A	C	K	M
-cV	1	-0.51	-0.56	-0.62	-0.44
		(1139)	(1410)	(1258)	(1334)
	2	0.57	0.94	0.45	1.69
		(44)	(22)	(34)	(16)
	3	-0.46	0.24	0.21	0.05
		(38)	(25)	(52)	(43)
-cVc	1	-0.60	-0.63	-0.66	-0.48
		(1252)	(1648)	(1317)	(1590)
	2	0.25	0.40	0.30	0.17
		(36)	(24)	(34)	(31)
	3	-0.23	0.08	0.15	-0.10
		(37)	(50)	(55)	(71)
-cVcv	1	-0.51	-0.38	-0.50	-0.48
		(1456)	(1727)	(1502)	(1745)
	2	0.17	0.34	0.55	0.29
		(46)	(30)	(41)	(30)
	3	-0.20	0.23	0.51	0.14
		(33)	(47)	(54)	(66)
-cVcvc	1	-0.58	-0.49	-0.61	-0.54
		(705)	(869)	(724)	(794)
	2	0.27	0.08	0.36	0.27
		(21)	(15)	(18)	(14)
	3	-0.04	0.22	0.48	-0.08
		(15)	(27)	(35)	(31)
-cVccv	1	-0.45	-0.40	-0.51	-0.35
		(742)	(891)	(814)	(886)
	2	0.11	-0.07	0.56	0.45
		(19)	(16)	(19)	(23)
	3	-0.20	-0.02	0.29	-0.20
		(26)	(16)	(28)	(31)

phrases than at ends of utterances, but this tendency rather weak for vowels in non-ultimate or closed syllables.

On the other hand, we might be observing at least two different strategies for signalling boundary depth. Some speakers tend to mark non-final intonational

Table 2. Mean normalized duration values for *post-stressed* vowels in clitic groups ending in different CV-patterns, for 3 positions: (1) IP-medial, (2) IP-final, (3) utterance-final (4 speakers). Sample sizes are given in brackets.

		speaker			
		A	C	K	M
-cVcv	1	-0.10	-0.15	-0.01	-0.02
		(1456)	(1727)	(1502)	(1745)
	2	0.35	0.99	0.12	0.34
		(46)	(30)	(41)	(30)
	3	-0.28	0.37	0.09	-0.10
		(33)	(47)	(54)	(66)
-cVcvc	1	-0.37	-0.31	-0.28	-0.28
		(705)	(869)	(724)	(794)
	2	-0.16	-0.05	-0.32	-0.39
		(21)	(15)	(18)	(14)
	3	-0.56	-0.45	-0.32	-0.42
		(15)	(27)	(35)	(31)
-cVccv	1	-0.11	-0.33	-0.19	-0.23
		(742)	(891)	(814)	(886)
	2	0.74	1.23	0.36	0.88
		(19)	(16)	(19)	(23)
	3	0.32	0.24	-0.16	-0.29
		(26)	(16)	(28)	(31)

phrases using durational characteristics, while others fail to reveal any durational differences; the latter group might be using other markers such as pitch, pausation or amplitude.

4.2 Utterance-Final Lengthening

In order to explore the nature of utterance-final lengthening (as opposed to IP-final lengthening), a comparison of duration values for IP-medial position and utterance-final position was performed. As shown in Table 1, there is a clear tendency that **stressed** vowel duration values are higher in utterance-final position (position 3) compared with IP-medial position (position 1). This difference is statistically significant ($p < 0.05$)

- for **3 out of 4** speakers for clitic groups ending in -cV;
- for **4 out of 4** speakers for clitic groups ending in -cVc;
- for **4 out of 4** speakers for clitic groups ending in -cVcv;
- for **4 out of 4** speakers for clitic groups ending in -cVcvc;
- for **4 out of 4** speakers for clitic groups ending in -cVccv.

Therefore, for stressed vowels this tendency is very strong in all cases.

However, the same general tendency does not seem to be observed for the **post-stressed** vowels (see Table 2). The difference between the two positions is statistically significant ($p<0.05$):

- for **1 out of 4** speakers for clitic groups ending in -cVcv;
- for **0 out of 4** speakers for clitic groups ending in -cVcvc;
- for **1 out of 4** speakers for clitic groups ending in -cVccv,

which demonstrates that post-stressed vowels are not lengthened much in utterance-final position.

This leads to a conclusion that the main carriers of utterance-final lengthening are stressed vowels and not post-stressed vowels. Similar results were obtained in our previous studies on phrase-final lengthening [4] for intonational phrases occurring non-utterance-finally.

5   Discussion

For Russian our data provide evidence that the deeper the boundary, the *weaker* the lengthening effect. In other words, the speaker marks the end of non-utterance-final intonational phrase better than the end of utterance-final intonational phrase—thus showing that the utterance is not finished and the listener is expected to wait for its ending. However, this relation might be speaker-specific, since not all the speakers show a statistically significant difference between these two boundary types.

The present study also enables us to draw conclusions on the locus of utterance-final lengthening. It is widely accepted that it is the final rhyme of the word where the phrase-final lengthening effect is most prominent (see, for example, [1, 3, 12] etc.). Some evidence that the stressed vowel is also involved, but to a lesser degree, can be found in [11] for English.

It is also known that phrase-final lengthening, although considered a universal phenomenon, can in some cases be language-specific. A well-known examples are Japanese [9] and Northern Finnish [7]—languages with phonemic vowel length—where the lengthening effect is limited so that the contrast between short and long vowels could be preserved. Russian, where the main correlate of lexical stress is vowel duration, might also differ from other European languages in terms of the locus of phrase-final lengthening.

Our data seem to support this hypothesis, showing that stressed vowels in penultimate syllables play a greater role in phrase-final lengthening than post-stressed vowels. However, it is worth noting here that final rhyme may not only consist of a vowel, but also include following consonants. Since the present study is focused on *vowel* duration, it does not provide a full answer to the question of how final rhyme is involved in this process. Our previous studies [5], though, have shown that absolute-final consonants do play a significant role in phrase-final lengthening.

Acknowledgements. The research is supported by the Russian Science Foundation (reserch grant # 14-18-01352).

References

1. Cambier-Langeveld, T.: Temporal marking of accents and boundaries. University of Amsterdam (2000)
2. Cooper, W.E., Danly, M.: Segmental and temporal aspects of utterance-final lengthening. Phonetica **38**(1–3), 106–115 (1981)
3. Flege, J.E., Brown Jr., W.S.: Effects of utterance position on English speech timing. Phonetica **39**(6), 337–357 (1982)
4. Kachkovskaia, T.: Phrase-final lengthening in Russian: pre-boundary or pre-pausal? In: Ronzhin, A., Potapova, R., Delic, V. (eds.) SPECOM 2014. LNCS, vol. 8773, pp. 353–359. Springer, Heidelberg (2014)
5. Kachkovskaia, T., Volskaya, N.: Phrase-final segment lengthening in Russian: preliminary results of a corpus-based study. In: Železný, M., Habernal, I., Ronzhin, A. (eds.) SPECOM 2013. LNCS, vol. 8113, pp. 257–263. Springer, Heidelberg (2013)
6. Kachkovskaia, T., Volskaya, N., Skrelin, P.: Final lengthening in Russian: a corpus-based study. In: Proceedings of INTERSPEECH-2013. ISCA, pp. 1438–1442 (2013)
7. Nakai, S., Kunnari, S., Turk, A., Suomi, K., Ylitalo, R.: Utterance-final lengthening and quantity in Northern Finnish. J. Phonetics **37**(1), 29–45 (2009)
8. Oller, D.K.: The effect of position in utterance on speech segment duration in English. J. Acoust. Soc. Am. **54**(5), 1235–1247 (1973)
9. Shepherd, M.A.: The scope and effects of preboundary prosodic lengthening in Japanese. USC Working Pap. Linguist. **4**, 1–14 (2008)
10. Skrelin, P.A., Volskaya, N.B., Kocharov, D., Evgrafova, K., Glotova, O., Evdokimova, V.: A fully annotated corpus of Russian speech. In: Proceedings of the Seventh conference on International Language Resources and Evaluation (LREC'10), pp. 109–112. European Language Resources Association (ELRA) (2010)
11. Turk, A., Shattuck-Hufnagel, S.: Multiple targets of phrase-final lengthening in American English words. J. Phonetics **35**, 445–472 (2007)
12. Wightman, C.W., Shattuck-Hufnagel, S., Ostendorf, M., Price, P.J.: Segmental durations in the vicinity of prosodic phrase boundaries. J. Acoust. Soc. Am. **91**(3), 1707–1717 (1992)

Automatic Detection of Voice Disorders

Ferenc Kazinczi[1](✉), Krisztina Mészáros[2], and Klára Vicsi[1]

[1] Laboratory of Speech Acoustics, Department of Telecommunications
and Media Informatics, Budapest University of Technology and Economics,
Budapest, Hungary
{kazinczi, vicsi}@tmit.bme.hu
[2] Department of Head and Neck Surgery, National Institute of Oncology,
Budapest, Hungary

Abstract. Speech and communication are the bases of our society and the quality of the speech can seriously affect any person's life. Besides the irregularities in the voice production can be caused by different diseases which can be treated better if they are diagnosed in an early stage. In this research we introduce a series of measurements based on continuous speech which could be a good base of developing a system that can automatically distinguish several voice disorders and identify the quality of the voice according to widely used RBH scale ((Rauhigkeit) (roughness) (Behauchtkeit) (breathiness) (Heiserkeit) (hoarseness)): 0 = normal voice quality, 3 = heavy huskiness). The different feature extraction and classification experiments presented in this research show how it is possible to separate healthy from pathological speech automatically and demonstrate possibilities to define the type of the pathological voice. The results suggest that the automatic classification into two or multiple classes can be improved by a multi-step pre-processing methodology, in which only the most significant acoustic features of a given class are extracted from the voice and used to train the SVM (Support Vector Machine) classifiers. We also present the importance of data acquisition, and how the selected features and the number of training samples can affect the accuracy and performance of the automatic voice disorder detection.

Keywords: Speech recognition · Voice disorder · Pathological voice · Principal component analysis · Support vector machine · RBH scale

1 Introduction

The accurate classification of pathological voices is still an important issue in medical decision support systems and speech processing. The performance depends on a variety of attributes ranging from the anatomy of voice production to the applied pre-processing algorithms and mathematical models used for machine learning [1–6]. There are differences from human to human in the bases of voice generation organs, and these differences could effect the measurable acoustic parameters (fundamental frequency, sound pressure, spectrum, etc.) of the generated sound of the speaker. This reflects the importance of a pathological and healthy speech databases.

© Springer International Publishing Switzerland 2015
A.-H. Dediu et al. (Eds.): SLSP 2015, LNAI 9449, pp. 143–152, 2015.
DOI: 10.1007/978-3-319-25789-1_14

In medical practice, the clinicians can classify the voice quality based on perceptual evaluation and mainly from continuous speech as part of a normal daily conversation. In previous researches we found that samples and vowels extracted from continuous speech could be a good basis of a system that can automatically characterize the quality of the patient's voice in a similar way to the clinicians do it practice [7]. Besides, there is a possibility in continuous speech for the observation of suprasegmental characteristics: intonation, and the duration of sonorants.

The acoustic characteristics of sustained voice and continuous speech for distinguishing a healthy from a pathological voice has been analyzed in many studies [1–15]. The most widely used acoustic parameters include: jitter, shimmer and harmonics-to-noise Ratio (HNR).

Taoi Li examined how the HNR can be used in case of 41 normal (healthy) and 111 pathological cases to classify them using an artificial neural network (ANN) [1]. The results indicate that in case of voice disorders the HNR could be a good bases of differentiating healthy from pathological voice and determine the degree of hoarseness from the patient's speech [1, 9].

There were also separate researches where jitter and shimmer parameters are used and others use the mel-frequency cepstral coefficients (MFCC) for characterizing the quality of the sound product and for separating healthy sound from sound affected by voices [3, 5, 6, 10–12].

In this paper we are going to focus on continuous speech based speech processing and classification where different combination of the jitter, shimmer, HNR and MFCC shall be examined. First, a detailed statistical analysis of acoustic parameters of vowels in continuous speech and sustained voice databases were examined, and the results were compared in healthy vs. pathological speech [7].

In this paper we present our feature extraction methodology and classification experiments on how it is possible to separate healthy from pathological speech automatically on the basis of continuous speech. In addition, we give solutions to differentiate healthy people from patients diagnosed with functional dysphonia or recurrent paresis and even how the quality of the voice can be objectively measured in comparison with the semi-subjective RBH scale ((Rauhigkeit) (roughness) (Behauchtkeit) (breathiness) (Heiserkeit) (hoarseness)): 0 = normal voice quality, 3 = heavy huskiness) applied by clinicians in daily routine. The research highlights the necessities of using proper acoustic features in case of the different classification problems and how the accuracy of the classification can be affected by the size of the database.

2   Method

2.1   Pathological and Healthy Speech Databases

The sound recordings were made in a consulting room at the Outpatients' Department of Head and Neck Surgery of the National Institute of Oncology. The following diseases occurred in the recorded database: functional dysphonia, recurrent paresis, tumors at various places of the vocal tract, gastro-oesophageal reflux disease, chronic inflammation of larynx, bulbar paresis, its symptoms (paralysis of lips, tongue, soft palate, pharynx and the muscles of larynx), amyotrophic lateral sclerosis, leukoplakia, spasmodic dysphonia and glossectomy. Recordings, for comparison, were also prepared with healthy patients who had attended for unrelated check-ups.

Speech samples were recorded by near field microphone (Monacor ECM-100), with Creative Soundblaster Audigy 2 NX: an outer USB sound card with 44,100 Hz sampling rate, at a 16-bit linear coding.

Each patient has to read out aloud the short story "The North Wind and the Sun", which is frequently used in phoniatric practice and has been translated into several languages.

The recorded sound samples were classified by a leading clinician according to the RBH sound perception evaluation scale, which is a popular scale in the practice among clinicians (In this research the evaluation was provided by one clinician and the results are being validated by other independent speech language specialists who are familiar with the RBH scale.). The scale classifies the voice samples into four classes on the basis of subjectively felt parameters provided by the RBH code. This scale was used to differentiate the degree of voice generation disorders in the database. Speech samples of the patients were labeled on the basis of this numerical scale [7–9].

Since we intended to process predefined vowel sequences of the continuous speech material, phoneme-level segmentation of speech was necessary. It was made in a semi-automatic way, using our own automatic speech recognizer.

This research focuses not only on differentiating healthy from pathological speech, but also defining the H parameter of RBH features in case of each patient, and even predicting which disease is the cause of the voice disorder. (The H parameter was chosen as the base of the classification because the hoarseness characterize better the quality of the voice and it is depending on the R and B parameters, which makes it more sensitive for the disorders [8, 9]). Those voices were considered as healthy which were given 0 score for the H parameter by the medical doctor. There are 63 healthy and 120 pathological speech samples in the database. The distribution of the samples according to the H parameter is 63 Healthy (H0), 64 H1, 24 H2 and 32 H3. According to diseases within the pathological voices there are 55 functional dysphonia (FD) and 33 recurrent paresis (RP) samples registered.

2.2 Pre-processing Methods

Previous research has confirmed that the selected acoustic parameters in the quasi-stationary part of the vowels in continuous speech could replicate the perceptual classification of experts much better than those in the traditionally used sustained sounds [7]. In this paper we also want to evaluate which parameters can be used to differentiate not just the healthy voices from the pathological, but even predict the Healthy (H0), H1, H2 and H3 status of the patients and identify the type of the voice disorder. To get the most significant features we established the following process shown in Fig. 1.

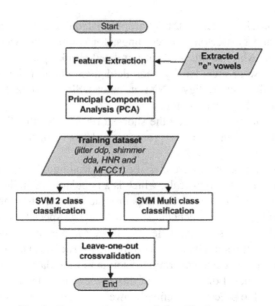

Fig. 1. Feature selection and classification process

As the base of this research the sound [e] was extracted from the reading tests. We chose the "e" vowel because in the Hungarian language this vowel is the most frequent one and the [e] sounds are supposed to occur approximately 50 times when reading the prompts. The other reason for selecting the vowel "e" is that we compared the result with measurements from other vowels and there were no significant differences accounted between the different vowels. During the extraction, samples shorter than 40 ms were discarded in order to eliminate sound containing less than 4 periods (We established this limit based on the sampling frequency 44.1 kHz and the minimum male fundamental frequency set to 100 Hz).

We used the Praat program in order to get the basic acoustic parameters for each patient [15]. At the middle of the [e] sound, the following acoustic parameters were measured: jitter, shimmer, HNR and mel-frequency cepstral coefficients (MFCC) [5, 11, 12]. This means an overall 37 features per sound per patient. The list of features includes all the parameters that can be calculated by Praat and the first twelve coefficients of MFCC. This also incorporates different calculation methods for the jitter and the shimmer (For example for the jitter it has jitter local, jitter relative average perturbation (rpa), jitter five-point period perturbation quotient (ppq5), jitter$_{ddp}$ difference between consecutive differences between consecutive periods etc.). The exact calculation methods, the list of used features and the equation can be found in accounts of earlier research [7, 9, 10, 15]. After the extraction, the sound "e" of each patient is represented by the mean of the measured acoustic parameters.

2.3   Parameter Selection

The raw parameter set, which we initially defined for the vowel [e], counted 37 parameters (see Sect. 2.2). For better recognition, first we tried to reduce the dimensionality of the input feature vector. To reach this we performed principal component analysis (PCA) on each class [13, 14]. In order to select the most significant parameters, we identified those principal components which can retain at least 90 % of the variation presented in the dataset: usually the first 10 principal components were selected. Then we selected those variables that had the highest eigenvector coefficients within the chosen principal components. In this examination PCA is used not only to select those parameters that are mandatory to characterize the classes, but to reduce the dimensionality of the input feature vector. Table 1 summarizes the results of the PCA-based feature selection method.

Based on the results in Table 1 we found that separate classes can be better characterized by different input features, but mostly the previously used jitter, shimmer and HNR parameters are the best. In addition, the first MFCC coefficient (mfcc1) could also be a useful parameter in the recognition of pathological voices.

Table 1. PCA selected features with the highest eigenvector coefficients within the first 10 principal components in each class in case of the sound "e"

Class	Highest eigenvector coefficient parameters of vowel "e" in the selected components	Explanatory power of the selected components
Healthy (H0)	$jitter_{ddp}$, $shimmer_{dda}$, mfcc1	93 %
H1	$jitter_{ddp}$, mfcc1	90 %
H2	$jitter_{ddp}$, $shimmer_{dda}$, HNR, mfcc1	93 %
H3	HNR	91 %
FD	$jitter_{ddp}$, mfcc1	92 %
RP	$shimmer_{dda}$, HNR, mfcc1	92 %

3   Classification Experiments

Based on the results of the PCA we selected the jitterddp, shimmerdda, HNR and mfcc1 features as the bases of the input vectors to train the classifier. In separate measurement sets we created different combinations of these features and measured the accuracy of the recognition. 2-class and multi-class support vector machine was used for the classification (LibSVM and Matlab SVM functions) [16, 17]. To validate the performance of the learning we used leave-one-out cross-validation (LOOCV). (The SVM was used with radial base function (RBF) kernel function, boxconstraint 0.2 and $\sigma = 1$.)

3.1 Classification of Healthy and Pathological Voices Based on RBH Scale

In this experiment we considered everyone with H0 diagnosis as healthy and everyone else with H1, H2 and H3 as pathological. The following measurements were taken into consideration and in each case the classification results were examined separately.

In Table 2, it can be seen, significantly, that the RBH scale semi-subjective behavior can seriously affect the accuracy of the classification. If we check the classification results of the neighboring H parameters separation (Healthy (H0) vs H1, H1 vs H2, H2 vs H3), we can see that the accuracy is below 70 %. On the other hand if we check the results of the non neighboring H parameter classifications (Healthy (H0) vs H2, Healthy (H0) vs H3, H1 vs H3) the accuracies are far higher, even reaching 91 %. The main cause behind this phenomenon is that the database is not big enough to clearly characterize these salient voices.

3.2 Accuracy of Healthy and Pathological Classification on Different Sized Training Sets

In case of classification problems it is always a question how the number of the samples can affect the performance of the classifier. Besides, the performance of the classifier is a good indicator whether the selected features are good enough for characterizing the given problem and it convergence to higher accuracy can be reached with a reasonable size of dataset. In order to demonstrate how the selected features affect the performance of the SVM based pathological and healthy classification, the size of the input dataset was iteratively increased and the accuracy was measured in each step. In each iteration we randomly selected a given number of samples from both the healthy and the pathological datasets and performed cross-validation. In order to have almost the same number of elements in each set at every step we added more healthy voices to the previously introduced database (These new healthy samples were recorded in the same way as the others, but these people were not examined by medical doctors). Table 3 represents the training sets and the accuracy reached in each step (Fig. 2).

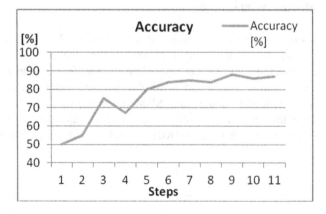

Fig. 2. Accuracy of the SVM classifier on different sized training sets

Table 2. Two class classification results with different combination of $jitter_{ddp}$, $shimmer_{dda}$, HNR and mfcc features of the vowel "e"

Parameters \classes	Accuracy		
	$jitter_{ddp}$, $shimmer_{dda}$, HNR	$jitter_{ddp}$, $shimmer_{dda}$, mfcc	$jitter_{ddp}$, $shimmer_{dda}$, HNR, mfcc
Healthy vs Pathological	73 %	73 %	83 %
Healthy vs H1	60 %	59 %	62 %
Healthy vs H2	76 %	76 %	78 %
Healthy vs H3	86 %	91 %	89 %
H1 vs H2	70 %	70 %	70 %
H1 vs H3	81 %	84 %	82 %
H2 vs H3	66 %	66 %	61 %

Table 3. Accuracy of the classifier based on the number of samples in each step used to train the SVM

Step	Num. of healthy samples	Num. of pathological samples	Accuracy [%]
1	10	10	50 %
2	10	10	55 %
3	10	10	75 %
4	15	15	67 %
5	30	30	80 %
6	41	45	84 %
7	50	60	85 %
8	69	75	84 %
9	80	90	88 %
10	90	105	86 %
11	100	120	87 %

In the first three iterations, selecting samples randomly from a given class, even with the same number of elements, results in big variances in accuracy if the data set is not big enough to characterize the given class. The following diagram shows how the accuracy develops in each step.

The classifier reached 87 % accuracy in the eleventh step, and based on the results we can clearly state that the bigger database we have to train the classifier, the better accuracy we can provide for such a classification.

3.3 Classification of Voice Disorder Based on Diseases

To observe the behaviour of the features in the case of multi-class classification environment we performed the following measurements (Table 4).

To separate the healthy voices from recurrent paresis voices the {jitter, shimmer, mfcc} can be used, but to separate functional dysphonia from recurrent paresis it is

Table 4. Classification of diseases based on jitter$_{ddp}$, shimmer$_{dda}$, HNR and mfcc features of the vowel "e"

Parameters \classes	Accuracy		
	jitter$_{ddp}$, shimmer$_{dda}$, HNR	jitter$_{ddp}$, shimmer$_{dda}$, mfcc	jitter$_{ddp}$, shimmer$_{dda}$, HNR, mfcc
Healthy vs FD	60 %	66 %	63 %
Healthy vs RP	70 %	85 %	82 %
FD vs RP	70 %	77 %	78 %

better to add the HNR to the parameter set. In the case of diseases it can be seen that healthy and FD can be significantly separated from the RP. This is natural, since with different types of diseases different speech parameters are distorted; it also has an anatomic root cause. Patients with RP have serious vocal cord dysfunctionality; this increases the power of noisy components in the speech and changes the spectrum of the sound. The separation of healthy from FD is challenging and the results show that the size of the dataset might not be large enough to train the SVM correctly.

3.4 Classification of Voice Disorders into Multiple Classes

We wanted to analyze this result further: in terms of how the samples can be separated into multiple classes using the same parameters, and how the supervised learning performs.

As we can see from the results in Table 5, the multi class classification has worse results than the 2 class problems. In order to see what happens during the classification, we examined the confusion matrix of classification (See Table 6) and we found that the classifier basically considered most of the H1 class elements as Healthy (H0), but the Healthy (H0) and H3 classes are separated with high accuracy. It could mean that the given input vector can only characterize the Healthy (H0) and H3 classes. The size of the training set could also affect the accuracy of the prediction (In these cases even the SVM settings can affect the learning performance).

Table 5. Multi-class classification results with different combination of jitter$_{ddp}$, shimmer$_{dda}$, hnr and mfcc1 features of vowel "e"

Parameters \classes	jitter$_{ddp}$, shimmer$_{dda}$, mfcc1	jitter$_{ddp}$, shimmer$_{dda}$, hnr	jitter$_{ddp}$, shimmer$_{dda}$, hnr, mfcc
Healthy (H0) – H3	48 %	47 %	51 %
Healthy, FD, RP	59 %	56 %	60 %

Table 6. Confusion matrix of jitter, shimmer and HNR based multi-class classification into Healthy (H0), H1, H2 and H3 with the number of each prediction

		Real Class			
		Healthy (H0)	H1	H2	H3
Predicted Class	Healthy (H0)	53	42	8	4
	H1	7	8	5	3
	H2	3	8	5	3
	H3	0	5	6	22

This confirms the results in Table 2 – that the neighbouring H parameters are hard to differentiate – and this result can also be linked to how the accuracy depends on the number of elements in each class.

4 Conclusions

In this research we examined how different classification problems can be applied on a pathological speech database. The results clearly demonstrate that an SVM based classifier can separate automatically the healthy and pathological voices with a relatively high (87 %) accuracy, and it can evaluate the RBH status of the voice with an overall 78–80 % accuracy. There are also possibilities to differentiate the type of the disorders but these results must be examined on bigger size of dataset. We also found that in the case of both two-class and multi-class classification the correct selection of the input features is essential and can significantly affect the performance of the machine learning. The other key point of the research is the importance of the size of the database and how the accuracy of the SVM can depend on the number of the elements in a given class. So in the future it is still important to concentrate not just on developing better pre-processing and machine learning methods, but on collecting more pathological data validated by clinicians.

References

1. Li, T., et al.: Classification of pathological voice including severely noisy cases. In: 8th International Conference on Spoken Language Processing, Interspeech-2004, pp. 77–80 (2004)
2. Peng, C., Chen, W., Zhu, X., Wan, B., Wei, D.: Pathological voice classification based on a single vowel's acoustic features. In: 7th International Conference on Computer and Information Technology, pp. 1106–1110. IEEE (2007)
3. Rabinov, C.R., Kreiman, J., Gerratt, B.R., Bielamowicz, S.: Comparing reliability of perceptual ratings of roughness and acoustic measures of jitter. J. Speech Hear. Res. **38**, 26–32 (1995)

4. Ritchings, R.T., McGillion, M., Moore, C.J.: Pathological voice quality assessment using artificial neural networks. Med. Eng. Phys. **24**, 561–564 (2002)
5. Grygiel, J., Strumiłło, P.: Application of mel cepstral representation of voice recordings for diagnosing vocal disorders. Przegląd Elektrotechniczny (Electr. Rev.), **88**(6), 8–11 (2012). ISSN 0033-2097
6. Sarria-Paja, M., Daza-Santacoloma, G., Godino-Llorente, J.I., Casellanos-Dominquez, G., Sáenz-Lechón, N.: Feature selection in pathological voice classification using dynamic of component analysis. In: Proceedings of the 4th International Symposium on Image/Video Communications over Fixed and Mobile Networks. Universidad de Deusto (2008)
7. Vicsi, K., Imre, V.: Voice disorder detection on the base of continuous speech. In: 4th Advanced Voice Function Assessment Workshop, COST Action 2103, New York (2010)
8. Mitrovic, S.: Characteristics of the voice in patients with glottic carcinoma evaluated with the RBH (roughness, breathiness, hoarseness) and GIRBAS (grade, instability, roughness, breathiness, asthenia, strain) scales. Med. Pregl. **56**, 337–340 (2003)
9. Yumotot, E.: Harmonics-to-noise ratio as an index of the degree of hoarseness. J. Acoust. Soc. Am. **71**(6) 1544–1550 (1982)
10. Farrús, M., Hernando, J., Ejarque, P.: Jitter and shimmer measurements for speaker recognition. TALP Research Center, Department of Signal Theory and Communications, Barcelona, Spain
11. Ravi Kumar, K.M., Ganesan, S.: Comparison of multidimensional MFCC feature vectors for objective assessment of stuttered disfluencies. Int. J. Adv. Netw. Appl. **02**(05), 854–860 (2011)
12. Vikram, C.M., Umarani, K.: Pathological voice analysis to detect neurological disorders using MFCC and SVM. Int. J. Adv. Electr. Electron. Eng. **2**(4), 87–91 (2013). ISSN (Print) 2278-8948
13. Jolliffe, I.T.: Principal Component Analysis. Department of Mathematical Sciences, King's College, London (2002)
14. Arjmandi, M.K., Pooyan, M., Mohammadnejad, H., Vali, M.: Voice disorders identification based on different feature reduction methodologies and support vector machine. In: Proceeding of ICEE2010. Isfahan University of Technology, 11–13 May 2010
15. Praat program. http://www.fon.hum.uva.nl/praat/manual/Voice.html. Accessed 30 June 2015
16. Multi SVM Matlab algorithm. http://www.mathworks.com/matlabcentral/fileexchange/33170-multi-class-support-vector-machine. Accessed 30 June 2015
17. LibSVM. https://www.csie.ntu.edu.tw/~cjlin/libsvm/. Accessed 30 June 2015

Semantic Features for Dialogue Act Recognition

Pavel Král[1,2]([✉]), Ladislav Lenc[1,2], and Christophe Cerisara[3]

[1] Department of Computer Science and Engineering Faculty of Applied Sciences,
University of West Bohemia, Plzeň, Czech Republic
{pkral,llenc}@kiv.zcu.cz
[2] NTIS - New Technologies for the Information Society Faculty of Applied Sciences,
University of West Bohemia, Plzeň, Czech Republic
[3] LORIA-UMR7503, Nancy, France
cerisara@loria.fr

Abstract. Dialogue act recognition commonly relies on lexical, syntactic, prosodic and/or dialogue history based features. However, few approaches exploit semantic information. The main goal of this paper is thus to propose semantic features and integrate them into a dialogue act recognition task to improve the recognition score. Three different feature computation approaches are proposed, evaluated and compared: Latent Dirichlet Allocation and the HAL and COALS semantic spaces. An interesting contribution is that all the features are created without any supervision. These approaches are evaluated on a Czech dialogue corpus. We experimentally show that all proposed approaches significantly improve the recognition accuracy.

Keywords: COALS · Dialogue act · HAL · Language model · LDA · Semantic spaces · Semantics · Speech act · Syntax

1 Introduction

Automatic Dialogue act (DA) recognition has received much attention in the last years, because this task is fundamental for many emerging dialogue systems. Many approaches have been proposed and evaluated on different corpora containing several dialogue acts. These methods use different type of information coming from the user input.

The features for dialogue act recognition are usually computed from lexical, syntactic, prosodic and/or dialogue history information. However, few approaches consider semantic features, while such features may bring additional information and prove useful to improve the accuracy of our dialogue act recognition system. For instance, because DA recognition systems are typically trained in a supervised way, a frequent cause of recognition errors are "unknown" words in the testing corpus that never occur in the training sentences. Lexical semantic similarity may partly address this issue by grouping words into coherent classes. Depending on how these semantic vectors are computed, these classes, or more generally "semantic distances", can also include some syntactic information, e.g.,

© Springer International Publishing Switzerland 2015
A.-H. Dediu et al. (Eds.): SLSP 2015, LNAI 9449, pp. 153–163, 2015.
DOI: 10.1007/978-3-319-25789-1_15

related to the relative position or degree of proximity of pairs of words within a sentence. This additional information can be used to improve DA recognition, in particular when the training and test conditions differ, or when the size of training corpus is relatively small.

The main goal of this paper is thus to propose semantic features for dialogue act recognition to improve DA recognition results. We describe next three different approaches, which respectively use Latent Dirichlet Allocation (LDA) [5], the Hyperspace Analogue to Language (HAL) [25] and Correlated Occurrence Analogue to Lexical Semantics (COALS) [33] semantic spaces to compute these features. The dialogue act recognition is further done by a supervised classification algorithm, which takes as input both the semantic and the baseline lexical features.

An interesting contribution is that all of these features are computed without any supervision. Another contribution is the proposal to use semantic space models (i.e. HAL and COALS), which, to the best of our knowledge, have never been used for dialogue act recognition. These models will further be compared. The last contribution consists in the evaluation of the proposed approaches on Czech, as a representative of morphologically rich language.

Our target application for the proposed dialogue act recognition approaches is a dialogue system that handles ticket reservation tasks. This system can exploit dialogue acts to better interpret the user's inputs. Our main interest is question (and order) detection, because these sentence modalities constitute an important clue for dialogue management. For example, when our system detects an explicit question (or an order), it has to treat it immediately and react accordingly.

The rest of the paper is organized as follows. Section 2 summarizes important DA recognition approaches with a particular focus on the recent methods using semantic features. Section 3 describes the different models we propose. Section 4 presents experimental results on the Czech Railways dialogue corpus. In the last section, we discuss the research results and we propose some future research directions.

2   Related Work

Relatively few studies on dialogue act modelling and automatic recognition have been published for the Czech language. Conversely, there are many works for other languages, especially for English and German.

Different sets of dialogue acts are defined in these works, depending on the target application and the available corpora. In [37], 42 dialogue act classes are defined for English, based on the Discourse Annotation and Markup System of Labeling (DAMSL) tag-set [1]. Switchboard-DAMSL tag-set [14] (SWBD-DAMSL) is an adaptation of DAMSL in the domain of telephone conversation. The Meeting Recorder DA (MRDA) tag-set [7] is another very popular tag-set, which is based on the SWBD-DAMSL taxonomy. MRDA contains 11 general DA labels and 39 specific labels. Jekat [11] defines 42 DAs for German and Japanese, with 18 DAs at the illocutionary level, in the context of the VERBMOBIL corpus.

These complete DA tag-sets are usually reduced for recognition into a few broad classes, because some classes occur rarely, or because other DAs are not useful for the target application. One typical regrouping may be [36]:

- statements
- questions
- backchannels
- incomplete utterance
- agreements
- appreciations
- other

Automatic recognition of dialogue acts is usually achieved using one of, or a combination of the following types of information:

1. lexical (and syntactic) information
2. prosodic information
3. context of each dialogue act

Lexical information (i.e. word sequence in the utterance) is useful for automatic DA recognition, because different DAs are usually composed from different word sequences. Some cue words and phrases can thus serve as explicit indicators of dialogue structure. For example, 88.4 % of the trigrams "<start> do you" occur in English in *yes/no questions* [15].

Several methods are used to represent lexical information [37]. Syntactic information is related to the *order* of the words in the utterance. For instance, in French and Czech, the relative order of the *subject* and *verb* occurrences might be used to discriminate between declarations and questions.

Words n-grams are often used to model some local syntactic information. Král et al. propose in [21] to represent word position in the utterance in order to take into account global syntactic information. Another type of syntactic information recently used for DA recognition are "cue phrases". These can be modelled with a subset of specific n-grams. The n value may vary from 1 to 4, which are selected based on their capacity to predict a specific DA and on their occurrence frequency [39]. A recent work in the dialogue act recognition field [20] also successfully uses a set of syntactic features derived from a deep parse tree.

Unfortunately, there are only few works that incorporate semantic features into the dialogue act recognition task. An interesting DA recognition approach using semantic information is presented in [24]. Sentence parse trees are computed on top of speech recognition output. Semantic information and the derivation rules of the partial trees are extracted and used to model the relationship between the DAs and the derivation rules. The resulting model is then used to generate a semantic score for dialogue act recognition when audio input is given.

The authors of [18] use for DA recognition syntactic and semantic relations acquired by information extraction methods. These features are successfully used as an input to a Bayesian network classifier. They use structured semantic features in the form of semantic predicate classes and semantic roles.

The authors of [28] study lexical semantics to recognize dialogue acts. They compare an unsupervised DA recognition approach based on the Latent Semantic Analysis (LSA) with another supervised one based on the Support Vector Machines (SVM). The authors show that the unsupervised method brings very good recognition results.

Ritter et al. propose in [32] another interesting unsupervised approach for dialogue act modelling and recognition. This method uses an LDA topic model together with clustering for DA recognition in twitter conversations. The LDA model is used to separate content words from dialogue indicators.

Prosodic information [36], particularly the melody of the utterance, is often used to provide additional clues to classify sentences in terms of DAs. The last useful information is the "dialogue history" which represents the sequence of recognized dialogue acts [37].

Dialogue act recognition is usually based on supervised machine learning as for instance Bayesian Networks [17], Discriminative Dynamic Bayesian Networks [13], BayesNet [29], Memory-based [22] and Transformation-based Learning [34], Decision Trees [26], Neural Networks [23], but also more advanced approaches such as Boosting [38], Latent Semantic Analysis [35], Hidden Backoff Models [4], Maximum Entropy Models [2], Conditional Random Fields [8,30] and Triangular-chain CRF [12].

3   Semantics for Dialogue Act Recognition

3.1   Latent Dirichlet Allocation

Latent Dirichlet Allocation (LDA) [5] is a popular topic model that assigns a topic to each word in the sentence. The semantically close words are usually represented by similar topics (e.g. synonyms). In this approach, we use a standard LDA model to compute a sentence topic for each word. These features will be used together with word labels for DA recognition.

3.2   Semantic Spaces

Semantic spaces are approaches that derive from word co-occurrence matrices a high dimensional semantic vector to represent every word in the vocabulary. The matrix is computed on a large unlabelled text corpus. Semantically close words are thus usually represented by similar vectors, according to some distance between vectors such as the cosine distance. Moreover, the vector space allows to create clusters of semantically close words by a clustering approach.

It has been shown in [6] that semantic spaces improve the results of language modelling. We assume in this work that these models can also bring relevant information for dialogue act recognition.

We use the HAL and COALS semantic spaces [25,33]. In the following, sentence-level semantic vectors are computed by additive composition of word-level semantic vectors, which are themselves computed with either the HAL or

COALS methods. The resulting sentence-level vector is then used as an additional semantic information for DA recognition. It is worth noting that these two semantic spaces have never been used for dialogue act recognition before.

Note that HAL and COALS are computed on relatively shorter context window, as compared to LDA, which takes in our case the full sentence as context. Hence, HAL and COALS will capture local dependencies between words while LDA will capture longer dependencies. Intuitively, the local word dependencies shall play a more important role than the distant ones for dialogue act recognition, because longer dependencies are more likely to capture information about the topic of the conversation rather than information about the possible substitutions of words within syntactic structures. Therefore, we expect HAL and COALS to give better results than LDA in the following experiments.

3.3 Dialogue Act Recognition

Let W be a sequence of n words w_i in the sentence, F be a sequence of semantic features f_i ($i \in [1; n]$) and C be a dialogue act class. We use two classifiers for dialogue act recognition:

- The first one is Naive Bayes [31] (also referred as an *unigram* when only word features available). This classifier which models simply $P(W|C)$ is used only as the first baseline.
- The second one is the Maximum Entropy (ME) [3] classifier. This classification algorithm is used to represent $P(C|W)$ or $P(C|W, F)$ in the semantic case. We use this classification approach as being very popular in the natural language processing field, because it has high recognition score. This approach is further also used (with lexical features only) as another baseline to show the impact of the proposed semantic features.

4 Evaluation

4.1 Corpus

The Czech Railways dialogue corpus, which contains human-human conversations, is used to validate the proposed approaches. The number of sentences of this corpus is shown in the second column of Table 1.

This corpus is divided into a training part, described in the first/top part of this table, and a testing part, described in the second/bottom part of the table. The training section of the corpus is used to train our LASER speech recognizer [9]. The testing part, composed of 2173 sentences pronounced by different speakers (see second part of Table 1) is used for testing the DA recognition systems. The sentences in the testing part of the corpus have been annotated manually with the following dialogue acts: statements (S), orders (O), yes/no questions (Q[y/n]) and other questions (Q). Note that in this corpus one utterance corresponds to one DA.

The automatic words transcription obtained with the LASER recognizer (1-best hypothesis) is used to compare the performances of our DA recognition systems on both manual and automatic speech transcriptions. Utterance recognition accuracy is 39.78 % and word recognition accuracy is 83.36 %.

Table 1. Composition of Czech Railways dialogue corpus

DA	No	Example	English translation
1. Training part			
Sent.	6234		
2. Testing part (annotated by DAs)			
S	566	Chtěl bych jet do Písku	I would like to go to Písek
O	125	Najdi další vlak do Plzně!	Look at for the next train to Plzeň!
Q[y/n]	282	Řekl byste nám další spojení?	Do you say next connection?
Q	1200	Jak se dostanu do Šumperka?	How can I go to Šumperk?
Sent.	2173		

4.2 Tools and Model Configuration

We use the LDA implementation from the MALLET [27] tool-kit for the following experiments. The LDA is trained with 1,000 iterations of the Gibbs sampling. The hyperparameters of the Dirichlet distributions are (as in [10]) initially set to $\alpha = 50/K$, where K is the number of topics and $\beta = 0.1$.

We use the S-Space package [16] for implementation of the HAL and COALS semantic space models and the whole above described corpus for training. For each semantic space, we use a four-words context window in both directions. Both semantic spaces use a matrix composed of 1,000 columns. Dimensionality reduction is not used in our experiments.

The LDA and both semantic space models are trained on the training part of the Railways corpus which is not annotated with the DAs (see first part of Table 1).

For DA recognition itself we use the Brainy [19] implementation of Naive Bayes and Maximum Entropy classifiers.

All experiments are realized using a cross-validation procedure, where 10 % of the corpus is reserved for the test. The resulting global accuracy has a confidence interval of ± 1 %.

4.3 Dialogue Act Recognition with Manual Speech Transcripts

Table 2 shows the results of a series of dialogue act recognition experiments with manual speech transcription. These experiments are realized in order to show the performance of the proposed approaches in "ideal" condition, i.e. without errors from Automatic Speech Recognition (ASR).

The first part of this table shows the results of our baseline approaches which use lexical word features with Naive Bayes (NB) and Maximum Entropy (ME) classifiers. The second part shows the results when the semantic features are combined with the baseline word features.

Only the maximum entropy classifier is used with semantic features, because of two main reasons: 1) The maximum entropy classifier has the best performances in terms of DA recognition accuracy; 2) More importantly, the Naive Bayes classifier assumes independence between the input features, and this assumption is clearly broken with the semantic features, because of their strong dependencies with the lexical features.

This table shows that every type of semantic features significantly improves the dialogue act recognition accuracy. We can also see that both semantic spaces outperform the LDA topic model. This result was expected, as already discussed in Sect. 3.2, because the semantic spaces model more local word dependencies than LDA, which are intuitively more important to characterize dialogue acts. The best recognition accuracy is obtained by the more sophisticated COALS model. Using the semantic features thus increases the recognition accuracy by 7.4 % over the baseline Naive Bayes approach and by 3.8 % over the discriminative Maximum Entropy model.

Table 2. Dialogue acts recognition accuracy for different approaches/classifiers and their combination with manual word transcription

Approach/ Classifier	Accuracy in [%]				
	S	O	Q[y/n]	Q	Glob
1. Lexical information (baselines)					
NB	93.5	77.6	96.5	89.9	**91.0**
ME	90.3	88.0	97.2	96.5	**94.6**
2. Semantic information					
LDA + ME	93.3	87.2	96.5	98.5	**96.4**
HAL + ME	95.1	96.0	97.9	97.9	**97.2**
COALS + ME	96.1	97.6	99.3	99.2	**98.4**

4.4 Dialogue Act Recognition with Automatic Speech Recognition

Table 3 shows the dialogue act recognition scores, when word transcriptions are estimated by the LASER speech recognizer. The results are obtained with a word class based trigram language model. The sentence speech recognition accuracy is 39.78 % and the word recognition accuracy is 83.36 %.

These experiments are done in order to show the robustness of our DA recognition approaches to the ASR errors. The structure of Table 3 is similar as in the previous case. This table shows that the DA recognition accuracy only slightly decreases, when word sequences are estimated automatically from the recognizer. The absolute decrease of the recognition score is very small and varies from 0.4 % to 3.9 % depending on the approach used. This confirms that our DA recognition system is quite robust to low and moderate ASR recognition errors.

Table 3. Dialogue acts recognition accuracy for different approaches/classifiers and their combination with word transcription by ASR

Approach/ Classifier	Accuracy in [%]				
	S	O	Q[y/n]	Q	Glob
1. Lexical information (baselines)					
NB	93.1	68.8	94.7	86.3	**88.2**
ME	87.5	77.6	89.7	95.2	**91.6**
2. Semantic information					
LDA + ME	88.3	80.8	89.0	96.3	**92.5**
HAL + ME	92.2	86.4	93.6	96.9	**94.8**
COALS + ME	95.9	96.8	97.5	99.0	**98.0**

5   Conclusions and Future Work

In this paper, we have shown the impact of semantic features for automatic dialogue act recognition. Three approaches to create semantic features have been proposed, implemented and evaluated on the Czech Railways dialogue act corpus. The dialogue act recognition model itself is a maximum entropy classifier, which has been trained in a supervised way. The experimental results confirm that the computed semantic features improve the dialogue act recognition score. We have further shown that, for this task, the semantic spaces significantly outperform the LDA model. This observation can be explained by the fact that these semantic spaces model more local dependencies between words than the LDA model.

Our approaches have been evaluated on the small Czech dialogue act corpus annotated with four dialogue acts. Our perspective thus consists in evaluation of the proposed methods on larger corpora and in other languages with more dialogue acts. The semantic features are particularly interesting for this task, because they can be computed without any supervision. Therefore, no additional annotation will be required when applying the proposed approach on another target language. We concentrated mainly on the creation of highly discriminative features. Another perspective thus consists in evaluation of the other classifiers for our task.

Acknowledgements. This work has been partly supported by the project LO1506 of the Czech Ministry of Education, Youth and Sports.

References

1. Allen, J., Core, M.: Draft of Damsl: Dialog Act Markup in Several Layers. In: http://www.cs.rochester.edu/research/cisd/resources/damsl/RevisedManual/RevisedManual.html (1997)
2. Ang, J., Liu, Y., Shriberg, E.: Automatic dialog act segmentation and classification in multiparty meetings. In: Proceedings of ICASSP. Philadelphia, USA, March 2005
3. Berger, A.L., Pietra, V.J.D., Pietra, S.A.D.: A maximum entropy approach to natural language processing. Comput. Linguist. **22**(1), 39–71 (1996)
4. Bilmes, J.: Backoff model training using partially observed data: application to dialog act tagging. Technical report UWEETR-2005-0008, Department of Electrical Engineering, University of Washington, August 2005
5. Blei, D.M., Ng, A.Y., Jordan, M.I., Lafferty, J.: Latent dirichlet allocation. J. Mach. Learn. Res. **3**, 2003 (2003)
6. Brychcín, T., Konopík, M.: Semantic spaces for improving language modeling. Comput. Speech Lang. **28**(1), 192–209 (2014)
7. Dhillon, R., Bhagat, S., Carvey, H., Shriberg, E.: Meeting recorder project: dialog act labeling guide. Technical report TR-04-002, International Computer Science Institute, February 2004
8. Dielmann, A., Renals, S.: Recognition of dialogue acts in multiparty meetings using a switching dbn. IEEE Trans. Audio Speech Lang. Process. **16**(7), 1303–1314 (2008)
9. Ekštein, K., Pavelka, T.: Lingvo/laser: prototyping concept of dialogue information system with spreading knowledge. In: NLUCS 2004, pp. 159–168. Porto, Portugal, April 2004
10. Griffiths, T.L., Steyvers, M.: Finding scientific topics. Proc. Nat. Acad. Sci. U.S.A. **101**(Suppl 1), 5228–5235 (2004)
11. Jekat, S., et al.: Dialogue acts in VERBMOBIL. In: Verbmobil Report 65 (1995)
12. Jeong, M., Lee, G.G.: Triangular-chain conditional random fields. IEEE Trans. Audio Speech Lang. Process. **16**(7), 1287–1302 (2008)
13. Ji, G., Bilmes, J.: Dialog act tagging using graphical models. In: Proceedings of ICASSP, vol. 1, pp. 33–36. Philadelphia, USA, March 2005
14. Jurafsky, D., Shriberg, E., Biasca, D.: Switchboard SWBD-DAMSL shallow-discourse-function annotation (coders manual, draft 13). Technical report 97–01, University of Colorado, Institute of Cognitive Science (1997)
15. Jurafsky, D., et al.: Automatic detection of discourse structure for speech recognition and understanding. In: IEEE Workshop on Speech Recognition and Understanding. Santa Barbara (1997)
16. Jurgens, D., Stevens, K.: The s-space package: an open source package for word space models. In: System Papers of the Association of Computational Linguistics (2010)
17. Keizer, S., Nijholt, A.: Dialogue act recognition with bayesian networks for dutch dialogues. In: 3rd ACL/SIGdial Workshop on Discourse and Dialogue, pp. 88–94. Philadelphia, USA, July 2002
18. Klüwer, T., Uszkoreit, H., Xu, F.: Using syntactic and semantic based relations for dialogue act recognition. In: Proceedings of the 23rd International Conference on Computational Linguistics: Posters, pp. 570–578. Association for Computational Linguistics (2010)
19. Konkol, M.: Brainy: a machine learning library. In: Rutkowski, L., Korytkowski, M., Scherer, R., Tadeusiewicz, R., Zadeh, L.A., Zurada, J.M. (eds.) ICAISC 2014, Part II. LNCS, vol. 8468, pp. 490–499. Springer, Heidelberg (2014)

20. Král, P., Cerisara, C.: Automatic dialogue act recognition with syntactic features. Lang. Resour. Eval. **48**(3), 419–441 (2014)
21. Král, P., Cerisara, C., Klečková, J.: Lexical structure for dialogue act recognition. J. multimedia (JMM) **2**(3), 1–8 (2007)
22. Lendvai, P., van den Bosch, A., Krahmer, E.: Machine learning for shallow interpretation of user utterances in spoken dialogue systems. In: EACL-03 Workshop on Dialogue Systems: Interaction, pp. 69–78. Adaptation and Styles Management, Budapest, Hungary (2003)
23. Levin, L., Langley, C., Lavie, A., Gates, D., Wallace, D., Peterson, K.: Domain Specific speech acts for spoken language translation. In: 4th SIGdial Workshop on Discourse and Dialogue. Sapporo, Japan (2003)
24. Liang, W.B., Wu, C.H., Chen, C.P.: Semantic information and derivation rules for robust dialogue act detection in a spoken dialogue system. In: Proceedings of the 49th Annual Meeting of the Association for Computational Linguistics: Human Language Technologies: short papers, vol. 2, pp. 603–608. Association for Computational Linguistics (2011)
25. Lund, K., Burgess, C.: Hyperspace analogue to language (hal): a general model semantic representation. In: Brain and Cognition. vol. 30, pp. 92101–4495. Academic Press INC JNL-Comp Subscriptions 525 B ST, STE 1900, San Diego, CA (1996)
26. Mast, M., et al.: Automatic classification of dialog acts with semantic classification trees and polygrams. In: Connectionist, Statistical and Symbolic Approaches to Learning for Natural Language Processing, pp. 217–229 (1996)
27. McCallum, A.K.: Mallet: a machine learning for language toolkit (2002). http:// mallet.cs.umass.edu
28. Novielli, N., Strapparava, C.: Exploring the lexical semantics of dialogue acts. J. Comput. Linguist. Appl. **1**(1–2), 9–26 (2010)
29. Petukhova, V., Bunt, H.: Incremental dialogue act understanding. In: Proceedings of the 9th International Conference on Computational Semantics (IWCS-9). Oxford, January 2011
30. Quarteroni, S., Ivanov, A.V., Riccardi, G.: Simultaneous dialog act segmentation and classification from human-human spoken conversations. In: Proceedings of ICASSP, Prague, Czech Republic, May 2011
31. Rish, I.: An empirical study of the naive bayes classifier. In: IJCAI 2001 Workshop on Empirical Methods in Artificial Intelligence, vol. 3, pp. 41–46. IBM, New York (2001)
32. Ritter, A., Cherry, C., Dolan, B.: Unsupervised modeling of twitter conversations. In: NAACL HLT 2010 - Human Language Technologies: Proceedings of the Main Conference of the 2010 Annual Conference of the North American Chapter of the Association for Computational Linguistics, pp. 172–180 (2010)
33. Rohde, D.L., Gonnerman, L.M., Plaut, D.C.: An improved model of semantic similarity based on lexical co-occurrence. Commun. ACM **8**, 627–633 (2006)
34. Samuel, K., Carberry, S., Vijay-Shanker, K.: dialogue act tagging with transformation-based learning. In: 17th international conference on Computational linguistics, vol. 2, pp. 1150–1156. Association for Computational Linguistics, Morristown, NJ, USA, Montreal, Quebec, Canada, 10–14 August 1998
35. Serafin, R., Di Eugenio, B.: LSA: extending latent semantic analysis with features for dialogue act classification. In: Proceedings of the 42nd Annual Meeting on Association for Computational Linguistics, Spain (2004)

36. Shriberg, E., et al.: Language and speech, special double issue on prosody and conversation. In: Can Prosody Aid the Automatic Classification of Dialog Acts in Conversational Speech?, vol. 41, pp. 439–487 (1998)
37. Stolcke, A., et al.: Dialog act modeling for automatic tagging and recognition of conversational speech. Comput. Linguist. **26**, 339–373 (2000)
38. Tur, G., Guz, U., Hakkani-Tur, D.: Model adaptation for dialogue act tagging. In: Proceedings of the IEEE Spoken Language Technology Workshop (2006)
39. Webb, N., Hepple, M., Wilks, Y.: Dialog act classification based on intra-utterance features. Technical report CS-05-01, Department of Computer Science, University of Sheffield (2005)

Conversational Telephone Speech Recognition
for Lithuanian

Rasa Lileikyte[(⊠)], Lori Lamel, and Jean-Luc Gauvain

CNRS/LIMSI, Spoken Language Processing Group, 91405 Orsay Cedex, France
{lileikyte,lamel,gauvain}@limsi.fr

Abstract. This paper presents a conversational telephone speech recognition system for the low-resourced Lithuanian language, developed in the context of IARPA-Babel program. Phoneme-based systems and grapheme-based systems are compared to establish whether or not it is necessary to use a phonemic lexicon. We explore the impact using Web data for language modeling and additional untranscribed data for semi-supervised training. Experimental results are reported for two conditions: Full Language Pack (FLP) and Very Limited Language Pack (VLLP), for which respectively 40 and 3 h of transcribed training data are available. Grapheme-based systems are shown to give comparable results to phoneme-based ones. Adding Web texts improves the performance of both the FLP and VLLP system. The best VLLP results are achieved using both Web texts and semi-supervised training.

Keywords: Conversational telephone speech · Lithuanian · KWS · STT

1   Introduction

The Lithuanian language is one of the least spoken European languages, with only about 3.5 million speakers. Lithuanian belongs to the Baltic subgroup of Indo-European languages. The language was standardized during the late 19th century and the early 20th century. Having preserved the majority of phonetic and morphological features [27], the Lithuanian language has rich inflection, a complex stress system, and flexible word order. It is based on the Latin alphabet with some additional language specific characters, as well as some characters borrowed from other languages.

There are two main dialects - Aukštaitian (High Lithuanian), and Samogitian (Žemaitian or Low Lithuanian), each with sub-dialects. The dominant dialect is Aukštaitian, spoken in the east and middle of Lithuania by 3 million speakers. Samogitian is spoken in the west of the country by about 0.5 millions speakers.

This paper presents the development of a conversational telephone speech (CTS) recognition system for the Lithuanian language. Today's speech recognition systems make use of statistical acoustic and language models which are typically trained on large data sets. Three main resources are needed: (1) telephone speech recordings with corresponding transcriptions for acoustic model

© Springer International Publishing Switzerland 2015
A.-H. Dediu et al. (Eds.): SLSP 2015, LNAI 9449, pp. 164–172, 2015.
DOI: 10.1007/978-3-319-25789-1_16

training, (2) written texts for language modeling, (3) and a pronunciation dictionary.

There have been few studies reporting on speech recognition for the Lithuanian language, in part due to the sparsity of linguistic resources. Systems for isolated digit recognition with some additional words are described in [19,21]. Isolated word recognizers trained on read-speech data are reported in [2,24,26]. In [18,25] Lithuanian broadcast speech recognition systems were trained on only 9 h of transcribed speech. A uni-code based graphemic system described in [5] reports on the transcription of conversational telephone speech in Lithuanian. In the context of the Quaero program, a broadcast news transcription system was jointly developed by LIMSI and Vocapia Research. An average word error rate (WER) of 27 % was obtained without any manually transcribed training data [16].

Our system was built in the context of IARPA-Babel project, which as for [5], provided training resources for two conditions: full language pack (FLP) with approximately 40 h of transcribed telephone speech and very limited language pack (VLLP) comprising only 3 h of transcribed data. An additional 40 h of untranscribed speech was available for semi-supervised training. Transcribing conversational telephone speech is a more complex task than transcribing broadcast news. Spontaneous telephone speech has a high variability of speaking rates and styles, grammar rules are not strictly followed, and there is limited frequency bandwidth and noisy audio channels. Additional text corpora were also provided for training. However, most of these texts are quite different from conversational speech [29]. We used the data prepared by our partner BBN, which contains texts collected from the Web such as Wikipedia, subtitles, and other sources. The text corpus consists of 26M words. Nevertheless, 40 h of transcribed data and 26M text words is a very small amount compared to the 2000 h of transcribed audio and over a billion words of language modeling text that are available for the English language [23].

The pronunciation dictionary is an important component of the system. To generate one, a grapheme based or phoneme based approach can be used. The advantage of using graphemes is that pronunciations are easily derived. Grapheme based systems have been shown to work well for various languages [12,15]. Yet, some languages, such as English, have a weak correspondence between graphemes and phonemes, and using graphemes leads to a degradation in system performance. Phoneme based systems usually provide good results as they better represent the speech production. However, designing the grapheme-to-phoneme rules requires the linguistic skills of an expert, making it an expensive process. For the Lithuanian language there is a quite strong dependency between the orthographic transcription and the phonetic form, and conversion rules are implemented easily in comparison to the English language that requires numerous exceptions.

In our study we addressed the following questions: (1) is a phoneme-based system better than grapheme-based one? (2) how can additional resources (untranscribed audio and Web texts) be used to improve the system performance, and what is their impact with respect to the original training conditions?

We describe the phonemic inventory of the Lithuanian language in Sect. 2. In Sect. 3 the experimental setup is defined. Section 4 presents results for different sets of graphemes and phonemes, when semi-supervised training and Web texts are used. Finally, in Sect. 5 we draw some conclusions.

2   Lithuanian Phonemic Inventory

The Lithuanian alphabet contains 32 letters. While most of them are Latin, there is also ė, and some borrowed letters: š, ž from Czech, and ą, ę from Polish [27]. Lithuanian is generally described as having 56 phonemes, consisting of 11 vowels and 45 consonants [1,22]. Consonants are differentiated as soft (palatalized) or hard (not palatalized). Consonants are always soft before certain vowels (i, į, y, e, ę, ė). There are 8 diphthongs that are composed of two vowels (ai, au, ei, ui, ou, oi, ie, uo) [13] and 16 mixed-diphthongs composed of a vowel (a, e, i, u), followed by sonorant (l, m, n, r), for example (al, am, an, ar). The language also has 4 affricates (c, č, dz, dž). The correspondence between the orthography and phonemes is provided in Table 1, where the International Phonetic Alphabet (IPA) is used to denote the phonemes.

Table 1. Lithuanian orthographic and phonemic correspondence

Vowels		Consonants			
a	/a/	p	/p/,/pʲ/	dz	/dz/,/dzʲ/
ą	/ɑ/	b	/b/,/bʲ/	č	/tʃ/,/tʃʲ/
e	/ɛ/	t	/t/,/tʲ/	dž	/dʒ/,/dʒʲ/
ę	/æ/	d	/d/,/dʲ/	m	/m/,/mʲ/
ė	/e:/	k	/k/,/kʲ/	n	/n/,/nʲ/
i	/i/	g	/g/,/gʲ/	l	/ɫ/,/lʲ/
į, y	/i:/	v	/v/,/vʲ/	r	/r/,/rʲ/
o	/o:/,/ɔ/	s	/s/,/sʲ/	j	/j/
u	/ʊ/	z	/z/,/zʲ/	f	/f/,/fʲ/
ų, ū	/u:/	š	/ʃ/,/ʃʲ/	ch	/x/,/xʲ/
		ž	/ʒ/,/ʒʲ/	h	/ɣ/,/ɣʲ/
		c	/ts/,/tsʲ/		

The following papers [8,22], inspired the grapheme to phoneme conversion rules used in this work. In total there are 43 rules of which 11 are contextual.

3   Experimental Setup

3.1   Data Set

All experiments use the data provided by the IARPA Babel program [10], more specifically the IARPA-babel304b-v1.0b dataset. The data is comprised of spontaneous telephone conversations. As mentioned earlier, two conditions are defined: (1) Full Language Pack (FLP) with about 40 h of transcribed speech for training, (2) Very Limited Language Pack (VLLP) is a subset of FLP comprising about 3 h of transcribed data. For this condition the remaining 37 h of FLP data is untranscribed and can be used only in an unsupervised manner [14,28]. In semi-unsupervised training a first set of models are built with the transcribed portion of the data. The recognizer is then used to generate transcripts for the untranscribed training data. An additional 40 h of untranscribed data is also available for semi-supervised training. According to the Babel evaluation conditions, for the FLP systems only the manual transcriptions of audio training data could be used for language modeling, whereas in the case of VLLP, the Web text corpora could also be used for training the language models. In both cases any available untranscribed data could be used for acoustic modeling.

Results are reported on the 10 h development data set. For the keyword spotting (KWS) experiments, the official 2015 list of development keywords provided by NIST was used. The FLP list has 4079 keywords. There are 412 keywords that do not appear in the training data and are considered to be out-of-vocabulary with respect to the FLP condition.

3.2   Baseline Recognition Systems

In our experiments the speech-to-text (STT) systems are built via a flat start training, where the initial segmentation is performed without any *a priori* information. The acoustic models are tied-state, left-to-right 3-state HMMs with Gaussian mixture observation densities [6]. The models are triphone-based and word position-dependent. The system uses acoustic features provided by BUT [9].[1]

The language model is trained with the LIMSI STK toolkit [17]. The models are built using the manual transcriptions and the Web texts. Prior to transcription, speech non-speech separation is carried out using the BLSTM algorithm as described in [7]. Then, for each speech segment, word lattices are generated using a 3-gram LM. The final word hypotheses are obtained via a consensus decoding [20]. The keyword spotting system is based on a consensus network, using both word and 7-gram sub-word units as described in [11].

3.3   Performance Metrics

The speech recognition performance is measured using the word error rate (WER). Actual term-weighted value (ATWV) is used to evaluate KWS [3]. The keyword specific ATWV for the keyword k at threshold t is computed as follows:

$$ATWV(k,t) = 1 - P_{FR}(k,t) - \beta P_{FA}(k,t) \tag{1}$$

---

[1] In order to abide by the evaluation rules, different features were provided by BUT for the FLP and VLLP conditions, in the latter case being trained on multilingual data.

where P_{FR} is the probability of a false reject and P_{PFA} of a false accept. The constant β is set to 999.9. This constant mediates the trade off between the false accepts and the false rejects. MTWV is the maximum value computed over all possible thresholds t. The words that are not in the system vocabulary are out-of-vocabulary words (oov), and the rest are in-vocabulary words (inv).

4 Experimental Results

We evaluated several phoneme-based and grapheme-based systems. Various sets of units and their combinations where investigated, including complex sounds as affricates, diphthongs, soft consonants and mappings for rarely seen units. The dictionary provided by Appen as part of the Babel language pack explicitly models stress and soft consonants, which are not used in our baseline phonemic dictionary.

Some of the various grapheme and phoneme sets are described in Table 2. Three units are used to model breath noise, hesitations and silence. The rare non Lithuanian characters appearing in the corpus are mapped to Lithuanian ones (x→ks, q→k, w→v).

Table 2. Grapheme and phoneme systems

System	#Units	Modification from baseline
FLP graph-baseline	35	graphs
FLP phone-baseline	32	phones
FLP phone	36	c→c, č→č, dž→D, dz→Z
FLP phone	38	diph, ou→o,u, oi→o,i
FLP phone	48	soft consonants, soft ch→x
VLLP graph-baseline	33	graphs, c→ts, f→v
VLLP graph	29	z→s, ch→ɣ, ę→ɛ, į,y→i:, ū,ų→u:
VLLP phone-baseline	31	phones, f→v
VLLP phone	29	z→s, ch→ɣ, ę→ɛ, į,y→i:, ū,ų→u:

In the grapheme lexicon each orthographic character is modeled as a separate grapheme. For the baseline phoneme set, affricates are split into a sequence of two phonemes (according IPA): c→ts, č→tʃ, dz→dz, dž→dʒ.

Linguists do not necessarily agree if affricates and diphthongs should be modeled as a single phoneme or a phoneme sequence. The FLP 36 phone set models affricates and the 38 phone set models diphthongs (in all cases the rare *ou, oi* were split into a sequence of vowels) as specific units. The soft consonants are included in the 48 phone set, but the rare *soft ch* is mapped to the hard one.

In case of VLLP, we investigated mappings for the rare units as they tend to be poorly modeled. For the baseline systems, only the two rarest units (c and f) are mapped. Furthermore, in the 29 graph set and the 29 phone set the number of units is reduced by mapping z, ch and $ę$ because they were rarely seen in the data. The $į$, y, $ū$ and $ų$ are also mapped as these units define the same sounds but have different representations due to grammar exceptions.

STT and KWS results for 5 FLP grapheme and phoneme systems are shown in Table 3. Only small differences in WER were observed, with the WER ranging from 44.4 % to 44.7 % for the FLP task. ATWV and MTWV are reported for the combined full-word and 7-gram sub-word keyword hits [11]. The column *Homoph* gives the number of homographs and the number of homophones for grapheme and phoneme-based systems, respectively.

Table 3. STT and KWS results for graphemic and phonemic FLP systems. ATWV and MTWV use both word and 7-gram sub-word units

System	#Units	Homoph	%WER	ATWV(all/inv/oov)	MTWV(all/inv/oov)
graph	35	522	44.6	0.578/0.592/0.450	0.579/0.592/0.472
phone	32	719	44.7	0.574/0.590/0.437	0.576/0.591/0.476
phone	36	718	44.6	**0.578/0.592/0.451**	**0.580/0.593/0.487**
phone	38	718	**44.4**	0.573/0.589/0.431	0.576/0.591/0.460
phone	48	717	44.6	0.570/0.585/0.436	0.573/0.587/0.472

Table 3 shows that by merging affricates a slight improvement in ATWV and MTWV is obtained with the phoneme based system (36 phones). In contrast, when diphthongs are modeled as separate units, there is a 0.3 % absolute increase in WER (44.7 % vs. 44.4 %). Note that the best phonemic system for STT does not have the best KWS performance. While the best WER result is obtained for the 38 phone system with units for diphthongs, the best ATWV/MTWV are obtained when affricates are modeled. The best phoneme based system gives a slightly lower WER (0.2 % absolute) than the grapheme based system (44.6 % vs. 44.4 %).

Table 4 presents the WER results for the VLLP systems. It can be observed that both the grapheme and phoneme based systems perform slightly better when the number of units is reduced. Comparing the best graphemic system with the best phonemic system, the phonemic system obtains an absolute WER gain of 0.2 % (52.2 % vs. 52.0 %) as was the case for FLP comparison. Since the mapping has increased the number of homophones from 1583 to 2418, the limited gain may be due to the increased lexical confusability.

Table 4. WER of VLLP systems

System	#Units	Homoph	%WER
graph	33	493	52.6
graph	29	1583	**52.2**
phone	31	1336	52.3
phone	29	2418	**52.0**

In the above FLP experiments we used only the manual transcriptions for language modeling. To build the VLLP systems, we also used Web data for training language models, and the remaining untranscribed data for semi-supervised acoustic model training. These extra resources help to reduce the performance difference between the two conditions. We assess the impact of the Web data and semi-supervised training for both the FLP and VLLP systems.

Table 5 summarizes the STT experiments when Web texts are used for language modeling, and lattice-based semi-supervised training of the acoustic models [4]. Comparing the top rows of each section (FLP 40 and VLLP 3), there is a large difference in the WER obtained by the two systems: the 37 h of audio with transcripts halves the OOV rate and reduces the relative WER by 25 %. As can be expected the Web texts have a much larger impact for the VLLP condition than the FLP one: for VLLP condition the WER is reduced by 6 % compared to just under 2 % for FLP.

Semi-supervised training did not improve the performance of the FLP system, but did improve the VLLP system performance both with and without Web data. Lattice-based semi-supervised training gave a 0.6 % absolute improvement over 1-best based training (51.1 % vs. 50.5 %) which is consistent with [4]. The best VLLP WER remains 10 % behind that of the FLP system. This difference carries over to word-based KWS: the overall ATWV with the VLLP system was 0.357 compared to 0.470 for the FLP one. However, using subword units can reduce this difference as reported in [11].

Table 5. WER results for different conditions: only manual transcriptions used for LM training, Web texts used, SST used for acoustic models

Set	Hours	Acoustic model	Language model	Lexicon	%OOV	%WER
FLP	40	trn	trn	30k	7.6	44.4
FLP	73	trn + SST	trn	30k	7.6	44.8
FLP	40	trn	trn + webtexts	60k	5.2	**42.4**
FLP	73	trn + SST	trn + webtexts	60k	5.2	42.4
VLLP	3	trn	trn	5.7k	16.7	59.3
VLLP	41	trn + SST	trn	5.7k	16.7	59.0
VLLP	3	trn	trn + webtexts	60k	6.0	53.3
VLLP	41	trn + SST	trn + webtexts	60k	6.0	**52.0**

5 Summary

We report on developing a conversational telephone speech recognition system for Lithuanian, a low-resourced language. We first analyzed the phonemic inventory to determine what set of units are best to use and to assess if a phoneme-based system outperforms a grapheme-based one. It is very interesting to observe that using phonemes was found to give only a slight improvement for the two training conditions (3 or 40 h of transcribed audio data). We attribute this small difference to the strong relationship between the orthographic and phonemic forms in Lithuanian.

Moreover, we explored the impact of using Web texts for training language models, and untranscribed data for semi-supervised training of the acoustic models. Adding Web texts to FLP system gave an improvement of almost 2 %. The VLLP system was improved over 7 % absolute using both Web data and semi-supervised training.

Acknowledgments. We would like to thank our IARPA-Babel partners for sharing resources (BUT for the bottle-neck features and BBN for the web data), and Grégory Gelly for providing the VADs.

This research was in part supported by the Intelligence Advanced Research Projects Activity (IARPA) via Department of Defense US Army Research Laboratory contract number W911NF-12-C-0013. The U.S. Government is authorized to reproduce and distribute reprints for Governmental purposes notwithstanding any copyright annotation thereon. Disclaimer: The views and conclusions contained herein are those of the authors and should not be interpreted as necessarily representing the official policies or endorsements, either expressed or implied, of IARPA, DoD/ARL, or the U.S. Government.

References

1. Ambrazas, V., Garšva, K., Girdenis, A.: Dabartinės lietuviu kalbos gramatika. (A Grammar of Modern Lithuanian), Vilnius, MELI (2006)
2. Filipovič, M., Lipeika, A.: Development of HMM/neural network-based mediumvocabularyisolated-word lithuanian speech recognition system. Informatica **15**(4), 465–474(2004)
3. Fiscus, J.G., Ajot, J., Garofolo, J.S., Doddington, G.: Results of the 2006 spoken term detection evaluation. In: Proceedings of ACM SIGIR, vol. 7, pp. 51–55 (2007)
4. Fraga-Silva, T., Gauvain, J.L., Lamel, L.: Lattice-based unsupervised acoustic model training. In ICASSP 2011, 36th International Conference on Acoustics, Speech and Signal Processing, Prague, Czech Republic, pp. 4656–4659 (2011)
5. Gales, M.J.F., Knill, K.M., Ragni, A.: Unicode-based graphemic systems for limited resource languages. In: ICASSP 2015 (2015)
6. Gauvain, J.L., Lamel, L., Adda, G.: The LIMSI broad-cast news transcription system. Speech Commun. **37**(1), 89–108 (2002)
7. Gelly, G. Gauvain, J.L.: Minimum word error training of RNN-based voice activity detection. In: Interspeech 2015, Dresden (2015)
8. Girdenis, A.: Teoriniai lietuviu fonologijos pagrindai (Theoretical Foundations of Lithuanian Phonology), 2nd edn. Mokslo ir enciklopediju leidybos inst., Vilnius (2003)
9. Grézl, F., Karafiát, M.: Semi-supervised bootstrapping approach for neural network feature extractor training. In: 2013 IEEE Workshop on Automatic Speech Recognition and Understanding (ASRU), pp. 470–475. IEEE (2013)
10. Harper, M.: "IARPA Babel Program." http://www.iarpa.gov/index.php/research-programs/babel
11. Hartmann, W., Le, V.B., Messaoudi, A., Lamel, L., Gauvain, J.L.: Comparing decoding strategies for subword-based keyword spotting in low-resourced languages. In: Proceedings of Interspeech, Singapore, pp. 2764–2768 (2014)
12. Kanthak, S., Ney, H.: Context-dependent acoustic modeling using graphemes for large vocabulary speech recognition. In: ICASSP, vol. 2, pp. 845–848 (2002)

13. Kazlauskienė, A., Raškinis, G.: Bendrinės lietuviu kalbos garsu dažnumas (The frequency of generic Lithuanian sounds). Respectus Philologicus **16**(21), 169–182 (2009)
14. Kemp, T., Waibel, A.: Unsupervised training of a speech recognizer: recent experiments. In: ESCA Eurospeech, pp. 2725–2728 (1999)
15. Killer, M.: Grapheme-based speech recognition. M.S. Thesis, Carnegie Mellon University (2003)
16. Lamel, L.: Unsupervised acoustic model training with limited linguistic resources. In: ASRU 2013 (2013)
17. Lamel, L., et al.: Speech recognition for machine translation in Quaero. In: IWSLT 2011, pp. 121–128 (2011)
18. Laurinčiukaitė, S., Lipeika, A.: Syllable-phoneme based continuous speech recognition. Elektronika ir Elektrotechnika **70**(6), 91–94 (2006)
19. Lipeika, A., Lipeikiene, J., Telksnys, L.: Development of isolated word speech recognition system. Informatica **13**(1), 37–46 (2002)
20. Mangu, L., Brill, E., Stolcke, A.: Finding consensus in speech recognition: word error minimization and other applications of confusion networks. Comput. Speech Lang. **14**(4), 373–400 (2000)
21. Maskeliūnas, R., Rudžionis, A., Ratkevičius, K., Rudžionis, V.: Investigation of foreign languages models for Lithuanian speech recognition. Elektronika ir Elektrotechnika **91**(3), 15–20 (2009)
22. Pakerys, A.: Lietuviu bendrinės kalbos fonetika. Enciklopedija, Valiulio Leidykla (The phonetics of generic Lithuanian language. Encyclopedia) (2003)
23. Prasad, R., et al.: The 2004 BBN/LIMSI 20xRT English Conversational Telephone Speech Recognition System. In: InterSpeech, pp. 1645–1648 (2005)
24. Raškinis, G., Raškinienė, D.: Building medium-vocabulary isolated-word Lithuanian HMM speech recognition system. Informatica **14**(1), 75–84 (2003)
25. Šilingas, D., Laurinčiukaitė, S., Telksnys, L.: Towards Acoustic Modeling of Lithuanian Speech. In: Proceedings of International Conference SPECOM 2004, pp. 326–333 (2004)
26. Vaiciunas, A., Raškinis, G.: Cache-based statistical language models of English and highly inflected Lithuanian. Informatica **17**(1), 111–124 (2006)
27. Vaišnienė, D., Zabarskaitė, J.: Lithuanuan Language in the Digital Age. Springer, White Paper Series (2012)
28. Zavaliagkos, G., Colthurst, T.: Utilizing untranscribed training data to improve performance. In: Proceedings of the DARPA Broadcast News Transcription and Understanding,Workshop, pp. 301–305 (1998)
29. Zhang, L., Karakos, D., Hartmann, W., Hsiao, R., Schwartz, R., Tsakalidis, S.: Enhancing low resource keyword spotting with automatically retrieved web documents. In: 2015 Interspeech (accepted, September 2015)

Long-Term Statistical Feature Extraction from Speech Signal and Its Application in Emotion Recognition

Erfan Loweimi, Mortaza Doulaty$^{(\boxtimes)}$, Jon Barker, and Thomas Hain

Speech and Hearing Research Group (SPandH),
University of Sheffield, Sheffield, UK
{e.loweimi,m.doulaty,j.barker,t.hain}@dcs.shef.ac.uk

Abstract. In this paper we propose a statistical-based parametrization framework for representing the speech through a fixed-length supervector which paves the way for capturing the long-term properties of this signal. Having a fixed-length representation for a variable-length pattern like speech which preserved the task-relevant information allows for using a wide range of powerful discriminative models which could not effectively handle the variability in the pattern length. In the proposed approach, a GMM is trained for each class and the posterior probabilities of the components of all the GMMs are computed for each data instance (frame), averaged over all utterance frames and finally stacked into a supervector. The main benefits of the proposed method are making the feature extraction task-specific, performing a remarkable dimensionality reduction and yet preserving the discriminative capability of the extracted features. This method leads to an 7.6 % absolute performance improvement in comparison with the baseline system which is a GMM-based classifier and results in 87.6 % accuracy in emotion recognition task. Human performance on the employed database (Berlin) is reportedly 84.3 %.

Keywords: Discriminative model · Emotion recognition · Feature extraction · Generative model · Speech signal

1   Introduction

Speech is the most natural way of human communication. It reflects many aspects of us and this turns it into a complicated signal which as well as its lingual content, encodes a wide variety of information including environmental and speaker-dependent information like identity, emotional state, accent, dialect, age and health condition. These components of the speech are combined through a complicated process and disentangling such a complex signal into the aforementioned underlying dimensions is a challenging yet interesting task from both signal processing and machine learning points of view.

© Springer International Publishing Switzerland 2015
A.-H. Dediu et al. (Eds.): SLSP 2015, LNAI 9449, pp. 173–184, 2015.
DOI: 10.1007/978-3-319-25789-1_17

In this regard, the main problem is that such attributes are subjective in essence and developing an objective model or method for capturing them is difficult. In fact, other than very general clues about the aforementioned properties in the time and frequency domains, we do not have any particular extra information to steer the hand-crafted deterministic parametrization algorithms to the right direction. As a result, in a wide range of applications in speech processing, MFCC serves as the swiss-knife army of this field and is used as the main feature representation despite the fact that it is basically proposed for speech recognition [5]. That is why the general tendency is to put the back-end at the center of attention for building a system in different applications.

Typically, pattern recognition systems consist of two main blocks, namely the front-end and back-end [6]. The front-end is tasked with extracting a representation of the data in which the task-pertinent attributes are preserved/enhanced and the irrelevant/misleading aspects of the data are filtered/weakened. This process, among other steps, requires data filtering in a very high-level domain where each attribute occupies a particular subspace. So, the front-end ideally should do *information filtering* in the *information space* and it turns out to be very challenging. The reason backs to the fact that such information space is categorically abstract and subjective. Therefore, mathematical underpinning of a mapping which takes the data from the low-level quantitative domain to such a high-level qualitative/subjective space is extremely complicated.

In this paper, we aim at enhancing the conventional feature extraction process with an interface which to some extent contributes toward conducting information filtering. This interface is a generative model which targets learning a task-dependent representation. As well, it affords further dimensionality reduction and renders a fixed-length representation for speech. This paves the way for the discriminative model employed at the back-end to return more accurate results because most of these models cannot effectively deal with the variable length patterns like speech. In the emotion recognition task, such coupling of the generative and discriminative models results in up to 7.6 % performance elevation in comparison with the GMM-based classifier and leads to 87.6 % accuracy which is higher than the reported human performance (84.3 %) on the this task and database [3].

The rest of this paper is organized as follows. In Sect. 2 the main difficulties and issues in extracting hand-crafted features are reviewed and discussed. The proposed parametrization method is introduced and explained in Sect. 3. Experimental results are presented and analyzed in Sects. 4 and 5 concludes the paper.

2   Feature Extraction

Feature extraction (also known as parametrization or front-end) bears the task of converting the sensory data into a sequence of numbers which should preserve the relevant information in a compact way and discards the irrelevant and misleading aspects of the data. As well, the front-end should present the data in an appropriate way. Since its output serves as the input of the back-end, parametrization process output should be in harmony with the assumptions that the back-end makes about its input. For instance, a back-end with a probabilistic basis makes some statistical assumptions about the distribution of its input. Coherency of the fed features with such expectations would substantially affect the overall performance and efficacy of the system.

Having a fixed back-end, different parametrization algorithms, could be assessed through three main criteria, namely discriminability, robustness and complexity. Discriminability is about the capability of the front-end in extracting features with both high intra-class similarity and inter-class dissimilarity. It could be evaluated by using the train data as the test data. Robustness relates to the ability of the feature in handling a reasonable amount of noise and/or mismatch between the test and train conditions. It is a challenging issue and could be assessed by using unseen/noisy data. Complexity is connected to the computational load of the feature extraction process and the lower the better.

Another important issue in feature engineering is that the output of the designed algorithm should be task-dependent because the target attribute and consequently the focus of the front-end for each application is different. That is, by setting the aim of a system to capture one of the speech properties, say speaker's emotional state, all the other elements like lingual content, speaker ID, etc. are turned into noise and should be suppressed. In practice, the ideal emotion recognition system should be able to recognize the emotion regardless of the speaker identity, lingual content and background noises.

Presence of irrelevant/misleading factors poses two main problems. First, learning the structure of the target attribute(s) under the existence of the other irrelevant attributes will be more difficult due to the clutter which they arise in the information space and it leads to hindering the learning process. Second, even if the system performs relatively well across the seen data during training, its accuracy over the unseen data would be strictly questionable. The reason backs to the fact that the misleading components would highly restrict the situation where the system performs well and the performance becomes oversensitive to any mismatch even in irrelevant traits. As a result, the generalization will be poor and any mismatch with the training condition, even in the irrelevant aspects, would noticeably degrade the accuracy and reliability of the system. To overcome these issues, the front-end should be able to filter out the misleading/irrelevant characteristics and only passes through the pertinent properties.

However, such filtering is not straightforward. As a matter of fact, it takes place in a conceptual space where each attribute presumably occupies a distinct subspace. This high-level information domain does exist based on what we subjectively perceive from the speech signal. However, expressing it in an efficient objective/mathematical way which allows for filtering nuisance characteristics and only permitting the relevant dimensions is highly challenging, if not impossible. That is why most of the researches in the pattern recognition field are focused on the back-end and less attention is paid to the front-end.

Dealing with these issues, researchers tended to develop techniques for *learning* the proper representation instead of using hand-crafted features. Currently, one of the very active branches of Machine Learning is Deep Learning via deep neural networks (DNN) which essentially solve the problem of data representation [1]. In other words, they learn a transform (or a set of transforms) that represents the input data in the most suitable way based on the task requirement. However, for efficient training of such models which have enormous parameters, a huge amount of data and computational power are required. Although the later is no longer an impeding factor, the former is still troublesome at some fields. For example, in the task of emotion detection from speech, most of the available databases are not sufficiently big and do not allow for employing models with too many variables.

3   Proposed Method

As mentioned, feature learning under having limited data is problematic. However, due to lack in practical clues for engineering the feature extraction process for capturing the most pertinent aspects of the signal, we need to carry out a feature learning to steer the parametrization process toward a right task-dependent direction and avoid passing through irrelevant dimensions. In this section, we introduce our proposed method which serves as an interface between the conventional front-end and the classifier.

3.1   Workflow

Figure 1 shows the main parts of the proposed approach. First, each speech waveform is converted into a feature matrix, X, a $D - by - N$ matrix where D is the dimension of the feature vector and N is the number of frames of the utterance. We have used MFCC, although any feature may be utilized at this phase. Then, all the available class data is pooled for training a GMM with M *Gaussians* in order to estimate the corresponding distribution. After training a GMM for each class, the posterior probability of each component, $p(m_k|X, \theta_c)$, is computed, where m_k and θ_c denote the kth Gaussian (component) and GMM's parameter set of class c, respectively.

In the next step, the posterior probabilities of each GMM are averaged over all frames of the utterance as follows

$$p(m_k|X, \theta_c) = \frac{1}{N} \sum_{n=1}^{N} log[p(m_k|x_n, \theta_c)] \tag{1}$$

where x_n represents the feature vector of the nth frame of the utterance. Posterior probability can be computed based on the Bayes' rule as follows

$$p(m_k|x_n, \theta_c) = \frac{p(x_n|m_k, \theta_c)\, p(m_k|\theta_c)}{p(x_n|\theta_c)} = \frac{p(x_n|m_k, \theta_c)p(m_k|\theta_c)}{\sum\limits_{k=1}^{M} p(x_n|m_k, \theta_c)p(m_k|\theta_c)} \tag{2}$$

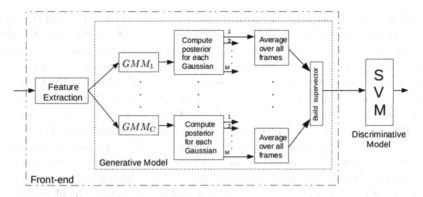

Fig. 1. Workflow of the proposed method. C and M denote number of classes and number of mixture components, respectively.

where $p(x_n|m_k, \theta_c)$ is the likelihood of the x_n (n^{th} frame) given the mth component of the GMM of the class c. The likelihood is computed based on the multivariate Gaussian distribution as follows

$$p(x_n|m_k, \theta_c) = \frac{\mathcal{H}_k^c}{\sqrt{(2\pi)^d}} exp(-\frac{1}{2}(x_n - \mu_k^c)^T \mathcal{H}_k^c (x_n - \mu_k^c)) \qquad (3)$$

where μ_k^c and $\mathcal{H}_k^c$ denote the mean vector and precision (also known as concentration) matrix of the kth component of class c, respectively.

After working out $p(m_k|x_n, \theta_c)$ for all components of all the GMMs and averaging over all utterance frames, the final feature vector is built by concatenating all the posterior probabilities in a supervector as follows

$$super\_vector = [\, p(m|X, \theta_{c_1})^T,\; p(m|X, \theta_{c_2})^T,\; ...,\; p(m|X, \theta_{c_C})^T\,] \qquad (4)$$

where C denotes number of classes and

$$p(m|X, \theta_{c_i})^T = [\, p(m_1|X, \theta_{c_i}),\; p(m_2|X, \theta_{c_i}),\; ...,\; p(m_M|X, \theta_{c_i})\,]. \qquad (5)$$

As a result, the speech signal will be represented by a fixed-length supervector which its length is C times M.

3.2 Advantages

As well as, representing the speech signal through a fixed-length pattern, the proposed theme has four main advantages

– First, it is no longer a general non-flexible feature extraction algorithm like MFCC. In comparison with the hand-engineered deterministic structure of the conventional front-ends, it has the spirit of the statistical feature learning paradigm. Such approach allows for learning some aspects of the data which are important for classification but we do not have any clear clue of them in order to somehow embed them in the parametrization workflow.

– Second, this approach provides an effective framework for capturing the long-term properties of the speech like emotion. From statistical standpoint, unlike the lingual content which changes on a short-term basis, the speaker-dependent attributes are fairly stationary during the utterance. As a result, it is more sensible to steer the front-end toward extracting features which reflect the long-term properties of the speech in tasks like emotion recognition. However, due to the non-stationarity of the speech and the Fourier transform limitation, we have to stick to the short-term processing. The proposed method paves the way for extracting the long-term properties of the speech from the short-term frame-based processing. This is due to the fact that the GMMs are trained based on all the frames of all the utterances of each class, without taking the timing issue into account. The underlying premise for validity of this argument is that the process to be modelled does not change in statistical term across the training data which holds with a reasonable approximation. In the second place, the supervector is the outcome of averaging over all the utterance frames. These two factors make the supervector highly correlated with the long-term properties of the signal which the GMMs are trained to capture them.

– Third benefit is further dimensionality reduction. In fact, instead of representing the speech via a matrix with D (typically 39) times N elements, the signal is represented with M times C elements which is by far more compact. As seen, the length of the supervector is no longer a function of neither the feature dimension nor the number of frames. As a result, a very lengthy and comprehensive feature set may be employed without increasing the computational load at the back-end.

– Since the supervector is built by stacking the posteriors of each class, it can be imagined that each class occupies a particular subspace in the feature space. This potentially enhances the discriminative capability of the extracted features and provides a better ground for the discriminative model to adjust the decision borders between the classes.

3.3   Comparison with UBM-GMM

Using universal background model (UBM) forms the status quo in the GMM-based feature extraction, in particular for speaker recognition [17]. In the UBM-based approach, at first, all the available training data is pooled and a shared universal model is trained for all the classes as the background. Then, for each utterance and through MAP adaptation [12], the parameters of the background model are modified and the supervector is built by stacking the mean vectors of all the Gaussians of the adapted model (Fig. 2). As a result, the length of the supervector would be M times D which in comparison with the proposed method (M times C) is noticeably higher for applications where number of classes are lower than the feature vector length. Emotion and environment detection are examples of such scenarios although in the task of speaker recognition it may not be the case. One solution for this issue is to do the classification on a hierarchical basis. As such at each level, number of the categories to be classified would be much less and this smooths the way for effective employment of the proposed method.

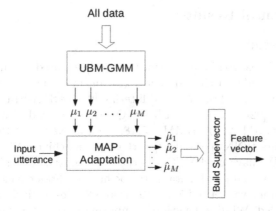

Fig. 2. Framework of the UBM-GMM method.

On the other hand, since the background model is trained with all the available data of all classes, the required number of gaussians for having a reasonable estimate of the corresponding distribution is expected to be higher than the components of a GMM which is trained for a single class. Another issue is that the adaptation process is done for each utterance, and the amount of the data provided by each signal is not enough for efficient model adaptation given the cardinality of the parameter set of the UBM. As well, the adaptation process has some hyperparameters such as *relevance factor* [17] which should be adjusted and there is no guarantee that the optimum value remains the same over all the classes and utterances. However, the proposed approach does not involve any adaptation and/or particular (hyper)parameter setting other than the number of mixture components.

3.4   Dimension of the Supervector and Curse of Dimensionality

Having a long feature vector along with a probabilistic model runs the risk of facing with the curse of dimensionality. However, the issue is more manageable as far as the classification is done with a discriminative model. This backs to the fact that in such models only subspaces around the decision borders are taken into account which has a rather low volume and lack of data for covering the space is less problematic. In fact, this issue is more serious in case of working with generative models which consider the region inside each class borders because it could have a massive volume in the high-dimensional space and is really difficult to be covered by the limited available training data. As a matter of fact, for the SVM classifier (which we have used as the back-end) only the borders and particularly support vectors are concerned. As a result, the volume that should be covered would be much smaller and lack of training data and working in the high-dimensional space due to using a supervector is less an issue.

4   Experimental Results

4.1   System Setup

A wide variety of features and classifiers have been used in emotion classification. On feature side, pitch and energy (mean, max, median, variance, etc.) [21], LPCC [20], wavelet [11], sub-band filtering [9], RASTA-PLP [19] and modulation spectrum [22] have been utilized. On the back-end also a wide range of methods like GMM [2,10], HMM [15,18], Neural Networks (NN) [8,14], K-nearest Neighbours (kNN) [7] and support vector machines (SVM) [4,20] have been employed. We have used MFCC as the feature extraction algorithm along with the log-energy, delta and delta-delta coefficients. Frame length, frame shift and number of filters was set to 25 ms, 10 ms and 25, respectively, and Hamming window was applied. Window length in computing both delta and delta-delta is set to 1. It was observed that this setting for computing the dynamic coefficients returns better results in comparison with 2 for delta and delta-delta which is typically used in speech recognition setup. For classification at the back-end, SVM with RBF (radial basis function) kernel is employed. Slack variable and gamma coefficient of the kernel were set to 12 and 2, respectively.

It should be noted that the SVM could not handle variable length patterns efficiently and removing the interface block noticeably degrades its performance. In fact, without a fixed-length representation, the SVMs should be trained by each individual frame. One problem is that the amount of emotion-correlated information within a frame is not enough for effective emotion discrimination. On the the other hand, for classification, the decision should be made on a frame-wise basis and the class with maximum bincount would be the output of the classifier. This strategy is called *Max-Wins voting* and as well as the problem of making decision based on short-term observations, it suffers from the issue that all the frames either discriminative or non-discriminative would have the same weight in the voting and consequently decision making. This could negatively affect the performance of the system, however, the proposed approach does not suffer form such issues.

The GMM and SVM were trained using Scikit-learn package [16]. A GMM classifier is used as the baseline system. It includes 25 Gaussians, trained with 5 iterations based on EM algorithm and the covariance matrix is full. Publicly available Berlin emotional database (Emo-DB) [3] has been used which includes 7 acted emotions, namely Anger (A), Boredom (B), Disgust (D), Happiness (H), Fear (F), Sadness(S) and Neutral (N). It consists of 535 signals and 10 speakers (5 male and 5 female) who are professional actors read the predefined sentences in an anechoic chamber, under supervised conditions. Sampling rate of the signals is 16 kHz with a 16-bit resolution. A human perception test to recognize various emotions with 20 participants resulted in a mean accuracy of 84.3 %.

4.2   Performance Evaluation

For evaluation, 5-fold cross-validation has been used. The confusion matrices of all folds were added together and from the resultant confusion matrix four performance metrics were computed which are *Accuracy, Recall Rate, Precision* and *F-measure* (also known as $F-score$ or $F_1 score$). Assuming the rows of the confusion matrix determine the actual class and its columns show the predicted class, these measures were computed as follows

$$conf\_mat = \sum_{fold=1}^{5} conf\_mat_{fold} \qquad (6)$$

$$Accuracy = \frac{\sum_{c=1}^{C} conf\_mat(c,c)}{\sum_{i=1}^{C} \sum_{j=1}^{C} conf\_mat(i,j)} \qquad (7)$$

$$Recall(c) = \frac{relevant\,class\,patterns\,\cap\,retrieved\,class\,patterns}{relevant\,class\,patterns}$$
$$= \frac{TruePositive}{TruePositive + FalseNegative} = \frac{conf\_mat(c,c)}{\sum_{j=1}^{C} conf\_mat(c,j)} \qquad (8)$$

$$Precision(c) = \frac{retrieved\,class\,patterns\,\cap\,relevant\,class\,patterns}{retrieved\,class\,patterns}$$
$$= \frac{TruePositive}{TruePositive + FalsePositive} = \frac{conf\_mat(c,c)}{\sum_{i=1}^{C} conf\_mat(i,c)} \qquad (9)$$

$$F-measure(c) = 2\frac{Recall(c).Precision(c)}{Recall(c) + Precision(c)} \qquad (10)$$

Accuracy is an overall performance measure and Recall Rate, Precision and consequently F-measure are defined for each class. In fact, they are originally designed for binary classification. For more details about these measures readers are referred to [13].

4.3 Results and Discussion

Tables 1 and 2 show the confusion matrices of the baseline and the proposed systems, respectively. As seen, the errors which the systems make are not similar which imply that by combining these two approaches via an appropriate framework better results may be achieved. In case of the proposed method, confusion is less and the main misclassification occurs between the Happiness and Anger. This error could be alleviated by doing a hierarchical classification.

Tables 3 and 4 show the performance metrics of the baseline and proposed method. As seen, the accuracy of the suggested system is noticeably higher than

Table 1. Confusion matrix of the baseline system.

	A	B	D	F	H	S	N
Anger	123	0	1	0	3	0	0
Boredom	0	61	1	0	0	5	14
Disgust	4	3	37	0	0	0	2
Fear	5	0	2	45	11	4	2
Happiness	29	0	0	2	39	0	1
Sad	0	1	0	0	0	60	1
Neutral	0	16	0	0	0	0	63

Table 2. Confusion matrix of the proposed method.

	A	B	D	F	H	S	N
Anger	123	0	1	0	3	0	0
Boredom	0	61	1	0	0	5	14
Disgust	4	3	37	0	0	0	2
Fear	5	0	2	45	11	4	2
Happiness	29	0	0	2	39	0	1
Sad	0	1	0	0	0	60	1
Neutral	0	16	0	0	0	0	63

the baseline. However, accuracy on its own is not enough for precise comparison of two systems and other factors such as precision and recall should be taken into account as they further clarify the type of errors which the system make. One important advantage of the proposed method is that the recall rate and precision are very closed to each other whereas in the baseline system (GMM-based classifier) the difference is noticeably higher. It should be noted that the recall and precision are inversely proportional and improving one would degrade the other. In an optimal setup they should be high and as close as possible which leads to having the maximum area under curve (AUC). This also results in having a higher F-score as it is the harmonic mean of the precision and recall. Harmonic mean is smaller than both geometric and arithmetic means and is close to the minimum of its inputs. So, as far as one of the metrics is too small, regardless of the goodness of the other one, the F-score would be poor. Therefore, the optimum performance in terms of the F-measure would be achieved if both recall and precision are high and almost equal. This is another point which shows the optimality of the proposed approach.

An important issue from practical standpoint is that how such approach can be extended to applications where there is a data stream. In such cases, the basic premise of the proposed method which was the stationarity of the speaker-dependent attribute across the signal is violated. In order to deal with this issue, one could decompose the stream into (overlapping) segments with an appropriate length. As such, instead of representing the whole utterance with a super-vector, each chunk is buffered and represented by a supervector and the decision is taken locally for each sub-utterance. Some segmentation algorithms may be used depending on the task which allow for having an adaptive variable-length buffers. As well, if the task allows a finite-state machine may be employed for handling the transitions between the outputs of the system (labels of segments) using the previous local decisions (history) and some prior knowledge. This could contribute toward alleviating the errors that potentially occur in chunking the data stream (having more than one class within a segment) and improving the accuracy of the system.

Table 3. Performance of the baseline system.

	Recall	Precision	F-measure
Anger	96.9	76.4	85.4
Boredom	75.3	75.3	75.3
Disgust	80.4	90.2	85.1
Fear	65.2	95.7	77.6
Happiness	54.9	73.6	62.9
Sadness	96.7	87.0	91.6
Neutral	79.8	75.9	77.8
Average	78.5	82.0	79.4
Accuracy	80.0		

Table 4. Performance of the proposed method.

	Recall	Precision	F-measure
Anger	89.0	85.0	86.9
Boredom	93.8	91.6	92.7
Disgust	78.3	92.3	84.7
Fear	91.3	90.0	90.7
Happiness	66.2	77.1	71.2
Sadness	96.8	95.2	96.0
Neutral	93.7	86.1	89.7
Average	87.0	88.2	87.4
Accuracy	87.6		

5   Conclusion

In this paper, we proposed an interface block between the front-end and the back-end which is based on a generative (GMM) model and leads to a fixed-length representation for speech, filtering out the unwanted attributes, extracting a long-term feature from the signal, further dimensionality reduction and yet preserving the discriminative information. It paves the way for efficient classification through discriminative models like SVMs and results in an 7.6 % absolute performance improvement. The main error which the proposed system makes during the classification is confusing the Anger and Happiness emotional states. Devising a hierarchical classification strategy could help in dealing with this issue. Using modulation spectrum, extending the feature vector with prosodic features and training the GMMs discriminatively could contribute toward improving the accuracy of the proposed method.

References

1. Bengio, Y.: Learning deep architectures for AI. Found. Trends Mach. Learn. **2**(1), 1–127 (2009). also published as a book. Now Publishers, 2009
2. Bozkurt, E., Erzin, E., Erdem, A.T.: Improving automatic emotion recognition from speech signals. In: Proceedings of the INTERSPEECH, pp. 324–327 (2009)
3. Burkhardt, F., Paeschke, A., Rolfes, M., Sendlmeier, W., Weiss, B.: A database of german emotional speech. In: Proceedings of Interspeech, Lissabon, pp. 1517–1520 (2005)
4. Chavhan, Y., Dhore, M.L., Yesaware, P.: Speech emotion recognition using support vector machine. Int. J. Comput. Appl. **1**(20), 6–9 (2010). published By Foundation of Computer
5. Davis, S., Mermelstein, P.: Comparison of parametric representations for monosyllabic word recognition in continuously spoken sentences. IEEE Trans. Acoustics, Speech Sig. Proces. **28**(4), 357–366 (1980)

6. Duda, R.O., Hart, P.E., Stork, D.G.: Pattern Classification, 2nd edn. WileyInterscience, New York (2000)

7. Feraru, M., Zbancioc, M.: Speech emotion recognition for srol database using weighted knn algorithm. In: 2013 International Conference on Electronics, Computers and Artificial Intelligence (ECAI), pp. 1–4, June 2013

8. Han, K., Yu, D., Tashev, I.: Speech emotion recognition using deep neural network and extreme learning machine. In: Interspeech 2014, September 2014

9. Hosseini, Z., Ahadi, S.: A front-end for emotional speech classification based on new sub-band filters. In: 2015 23rd Iranian Conference on Electrical Engineering (ICEE), pp. 421–425, May 2015

10. Hosseini, Z., Ahadi, S., Faraji, N.: Speech emotion classification via a modified gaussian mixture model approach. In: 2014 7th International Symposium on Telecommunications (IST), pp. 487–491, September 2014

11. Krishna Kishore, K., Krishna Satish, P.: Emotion recognition in speech using mfcc and wavelet features. In: 2013 IEEE 3rd International Advance Computing Conference (IACC), pp. 842–847, February 2013

12. Lee, C.H., Gauvain, J.L.: Speaker adaptation based on map estimation of hmm parameters. In: Proceedings of the 1993 IEEE International Conference on Acoustics, Speech, and Signal Processing: Speech Processing, ICASSP 1993, vol. II, pp. 558–561. IEEE Computer Society, Washington, DC(1993)

13. Murphy, K.P.: Machine Learning: A Probabilistic Perspective. The MIT Press, Cabridge (2012)

14. Nicholson, J., Takahashi, K., Nakatsu, R.: Emotion recognition in speech using neural networks. In: 6th International Conference on Neural Information Processing, Proceedings. ICONIP 1999, vol. 2, pp. 495–501 (1999)

15. Nwe, T.L., Foo, S.W., De Silva, L.C.: Speech emotion recognition using hidden Markov models. Speech Commun. 41(4), 603–623 (2003)

16. Pedregosa, F., et al.: Scikit-learn: Machine learning in Python. J. Mach. Learn. Res. 12, 2825–2830 (2011)

17. Reynolds, D.A., Quatieri, T.F., Dunn, R.B.: Speaker verification using adapted gaussian mixture models. In: Digital Signal Processing, p. 2000 (2000)

18. Schuller, B., Rigoll, G., Lang, M.: Hidden markov model-based speech emotion recognition. In: Proceedings of the 2003 International Conference on Multimedia and Expo, ICME 2003, vol. 2. pp. 401–404. IEEE Computer Society, Washington, DC (2003)

19. Schwenker, F., Scherer, S., Magdi, Y.M., Palm, G.: The GMM-SVM supervector approach for the recognition of the emotional status from speech. In: Alippi, C., Polycarpou, M., Panayiotou, C., Ellinas, G. (eds.) ICANN 2009, Part I. LNCS, vol. 5768, pp. 894–903. Springer, Heidelberg (2009)

20. Shen, P., Changjun, Z., Chen, X.: Automatic speech emotion recognition using support vector machine. In: 2011 International Conference on Electronic and Mechanical Engineering and Information Technology (EMEIT), vol. 2, pp. 621–625, August 2011

21. Ververidis, D., Kotropoulos, C., Pitas, I.: Automatic emotional speech classification. In: IEEE International Conference on Acoustics, Speech, and Signal Processing, Proceedings. (ICASSP 2004), vol. 1, pp. I-593-6, May 2004

22. Wu, S., Falk, T.H., Chan, W.Y.: Automatic speech emotion recognition using modulation spectral features. Speech Communication 53(5), 768–785 (2011)

Rhythm-Based Syllabic Stress Learning Without Labelled Data

Bogdan Ludusan[1]([✉]), Antonio Origlia[2], and Emmanuel Dupoux[1]

[1] LSCP, EHESS/ENS/CNRS, Paris, France
bogdan.ludusan@ens.fr
[2] PRISCA-Lab, Federico II University, Naples, Italy

Abstract. We propose a method for syllabic stress annotation which does not require manual labels for the learning process, but uses stress labels automatically generated from a multiscale model of rhythm perception. The model outputs a sequence of events, corresponding to the sequences of strong-weak syllables present in speech, based on which a stressed/unstressed decision is taken. We tested our approach on two languages, Catalan and Spanish, and we found that a classifier employing the automatic labels for learning improves performance over the baseline for both languages. We also compared the results of this system with those of an identical learning algorithm, but which employs manual labels for stress, as well as to the results of a clustering algorithm using the same features. It showed that the system employing automatic labels has a performance close to the one using manual labels, with both classifiers outperforming the clustering algorithm.

Keywords: Prosody learning · Rhythm · Stress annotation

1 Introduction

With the development of speech-based applications most of which use supervised learning principles, there is a growing need for annotations in large quantities. However, manual speech annotation is a long and demanding process. In order to overcome this problem, researchers can employ automatic labelling methods, but even those generally need labelled data to build their model. This becomes more important when the annotation needed is a prosodic one, as there are only a few large corpora annotated for prosody.

The majority of systems proposed for prominence detection use supervised learning and, thus, require manual labels (e.g. [3,9,22]), but there are also approaches that do not demand any labelled data, or only partially labelled data. They include methods employing semi-supervised learning (e.g. [12,15]), using unsupervised learning (e.g. [2,4]), or rule-based systems (e.g. [1,24]). Still, many of the above-mentioned studies that employ learning paradigms rely on lexical or syntactic information and, thus, require additional annotations. For this reason, it is desirable to use systems based on acoustic features, as these features

© Springer International Publishing Switzerland 2015
A.-H. Dediu et al. (Eds.): SLSP 2015, LNAI 9449, pp. 185–196, 2015.
DOI: 10.1007/978-3-319-25789-1_18

either do not require any annotations or, if segmental information is needed for their extraction, it can be obtained relatively cheaply. Thus, we envisage here an annotation system that exploits the following characteristics: a learning-based approach requiring no labelled examples and using only acoustic features.

Since speech rhythm can be seen as an alternating pattern of strong-weak events, one can employ rhythm information towards learning the stressed/unstressed discrimination. As one of the main prosody components, rhythm has seen an increasing applicative interest recently. Rhythm information has been used for language discrimination/identification [14,26], syllable nuclei detection [30], speech rate estimation [10], prominence detection [16] or even emotion recognition [21].

We propose here a stress annotation method which takes advantage of the information given by a multiscale model of rhythm perception [28]. The output of the rhythm model has been used already for prominence detection, in conjunction with pitch information [16], and in this paper we employ labels derived from the aforementioned model, to train classifiers. We will show that automatically derived rhythm-based labels are useful for the stress annotation task and that they give a similar performance to when manual stress labels are employed, eliminating the need for prior prosodic annotation.

The paper is structured as follows: in Sect. 2 we introduce the rhythm perception model employed in this study and we explain how stress labels are obtained from this model. After presenting the materials used and the acoustic features extracted from them, we illustrate, in Sect. 3, the results obtained with the proposed approach. Section 4 includes an extended discussion of the results and draws some conclusions about the implications of this work.

2   Methods

For the conducted experiments, we used syllable stress labels coming from a model of rhythm perception and we performed classification based on them. This was compared with the results obtained with the same classifier, but using manual stress labels and with a clustering method, on two different languages: Catalan and Spanish. The acoustic features extracted have been previously employed for the task of prominence detection [5], being the best set of features obtained, among the ones considered in that study.

2.1   Rhythm-Based Automatic Label Generation

The stress labels were automatically generated from a model of rhythm perception [28]. The original model takes into account the effect of the peripheral auditory system on the speech signal, by approximating the transformations that the signal undergoes in the human ear. At the output of this pre-processing step the sum of the response of the auditory nerves is obtained, which is further used in the model. The simulation of the auditory nerve response is then processed with a bank of time-domain Gaussian filters having a range of filter widths. For each filter output, the peaks of the function are determined and plotted in a two-dimensional space (time where they occur versus the filter width). By plotting the peaks for all the filter outputs in this 2D space we would obtain a hierarchical representation, called the rhythmogram, which represents the stronger acoustic events present in speech by peaks over more filter widths.

We employ in this paper an approximation of the original model, which was previously used for prominence detection [16]. The approximation uses the signal energy rather than the output of the nerve response, the former being more convenient for computing the rhythmogram, in the case of large scale events [29]. Furthermore, the output of the model was quantized, the rhythmogram being represented by a sequence of events and their heights (called P-values). The P-value of each event is obtained by summing all peak values of the event, across all filter widths (see [14] for more details).

The parameters required by the models (minimum and maximum filter width, and the total number of filters in the filter bank) were the ones obtained in [16], on a small corpus of English speech. The following steps were adopted for obtaining the rhythmogram: the speech signal was resampled at 500 Hz and full wave rectification was performed on it. The ear's loudness function was then modelled, by taking the cubic root of the signal, and one hundred logarithmically-distanced Gaussian filters were applied. The resulting representation is illustrated in Fig. 1 for one of the sentences used in the experiments. The upper part shows the waveform of the speech signal, the middle part the obtained rhythmogram, while the lower part displays the syllable-level segmentation along with the oracle syllabic stress annotation.

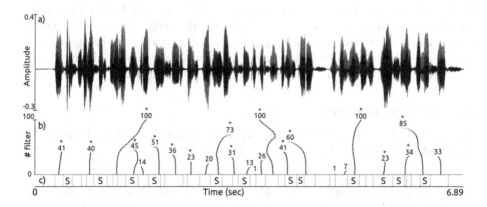

Fig. 1. (a) Waveform of the phrase "El precio de la cesta de la compra puede ser hasta un veinticinco por ciento más caro, dependiendo de cuál sea el comercio en el que compremos"; (b) rhythmogram of the phrase, along with the corresponding P-values (values marked with an asterisk represent local maxima of the stress function); and (c) oracle syllable segmentation and stress annotation (label S represents stressed syllables, while no label corresponds to unstresssed syllables).

Taking a closer look at the rhythmogram in Fig. 1 we can see that it contains 24 events (the lines in the middle panel), with P-values ranging from 1 to 100. The points plotted at the bottom level represent the peaks obtained with the narrowest filter from the bank of filter (filter index 1), thus the output function contained many peaks (24). As the filter index increases, the applied filter becomes wider and smooths more the signal energy, thus, when reaching index 60 the output of the filter would have only 6 peaks. The maximum index, 100, corresponds to the widest filter applied and, for this particular example, it gave 3 peaks in its output.

In order to obtain syllable-level stress labels, we first compute the P-values for all the events of the rhythmogram. Then, we determine the events between the boundaries of each syllable and take their P-value. If multiple events are found, only the largest event (in P-value) is considered. If no event was found, the value 0 is assigned to that particular syllable. Next, we define a stress function with the obtained P-values for all the syllables and we consider the local maxima of this function as being the stressed syllables, with the rest of the syllables belonging to the unstressed class. The local maximum is computed in a three-syllable window, centered on the current syllable. For this study we used the syllable boundaries available with the corpus, the syllable segmentation being obtained from automatically aligned phonetic transcription.

Going back to the example illustrated in Fig. 1, the resulting stress function will have its first value equal to 40, as the P-value of the event found within the boundaries of the first syllable was 40. The following three values will be equal to 0, since no rhythmogram event was found inside the following three syllables, followed by 38 (the height of the second event, which falls inside syllable number five). Next, we have two other values equal to 0, followed by 100 (the height of the third rhythmogram event, which falls inside the eight syllable of the utterance), and so on. Please note that in order to assign an event as belonging to a particular syllable it has to 'begin' inside that syllable (the peak obtained with the narrowest width should fall inside the syllable boundaries).

2.2   Acoustic Features

The acoustic features set used in the experiments is the one that obtained the best performance in [5]. The considered features are:

- syllable length (*sylLen*)
- nucleus length, estimated as the -3dB band of the energy maximum inside the syllable (*nucLen*)
- average energy inside the syllable nucleus (*avgEne*)
- ratio between the voiced time in the syllable and the total syllable length (*V_L*)
- likelihood of hearing the pitch movement passing through the nucleus as a dynamic tone (glissando) (Γ)

This last parameter is motivated by previous studies on tonal movement perception [11, 18, 23, 27] and has also been used in studies on pitch stylization [19, 20].

After producing a linear stylization of the pitch profile with the approach presented in [20] and considering the segment $[s_1, s_2]$ passing through the syllable nucleus, the Γ_{s_1,s_2} parameter is then computed as follows:

$$\Gamma_{s_1,s_2} = \begin{cases} 1 & if\ V_{s_1,s_2} > \frac{0.32}{T_e^2} \\[2ex] \frac{V_{s_1,s_2} T_e^2}{0.32}\ otherwise \end{cases} \tag{1}$$

where V_{s_1,s_2} is the absolute rate of change of the pitch excursion (in semi-tones/second). The glissando perception threshold $\frac{0.32}{T_e^2}$ used as reference in this formula is the same as the one used in [18]. While a tonal movement exceeding the threshold will have value 1, movements below it will be given a glissando likelihood value between 0 and 1. The Γ_{s_1,s_2} parameter reacts both to rising and descending pitch movements.

2.3   Learning Algorithms

For stress classification we have chosen a simple learning algorithm, naive Bayes. It is a probabilistic linear classifier which assumes that feature values are inde-pendent of other features, given the class. Since part of the features employed are not normally distributed (especially V_L and Γ), we decided to employ kernel-density estimate using Gaussian kernels [13]. The implementation of the algo-rithm is the one given by the Weka toolbox [8].

In order to compare the performance of the classifier that uses automatic labels with that of a system employing no annotated labels, we have chosen to use an unsupervised method based on the expectation-maximization (EM) algorithm. The EM clustering models the features by means of a mixture of Gaussians, the particular implementation used here [8] considering a similar feature independence assumption to the naive Bayes classifier.

2.4   Materials

We use for the experiments presented here the news sub-part of the Glissando corpus [7]. The chosen part contains broadcast news recorded by professional speakers in two languages: Catalan and Spanish, totalling over 12 h of speech (equally divided between the two languages). For each language the same number of speaker, 8, was employed, 4 females and 4 males.

The corpus was entirely transcribed and annotated, including segmental and prosodic levels. At the segmental level, both phonetic and syllabic segmentations are available, while the prosodic level contains prosodic boundary annotations. Besides prosodic phrasing, the corpus is also marked for syllabic stress, which we will use in our experiments as oracle labels.

3   Experiments

The experiments were conducted in a leave-one-speaker-out cross-validation setting. By training the model on all the speakers except one and testing on the latter, we ensure a speaker-independent model that does not capture any individual speech particularities.

For the evaluation of the results we employed the area under the receiver operating characteristic (ROC) curve (AUC). The ROC curve is obtained by varying the class decision threshold and plotting the obtained true positive rate (TPR) versus false positive rate (FPR), for each value of the threshold. We use the AUC due to its straightforward way in interpreting the results, as well as for being a measure independent of the size and the distribution of the label classes. Given a random pair of syllables, one representing a stressed syllable, and the other an unstressed syllable, the AUC can be described as the probability of making a correct choice in a forced choice task and thus the chance level for the AUC is always equal to 0.5. Furthermore, as we are interested in the overall learning quality, the AUC provides a single value for both classes (stressed/unstressed), whereas other measures, like precision, recall or F-score, would need two values to characterize the system, one for each class. While both oracle and automatically generated labels were used in the learning process, the evaluation was always performed using the oracle stress labels as the reference.

3.1   Learning with Automatic Labels

The rhythm-based procedure presented in Sect. 2.1 was applied in order to obtain the automatic labels used in the learning process and it was also considered as the baseline (further called *baselineRhy* system). The results obtained with the *baselineRhy* and the supervised algorithm that employs automatic labels for learning (referred to as the *rhyLabel* system) are displayed in Fig. 2. They are illustrated for each speaker of the two languages investigated (Catalan in the upper panel and Spanish in the lower panel), with female speakers on the left and male speakers on the right side. Each speaker is represented by a point in the FPR-TPR space, in the case of the *baselineRhy* and by a curve in the same space, for the *rhyLabel* system.

They show that stress learning based on labels derived from a rhythm model improves over the baseline (the model used to generate the labels, in the first place). It appears though that there are some differences between languages: while for Catalan we obtained important improvements for all speakers, we had lower improvements for Spanish, with one speaker seeing no improvement over the baseline. If we are to compare the *baselineRhy* results with those of the *rhyLabel* system, by taking, for each speaker, the point on the curve which corresponds to the same FPR as the baseline, we would obtain: an increase in TPR, for Catalan, ranging between 2.2 % and 21.8 % (14.0 %, on average), and a change in TPR, for Spanish, ranging between -0.1 % and 15.1 % (5.0 %, on average).

Next, we compared the results obtained when automatic labels are used to learn a classifier (*rhyLabel*) to those obtained when manual stress annotations

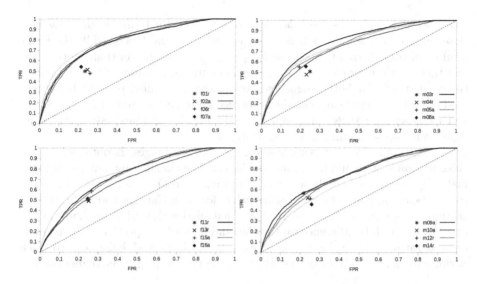

Fig. 2. Classification results obtained when labels derived from the rhythmogram are used for learning a naive Bayes classifier on acoustic features (curves in the ROC space), compared with the results given by the baseline (points in the ROC space), for each speaker. The upper panel shows the results for Catalan, while the lower panel the ones for Spanish. For the two languages, the results corresponding to the female speakers are on the left and those for the male speakers on the right. Chance level is represented by a dotted line.

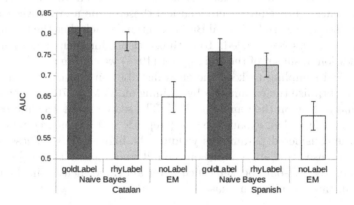

Fig. 3. Area Under the ROC Curve classification results obtained for Catalan and Spanish, when using the following systems: naive Bayes employing oracle labels (*gold-Label*), naive Bayes employing labels obtained from the rhythmogram (*rhyLabel*) and EM clustering (*noLabel*). The error bars represent the standard deviation across speakers. The three systems use the same set of acoustic features.

are employed with the same learning algorithm and the same acoustic features (*goldLabel*). These two cases were then compared against a clustering algorithm that takes in input the same features (*noLabel*). Both comparisons (*rhyLabel* ↔

goldLabel and *rhyLabel* ↔ *noLabel*) are important, as the former would show how close the use of automatic labels is to the use of manual labels, while the latter would indicate whether the proposed approach performs better than another system that does not employ manually labelled data for learning. Figure 3 displays the comparison, for both languages. Per-speaker two-tailed paired t-tests were used to test the significance of the difference between the *rhyLabel* ↔ *goldLabel* and *rhyLabel* ↔ *noLabel* and they showed that all differences are statistically significant, for both languages ($p < 0.001$).

As expected, the best performance, on both languages, is obtained by the classification system employing manual labels. The classifier making use of automatically determined labels performs well, with the difference between it and the topline being only about 3 % absolute value, for both Catalan and Spanish. Furthermore, its performance is well above that of the clustering method (13.3 % and 12.2 % absolute difference for Catalan and Spanish, respectively). This result demonstrates that learning without labels can be improved, being able to reach performances relatively close to that of a system employing manual annotations.

3.2 Learning with Different Classifiers

In order to test the generalizability of the results obtained, we trained two additional classifiers, one based on logistic regression and the other one based on support vector machines (SVMs), using the same setting as in the previous experiment (same acoustic features, trained with both manual and automatic labels). The logistic regression classifier had no parameters to set, so the training and evaluation procedures were identical those of the naive Bayes classifier. For SVMs instead, we used a Radial Basis Function kernel, thus the tuning of the C and γ parameters was needed. To optimize them, for each speaker independent test session, a subset of the training set (10 %) was automatically extracted using stratified sampling to keep the class distribution consistent. This subset was used to optimize the parameters by training an SVM on 70 % of the selected material and testing on the remaining 30 %. The search space was limited to the interval $[0.5 - 1.5]$ and an exhaustive grid (step size $= 0.1$) was used to select the values of the considered parameters yielding the highest AUC. These parameters were then used to train a model on the full training set. The evaluation was performed on the left-out speaker which never appeared, neither in the training phase nor in the optimization phase.

Table 1. Average improvement in TPR, for an equivalent FPR level, between the baselineRhy and rhyLabel systems, for the three classifiers used in this study.

Learning algorithm	Catalan	Spanish
Naive Bayes	.140	.050
Logistic Regression	.084	.006
Support Vector Machines	.102	.047

Table 2. AUC results obtained with the rhyLabel and goldLabel system when employing different learning algorithms. For comparison, we report also the results of the noLabel system, which uses EM clustering.

Learning algorithm	Catalan			Spanish		
	goldLabel	rhyLabel	noLabel	goldLabel	rhyLabel	noLabel
Naive Bayes	.815	.782	-	.757	.726	-
Logistic Regression	.819	.766	-	.758	.711	-
Support Vector Machines	.798	.734	-	.719	.690	-
EM clustering	-	-	.649	-	-	.604

When we compare the results obtained with the classifiers trained with the rhythmogram-derived labels with those of the rhythmogram system alone we observe significant improvements (see Table 1). The TPR increase when using logistic regression and SVMs, for an equivalent FPR, reaches 8.4 % and 10.2 % for Catalan and 0.6 % and 4.7 % for Spanish. The *rhyLabel* ↔ *goldLabel* comparison (in Table 2) shows an advantage of 5.3 % and 6.4 % (Catalan) or 4.7 % and 2.9 % (Spanish), for the latter. The *rhyLabel* ↔ *noLabel* comparison shows the same trend as the naive Bayes classifier: an improvement over the unsupervised learner of 11.7 % and 8.5 %, for Catalan, or 10.7 % and 8.6 %, for Spanish. SVMs appear to perform slightly worse than naive Bayes and logistic regression: a possible explanation for this is that SVMs are more sensitive to low-quality annotations than the other considered approaches [6]. The performance difference between the learning algorithms and the two languages will need to be investigated further.

4   Discussion and Conclusions

We presented here an approach for stress annotation that does not employ any manual labels for the learning process. We have shown that the labels obtained from a rhythm-based system, based only on the speech signal, can be successfully employed for learning. The trained classifier outperformed both the baseline and a clustering system that used the same acoustic features, while giving a performance close to that of a classifier that employs manual prosodic labels and identical features. As seen also in [16], the information given by the rhythmogram seems to be quite robust and language independent. In the aforementioned paper, the parameters of the system were determined on English and then tested on Italian and French, with good results. We applied here the same parameter settings on two new languages, Spanish and Catalan, with equally good performance.

The current study can be seen as one step further in improving the automatic annotation of prosody, without the use of labelled data in the learning process. With more improvements in this direction we can envisage more prosodic studies and better analyses, as the access to annotated data increases. This could be especially important as a starting point for languages which, for the moment, have no prosodic annotations. At the same time, an increase in available prosodic annotation might lead to more speech applications taking advantage of prosodic knowledge or to improvements in systems which use prosody determined through unsupervised or rule-based methods [17].

As mentioned before, the approach proposed here employs no labelled data in the learning process. Still, it made use of a small prominence annotated dataset for the setting of the parameters of the rhythmogram. But since those parameters were determined on English and our approach tested on two different languages, Catalan and Spanish, we could assume that the same procedure can be applied to other languages also. By making use of less than 20 min of annotation from a highly resourced language like English, we were able to annotate with reasonable performance more than 12 h of data from two previously unseen languages.

One significant issue that remains to be investigated is the role that syllable segmentation plays on the stress annotation performance. Since none of the current signal-based methods for syllable segmentation give excellent results, it is important to know to what extent the segmentation errors affect the prosodic annotation learning. One alternative would be to derive syllables from forced-aligned phonemic annotations. While this would greatly reduce the number of languages on which the approach could be applied, there are many more languages that have enough data to train acoustic models for forced alignment than languages in which extensive prosodic annotations exist. If this is not an option and a manual annotation has to be performed, it is nevertheless cheaper and easier to perform the segmentation stage manually (since this would also be needed to further add the prosodic annotation) and to have a system that would do a first pass prosodic annotation that needs to be corrected, than to do the whole prosodic annotation from scratch.

Our investigation was limited to the annotation of stressed syllables because of the lack of large corpora annotated for prominence. Once such corpora become available we plan to extend our study to the annotation of the more general phenomenon of acoustic prominence. Since the system employed in this study for the automatic generation of stress labels has been used successfully for prominence detection, we expect the procedure proposed here to perform well also in the case of acoustic prominence annotation.

We considered the present investigation as a case study on the use of automatically generated labels for learning stress and, thus, the emphasis was not put on the learning algorithms. In order to obtain better performance we would like to explore the use of state-of-the-art learning paradigm for prominence labelling, latent dynamic conditional neural fields [25], as well as learning algorithms which are robust to low quality labels [6].

Acknowledgments. BL and ED's work was funded by the European Research Council (ERC-2011-AdG-295810 BOOTPHON). It was also supported by the Agence Nationale pour la Recherche (ANR-10-LABX-0087 IEC, ANR-10-IDEX-0001-02 PSL*), the Fondation de France, the École des Neurosciences de Paris, and the Région

Île-de-France (DIM cerveau et pensée). AO's work was supported by the Italian Ministry of University and Research and EU under the PON OR.C.HE.S.T.R.A. project.

References

1. Abete, G., Cutugno, F., Ludusan, B., Origlia, A.: Pitch behavior detection for automatic prominence recognition. In: Proceedings of Speech Prosody (2010)
2. Ananthakrishnan, S., Narayanan, S.: Combining acoustic, lexical, and syntactic evidence for automatic unsupervised prosody labeling. In: Proceedings of INTER-SPEECH, pp. 297–300 (2006)
3. Ananthakrishnan, S., Narayanan, S.: Automatic prosodic event detection using acoustic, lexical, and syntactic evidence. IEEE Trans. Audio Speech Lang. Process. **16**(1), 216–228 (2008)
4. Chiang, C.Y., Chen, S.H., Yu, H.M., Wang, Y.R.: Unsupervised joint prosody labeling and modeling for Mandarin speech. J. Acous. Soc. Am. **125**(2), 1164–1183 (2009)
5. Cutugno, F., Leone, E., Ludusan, B., Origlia, A.: Investigating syllabic prominence with conditional random fields and latent-dynamic conditional random fields. In: Proceedings of INTERSPEECH, pp. 2402–2405 (2012)
6. Folleco, A., Khoshgoftaar, T., Van Hulse, J., Bullard, L.: Identifying learners robust to low quality data. In: Proceedings of the IEEE International Conference on Information Reuse and Integration, pp. 190–195 (2008)
7. Garrido, J.M., Escudero, D., Aguilar, L., Cardeñoso, V., Rodero, E., De-La-Mota, C., González, C., Vivaracho, C., Rustullet, S., Larrea, O., et al.: Glissando: a corpus for multidisciplinary prosodic studies in Spanish and Catalan. Lang. Resour. Eval. **47**(4), 945–971 (2013)
8. Hall, M., Frank, E., Holmes, G., Pfahringer, B., Reutemann, P., Witten, I.H.: The WEKA data mining software: an update. SIGKDD Explor. **11**(1), 10–18 (2009)
9. Hasegawa-Johnson, M., Chen, K., Cole, J., Borys, S., Kim, S.S., Cohen, A., Zhang, T., Choi, J.Y., Kim, H., Yoon, T., et al.: Simultaneous recognition of words and prosody in the Boston University Radio Speech Corpus. Speech Commun. **46**(3), 418–439 (2005)
10. Heinrich, C., Schiel, F.: Estimating speaking rate by means of rhythmicity parameters. In: Proceedings of INTERSPEECH, pp. 1873–1876 (2011)
11. House, D.: Differential perception of tonal contours through the syllable. In: Proceedings of ICSLP, pp. 2048–2051 (1996)
12. Jeon, J.H., Liu, Y.: Semi-supervised learning for automatic prosodic event detection using co-training algorithm. In: Proceedings of ACL-IJCNLP, pp. 540–548 (2009)
13. John, G., Langley, P.: Estimating continuous distributions in Bayesian classifiers. In: Proceedings of the Eleventh conference on Uncertainty in Artificial Intelligence, pp. 338–345 (1995)
14. Lee, C., Todd, N.M.: Towards an auditory account of speech rhythm: application of a model of the auditory 'primal sketch' to two multi-language corpora. Cognition **93**(3), 225–254 (2004)
15. Levow, G.A.: Unsupervised and semi-supervised learning of tone and pitch accent. In: Proceedings of NAACL-HLT, pp. 224–231 (2006)

16. Ludusan, B., Origlia, A., Cutugno, F.: On the use of the rhythmogram for automatic syllabic prominence detection. In: Proceedings of INTERSPEECH, pp. 2413–2416 (2011)

17. Ludusan, B., Ziegler, S., Gravier, G.: Integrating stress information in large vocabulary continuous speech recognition. In: Proceedings of INTERSPEECH, pp. 2642–2645 (2012)

18. Mertens, P.: The prosogram: semi-automatic transcription of prosody based on a tonal perception model. In: Proceedings of Speech Prosody (2004)

19. Origlia, A., Abete, G., Cutugno, F.: A dynamic tonal perception model for optimal pitch stylization. Comput. Speech Lang. **27**(1), 190–208 (2013)

20. Origlia, A., Cutugno, F.: A simplified version of the OpS algorithm for pitch stylization. In: Proceedings of Speech Prosody, pp. 992–996 (2014)

21. Ringeval, F., Chetouani, M., Schuller, B.W.: Novel metrics of speech rhythm for the assessment of emotion. In: Proceedings of INTERSPEECH, pp. 346–349 (2012)

22. Rosenberg, A.: AutoBI - a tool for automatic ToBI annotation. In: Proceedings of INTERSPEECH, pp. 146–149 (2010)

23. Rossi, M.: Interactions of intensity glides and frequency glissandos. Lang. Speech **21**(4), 384–396 (1978)

24. Tamburini, F.: Automatic prominence identification and prosodic typology. In: Proceedings of INTERSPEECH, pp. 1813–1816 (2005)

25. Tamburini, F., Bertini, C., Bertinetto, P.M.: Prosodic prominence detection in Italian continuous speech using probabilistic graphical models. In: Proceedings of Speech Prosody, pp. 285–289 (2014)

26. Tepperman, J., Nava, E.: Long-distance rhythmic dependencies and their application to automatic language identification. In: Proceedings of INTERSPEECH, pp. 1061–1064 (2011)

27. t'Hart, J., Collier, R., Cohen, A.: A perceptual study of intonation: an experimental-phonetic approach. Cambridge University Press, Cambridge (1990)

28. Todd, N.M.: The auditory "Primal Sketch": a multiscale model of rhythmic grouping. J. New Music Res. **23**(1), 25–70 (1994)

29. Todd, N.M., Brown, G.: Visualization of rhythm, time and metre. Artif. Intell. Rev. **10**, 253–273 (1996)

30. Zhang, Y., Glass, J.R.: Speech rhythm guided syllable nuclei detection. In: Proceedings of ICASSP, pp. 3797–3800 (2009)

Unsupervised and User Feedback Based Lexicon Adaptation for Foreign Names and Acronyms

André Mansikkaniemi$^{(\boxtimes)}$ and Mikko Kurimo

Department of Signal Processing and Acoustics, School of Electrical Engineering,
Aalto University, Aalto, Finland
{andre.mansikkaniemi,mikko.kurimo}@aalto.fi

Abstract. In this work we evaluate a set of lexicon adaptation methods for improving the recognition of foreign names and acronyms in automatic speech recognition (ASR). The most likely foreign names and acronyms are selected from the LM training corpus based on typographic information and letter-ngram perplexity. Adapted pronunciation rules are generated for the selected foreign name candidates using a statistical grapheme-to-phoneme (G2P) model. A rule-based method is used for pronunciation adaptation of acronym candidates. In addition to unsupervised lexicon adaptation, we also evaluate an adaptation method based on speech data and user corrected ASR transcripts. Pronunciation variants for foreign name candidates are retrieved using forced alignment and second-pass decoding over partial audio segments. Optimal pronunciation variants are collected and used for future pronunciation adaptation of foreign names.

Keywords: Speech recognition · Lexicon adaptation · Unsupervised pronunciation adaptation · Forced alignment · User feedback

1 Introduction

The lexicon, or pronunciation dictionary is a vital component in a modern automatic speech recognition system (ASR). The correct recognition of words relies on having accurate phoneme-level pronunciation rules stored in the lexicon. The pronunciation rules are either automatically generated or handcrafted. The ease of adding pronunciation rules for new words depends on the language. In Finnish ASR, which we focus on in this work, accurate pronunciation rules can easily be generated for new native words due to the one-to-one relationship between letters and phonemes. In this work, we focus lexicon adaptation efforts on words that are not covered by standard Finnish pronunciation rules, namely foreign proper names (FPNs) and acronymns (ACRs).

The occurrence rate of foreign names and acronyms is usually quite low in speech, in the order of a 1–5 %. Devoting effort to recognise them is motivated by the high information value they carry. The improved recognition of FPNs and ACRs can enhance the overall intelligibility of ASR output and also improve performance of spoken document retrieval.

© Springer International Publishing Switzerland 2015
A.-H. Dediu et al. (Eds.): SLSP 2015, LNAI 9449, pp. 197–206, 2015.
DOI: 10.1007/978-3-319-25789-1_19

Relatively few works have been devoted exclusively to improving recognition of foreign names and acronyms. Many ASR systems focus adaptation efforts on native words and use manually crafted pronunciation rules for the most common FPNs. Also in many languages FPNs are often pronounced more or less incorrectly with native pronunciation rules. The problem of recognizing foreign words is usually more of a problem for smaller languages where influence from other languages is bigger and FPN occurrence more frequent.

In [9], recognition of foreign names is improved in a US name recognition task by language identification and pronunciation adaptation. The linguistic origins of names in the training data are identified using language-specific letter-ngram models. Pronunciation rules are generated using language-specific grapheme-to-phoneme (G2P) conversion models. In [8], an ASR system is adapted to improve recognition of English, German, and Latin words in continuous Czech speech. Foreign OOV words in Czech training texts are detected using language-specific letter-ngram models. Pronunciation rules for the detected FPNs are generated using a G2P converter. In [1], a discriminative tree search approach is used to the find an optimal set of pronunciation variants for non-native proper names based on spoken isolated word utterances. Acronym adaptation was performed in [5], by identifying words in the LM training texts written in all capital letters as ACR candidates and generating pronunciation variants for them.

Pronunciation modeling of proper names in general has also been given some attention in the ASR community, especially in languages where letter-to-sound (L2S) rules are less straightforward than in Finnish. In [6,13], a combination of grapheme-to-phoneme (G2P) and phoneme-to-phoneme (P2P) models trained on spoken data were used to improve recognition of proper names in Dutch ASR.

This paper builds on our previous works related to unsupervised LM and vocabulary adaptation [10,11]. In this work, we explore an adaptation setting illustrated in Fig. 1. Different from our previous two works, which were

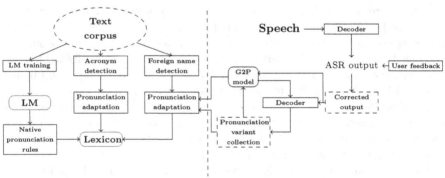

Fig. 1. Unsupervised lexicon adaptation is illustrated on the left side of the figure. User feedback based adaptation and its' connection to unsupervised lexicon adaptation is illustrated on the right side of the figure.

implemented in an unsupervised LM adaptation environment, this time we apply our unsupervised lexicon adaptation methods in parallel with a normal LM training phase. Foreign name and acronym candidates are detected in the training text corpus. Adapted pronunciation rules are generated for the most likely candidate words. A rule-based method is used for acronyms and a G2P model is used for foreign names. The pronunciation rules are finally added to the lexicon. We also evaluate a new user feedback based approach to pronunciation adaptation. The idea is to learn pronunciation rules from speech directly. The first-pass output from the ASR decoder is assumed to be manually corrected by a user. Using the G2P model multiple pronunciation variants are generated for all the words. The most likely pronunciation variants for FPN candidates in the corrected ASR transcript are extracted using forced alignment decoding. This procedure is run over a significant amount of speech data and the discovered pronunciation variants are collected and used when performing pronunciation adaptation for new unseen data.

In Sects. 2-3 we describe the methods related to unsupervised lexicon adaptation. In Sect. 4 we give a detailed description of user feedback based lexicon adaptation. In Sect. 5 the test data and experimental settings are described. In Sect. 6 we present and discuss the results. In Sect. 7 the conclusions are presented.

2   Detecting Special Vocabulary Units

In order to adapt the lexicon for improved recognition of FPNs and ACRs, these vocabulary units must first be detected in the LM training text. A score needs to be calculated for each detected word expressing how likely the chosen word is either an FPN or ACR. The most likely candidates are then chosen for pronunciation adaptation.

Scoring how likely a single word is of foreign origin has previously been implemented using language-specific letter-ngram models [2,8]. We implement a detection algorithm based on two different selection methods on how to distinguish foreign words from native words. Words are first converted into their base forms ("shanghaissa" → "shanghai") using a stemming algorithm[1]. This is to ensure that native morphology such as inflections are removed from the FPN vocabulary units before G2P conversion

1. Select all words starting in an uppercase letter as FPN candidates. Words composed entirely of uppercase letters are excluded however. In languages written in Latin characters names are by convention written starting with an uppercase letter. We can use this typographic information to select the most likely FPNs.
2. Calculate the letter perplexity for each FPN candidate using a letter-ngram model trained on native Finnish words. Words with higher letter perplexity are more likely to be foreign words.

---

[1] *Snowball* - http://snowball.tartarus.org/.

$$ppl(word) = \sqrt[n]{\prod_{i=1}^{n} \frac{1}{P(l_i|h)}} \tag{1}$$

3. Calculate the case frequency for the stem of the word. Words starting more often in uppercase letter than lowercase are more likely to be foreign words. Case frequency is calculated over the retrieved in-domain articles.

$$case(word) = \frac{\sum uppercase(word)}{\sum uppercase(word) + \sum lowercase(word)} \tag{2}$$

The selection scores, letter perplexity (*ppl*) and case frequency (*case*), are normalized and a final score is calculated by multiplying the two scores (3). The scoring function is quite rudimentary but it gives an easy method to select the most likely FPNs.

$$score(word) = ppl(word) * case(word) \tag{3}$$

When selection scores for all FPN candidates have been calculated they are sorted in descending order. Words at the top of the list can then be considered to be more likely of foreign origin.

Acronyms can be detected in a similar manner to foreign proper names based on typographic information. We can select all words composed of only uppercase letters as ACR candidates. The ACR candidates are sorted according to occurance rate. The most frequently used acronyms are selected for adaptation.

In ASR experiments we need to limit the number of new vocabulary units that are added to the lexicon. Recognition accuracy is degraded if too many new many pronunciation variants are introduced. We set a percentage threshold value F, how many new FPN pronunciation variants are added in relation to the original lexicon size. A similar value A is set for the relative share of new ACR pronunciation variants.

A more formal way of classifying words into native, foreign, and acronym classes, could be done with a posterior probability function $P(C|X)$ (C = class, X = features). This would however require large amounts of annotated training data, which is not always available. In terms of speech recognition accuracy, it's also doutbtful if this method would perform significantly any better than our scoring functions introduced earlier. We might however evaluate this type of posterior probability based classification method in the future.

3   Pronunciation Adaptation

Once the most likely FPNs and ACRs have been selected the first step is to generate adapted pronunciation rules for them.

For FPNs we have to generate adapted pronunciation rules since they usually differ from native Finnish one-to-one mapping between letters and phonemes. To automatically generate pronunciation variants we use *Sequitur G2P*, a data-driven grapheme-to-phoneme converter based on joint-sequence models [3].

A pronunciation model is trained on a lexicon consisting of 2000 FPNs with a manually given pronunciation hand-picked from a Finnish newswire text collection. The FPNs picked for training the pronunciation model are of mixed linguistic origin, including English, German, Russian, and Swedish names. The pronunciation model is used to generate the most probable pronunciation variants for the FPN candidates. In this work we generate four pronunciation variants for each selected FPN word and give all pronunciations equal probabilities.

Pronunciation variants for ACRs can be divided into two classes, alphabetic and phonetic pronunciations. In alphabetic pronunciations each initial letter is pronounced separately as an alphabetic unit. In phonetic pronunciation the acronym is pronounced with standard Finnish one-to-one G2P mapping, as a native subword. In this work both the alphabetic and phonetic variant are generated for each ACR candidate.

4   User Feedback Based Adaptation

A different approach compared to automatically generating pronunciation variants from a static G2P model is to gradually learn new pronunciation variants from speech directly. We use our ASR system [7] and assume there is a user who is willing to correct all mistakes made by the recognizer. We run the pronunciation discovery process over all FPN candidate words.

We use the same G2P model trained for unsupervised pronunciation adaptation, to generate multiple pronunciation variants (n=20) for one word at a time. An LM and lexicon are generated for discovering the most likely pronunciation variant. The lexicon holds all the generated pronunciation variants. The LM holds the word with different indices representing the different pronunciation variants, all having equal probabilities. Forced alignment is used to look for the start and ending times for the word in the speech data. By extracting this segment of the speech and running it through the recognizer with the constrained LM and lexicon we can assume that the Viterbi algorithm used in our ASR decoder finds the most likely pronunciation variant. The word and pronunciation variant are added into an extra lexicon where we collect all discovered pronunciation variants.

We can use the extra lexicon when performing pronunciation adaptation on new speech data. We can retrieve pronunciation variants from the extra lexicon that are not included in the four pronunciation variants that are G2P-generated for each word.

An additional method to use the discovered pronunciation variants, is to use them to re-train the G2P model. The idea of re-training the G2P model is to also improve pronunciation modeling for words which are not discovered in the training data. We also utilise this option when applying user feedback based adaptation.

5 Experiments

Experiments were run to test the adaptation framework illustrated in Fig. 1. Initial experiments were run on a development set. Final experiments were conducted on an evaluation set.

5.1 System

Speech recognition experiments were run on the Aalto ASR system [7]. The Finnish background language model was trained on the Kielipankki corpus (140 million words) and a smaller Web corpus consisting of online news articles (7 million words). A model for morpheme segmentation and a 46k morph lexicon were trained on the LM training corpora using Morfessor [4]. A Kneser-Ney smoothed varigram LM (max n=10 for n-grams) was trained on the morph-segmented background corpus with the variKN language modeling toolkit [12]. A back-off letter ngram model ($n=2$) was trained on the Kielipankki word list, composed mainly of native Finnish words but also a small number of foreign names, for the FPN detection algorithm.

5.2 Data

The main experimental data consists of Finnish radio and TV news segments in 16 kHz audio. The recordings were collected in 2011–2012 from YLE news and sports programs. The data is divided into a 5 h (35,056 words) development set and a 5 h (36,812 words) evaluation set. In the development set 1,621 of the words (4.6 %) were categorized as foreign words and 465 as acronyms (1.3 %). In the evaluation set 1,381 of the words (3.8 %) were categorized as foreign words and 411 as acronyms (1.1 %).

6 Results

Performance of the adapted ASR system was measured using three different values: average word error rate (WER), foreign proper name error rate (FER), and acronym error rate (ACER).

6.1 Unsupervised Lexicon Adaptation

First experiments were run on the development set. In the first experiment we evaluated foreign proper name adaptation using the different selection scores (*ppl*, *case*, *ppl*case*) and with different values for the threshold parameter F (0–24 %). Results of this experiment are in Table 1.

Gradual improvement in recognition accuracy is quite similar for all the selection scores. Foreign name error rate drops a little faster for *ppl*case* compared to using the standalone *ppl-* and *case*-scores. It seems to be the optimal score for selecting the most likely foreign name candidates while at the same time also

Table 1. Results of foreign proper name adaptation on the development set.

F[%]	ppl		case		ppl*case	
	WER[%]	FER[%]	WER[%]	FER[%]	WER[%]	FER[%]
0	**27.6**	**60.6**	**27.6**	**60.6**	**27.6**	**60.6**
4	27.5	58.3	27.6	60.4	27.5	58.1
8	27.5	57.2	27.7	60.1	27.5	56.0
12	27.5	56.0	27.7	58.7	**27.4**	55.3
16	**27.4**	55.2	**27.5**	56.4	27.4	54.9
20	27.5	55.1	27.5	55.0	27.4	**54.0**
24	27.5	**54.2**	27.5	**54.2**	27.5	54.3

filtering out native words. At most, when F is set to 20 %, ppl*case drops FER from 60.6 % to 54.0 % (11 % rel. change), and WER is dropped from 27.6 % to 27.4 % (1 % rel. change).

In the second experiment, we evaluated acronym adaptation with different values for the threshold parameter A (0–4 %). Results are in Table 2. Optimal performance is reached quite quickly for acronym adaptation. For a relatively small threshold value (2 %) ACER is reduced from 85.6 % to 40.0 % (53 % rel. change). Acronym adaptation has a more significant effect on WER compared to foreign name adaptation. The relative WER reduction is over 3 % for acronym adaptation.

Table 2. Results of acronym adaptation on the development set.

A[%]	freq	
	WER[%]	ACER[%]
0	27.6	85.6
1	26.7	40.9
2	26.8	40.0
3	26.8	40.0
4	26.8	40.0

In the final unsupervised lexicon adaptation experiment on the development set, we combined acronym and foreign proper name adaptation, using optimal threshold value settings (F=20 %, A=2 %). Results are in Table 3. Average WER drops from 27.6 % to 26.5 % (4 % rel. change), which is pretty much the expected cumulative reduction when combining both ACR and FPN adaptation. Compared to standalone FPN and ACR adaptation the error rates for foreign names and acronyms remain pretty much unchanged with a slight decrease of FER (from 54.0 % to 53.7 %) and slight increase of ACER (40.0 % to 40.4 %).

Table 3. Results of combined foreign name and acronym adaptation on the development set.

F[%]	A[%]	ppl*sim,freq		
		WER[%]	FER[%]	ACER[%]
0	0	27.6	60.6	85.6
20	2	26.5	53.7	40.4

6.2 User Feedback Based Adaptation

In the next set of experiments, we evaluated user feedback based pronunciation adaptation. We implemented it as was described in Sect. 4 and illustrated in Fig. 1. The idea is to discover pronunciation variants from speech directly by running a constrained second-pass ASR for the user corrected first-pass ASR outputs. To test the method, the discovered pronunciation variants are used on new unseen speech data. To maximize the number of discovered pronunciation variants we applied a type of cross-evaluation for each speech segment by using all the other segments in the development and evaluation sets as training data in user feedback adaptation process.

We compare the performance of user feedback based adaptation (*Feedback*) against the baseline and the best performing unsupervised lexicon adaptation system (*Lexicon*, $F=20\%$, $A=2\%$). We also evaluate the method of using user feedback based adaptation to retrain the G2P model (*G2P-retrain*). Statistical significance is measured using *Matched Pair Sentence Segment testing* (*MAPSSWE*, $p=0.01$). Results of the experiments on the development set are in Table 4.

Comparing the performance between lexicon and user feedback based adaptation we can observe that there is a small, 3% relative reduction in error rate for foreign names. Average WER remains unchanged however. Retraining the G2P model does not give any further improvements in FER.

Results of the experiments on the evaluation set are in Table 5, and they are quite similar to the development set results in terms of relative changes. Unsupervised lexicon adaptation lowers FER with 10%. Average WER is lowered with 3% and ACER with over 60%. User based feedback adaptation only

Table 4. Results of lexicon and user feedback based adaptation on the development set. Statistically significant improvement compared to the baseline system (*) and lexicon adaptation (**) is marked.

Adaptation	WER[%]	FER[%]	ACER[%]
-	**27.6**	**60.6**	**85.6**
Lexicon	26.5*	53.7*	40.4*
Lexicon + Feedback	26.5*	51.9**	40.9*
Lexicon + Feedback + G2P-retrain	**26.4***	**51.9***	**41.1***

Table 5. Results of lexicon and user feedback based adaptation on the evaluation set. Statistically significant improvement compared to the baseline system (*) and lexicon adaptation (**) is marked.

Adaptation	WER[%]	FER[%]	ACER[%]
-	**29.2**	**59.9**	**79.1**
Lexicon	28.2*	53.9*	31.9*
Lexicon + Feedback	28.2*	**53.4***	31.9*
Lexicon + Feedback + G2P-retrain	**28.1***	53.8*	**31.6***

manages to lower FER with a little over 1 %. Similar to the development set, retraining the G2P model doesn't give any additional improvement.

Even though the training set is large, over 10h for each speech segment, the reductions in error rate are not that convincing. it is of course important to remember is that the correct recognition of foreign names is also related to the language model and how well it is able to estimate their occurrence.

7   Conclusions and Discussion

In this work, the goal was to reduce the error rates of foreign names and acronyms in Finnish speech recognition. We implemented an unsupervised lexicon adaptation method, based on detecting foreign name and acronym vocabulary units in the LM training corpus and generating new pronunciation rules for the most likely candidates. Alongside unsupervised lexicon adaptation, we implemented a pronunciation adaptation process based on speech data and user corrected ASR transcripts. Optimal pronunciation variants for foreign names were extracted using a G2P model to generate pronunciation rule candidates and a Viterbi ASR decoder to choose the most likely pronunciation variant. Pronunciation variants were gathered from speech data and stored in an extra lexicon which was later used in pronunciation adaptation for new unseen data.

Unsupervised lexicon adaptation is successful in lowering error rates for both foreign names and acronyms, while at the same time also lowering the average word rate significantly. Recognition accuracy of acronyms is improved the most. Recognition accuracy of foreign names is also improved but the change is less radical than for acronyms. The variation in pronunciation of foreign names is greater compared to acronyms. it is also possible that the number of different acronyms is smaller than the number of different foreign names and therefore they are more easily estimated by the LM.

User feedback based adaptation only manages to improve recognition of foreign names slightly. It remains to be seen if this method can be improved further. One approach would be to combine this with a discriminative method that would also filter away pronunciation variants that are harmful for the overall recognition accuracy.

For future work, we will try to improve the performance of unsupervised lexicon adaptation. Most room for improvement lies in the recognition of foreign proper names. A more thorough analysis is needed of what type of foreign names are mis-recognized and if the errors stem from poor language or pronunciation modeling. We also intend to integrate the existing lexicon adaptation framework with more sophisticated LM training methods, such as neural network language models.

Acknowledgements. This work has been funded by the Academy of Finland under the Finnish Centre of Excellence in Computational Inference programme. The experiments were performed using computational resources provided by the Aalto Science-IT project.

References

1. Adde, L., Svendsen, T.: Pronunciation variation modeling of non-native proper names by discriminative tree search. In: Proceedings of IEEE International Conference on Acoustics, Speech and Signal, ICASSP (2011)
2. Ahmed, B., Cha, S.H., Tappert, C.: Detection of foreign entities in native text using n-gram based cumulative frequency addition. In: Proceedings of the CSIS Research Day, Pace University (2005)
3. Bisani, M., Ney, H.: Joint-sequence models for grapheme-to-phoneme conversion. Speech Commun **50**(5), 434–451 (2008)
4. Creutz, M., Lagus, K.: Unsupervised morpheme segmentation and morphology induction from text corpora using morfessor 1.0. Technical report, Helsinki University of Technology, technical Report A81, Publications in Computer and Information Science (2005)
5. Driesen, J., Bell, P., Sinclair, M., Renals, S.: Description of the uedin system for German asr. In: Proceedings of the International Workshop on Spoken Language Translation, IWSLT (2013)
6. Heuvel, H., Reveil, B., Martens, J.P.: Pronunciation-based asr for names. In: Proceedings of Interspeech (2009)
7. Hirsimäki, T., Pylkkönen, J., Kurimo, M.: Importance of high-order n-gram models in morph-based speech recognition. IEEE Trans. Audio Speech Lang. Process. **17**, 724–732 (2009)
8. Lehečka, J., Švec, J.: Improving speech recognition by detecting foreign inclusions and generating pronunciations. In: Habernal, I. (ed.) TSD 2013. LNCS, vol. 8082, pp. 295–302. Springer, Heidelberg (2013)
9. Maison, B., Chen, S., Cohen, P.S.: Pronunciation modeling for names of foreign origin. In: IEEE Workshop on Automatic Speech Recognition and Understanding, ASRU (2003)
10. Mansikkaniemi, A., Kurimo, M.: Unsupervised topic adaptation for morph-based speech recognition. In: Proceedings of Interspeech (2013)
11. Mansikkaniemi, A., Kurimo, M.: Adaptation of morph-based speech recognition for foreign names and acronyms. IEEE Trans. Audio Speech Lang. Process. **23**(5), 941–950 (2015)
12. Siivola, V., Hirsimäki, T., Virpioja, S.: On growing and pruning kneser-ney smoothed n-gram models. IEEE Trans. Audio Speech Lang. Process. **15**(5), 1617–1624 (2007)
13. Yang, Q., Martens, J.P., Konings, N., Heuvel, H.: Development of a phoneme-to-phoneme (p2p) converter to improve the grapheme-to-phoneme (g2p) conversion of names. In: Proceedings of the Fifth International Conference on Language Resources and Evaluation, LREC (2006)

Combining Lexical and Prosodic Features for Automatic Detection of Sentence Modality in French

Luiza Orosanu[1,2,3]($\boxtimes$) and Denis Jouvet[1,2,3]

[1] Speech Group, Inria, LORIA, 54600 Villers-lès-Nancy, France
{luiza.orosanu,denis.jouvet}@loria.fr
[2] Speech Group, LORIA, Université de Lorraine, LORIA,
UMR 7503, 54600 Villers-lès-Nancy, France
[3] Speech Group, LORIA, CNRS, LORIA, UMR 7503,
54600 Villers-lès-Nancy, France

Abstract. This article analyzes the automatic detection of sentence modality in French using both prosodic and linguistic information. The goal is to later use such an approach as a support for helping communication with deaf people. Two sentence modalities are evaluated: questions and statements. As linguistic features, we considered the presence of discriminative interrogative patterns and two log-likelihood ratios of the sentence being a question rather than a statement: one based on words and the other one based on part-of-speech tags. The prosodic features are based on duration, energy and pitch features estimated over the last prosodic group of the sentence. The evaluations consider using linguistic features stemming from manual transcriptions or from an automatic speech transcription system. The behavior of various sets of features are analyzed and compared. The combination of linguistic and prosodic features gives a slight improvement on automatic transcriptions, where the correct classification performance reaches 72 %.

Keywords: Speech-to-text transcriptions · Question detection · Prosody · Likelihood ratio · Part-of-speech tags

1 Introduction

The automatic detection of sentence modality has been studied in the past decades with different objectives: to model and detect the speech structure [1], to distinguish questions from statements [2–7], to create the summary of documents or meetings [4], to enrich an automatic transcription with punctuation marks [8], etc.

The most useful cues for the detection of sentence modality are the prosodic features (computed over the speech signal) and the linguistic features (computed over the word transcription). There are two scenarios for the linguistic features:

© Springer International Publishing Switzerland 2015
A.-H. Dediu et al. (Eds.): SLSP 2015, LNAI 9449, pp. 207–218, 2015.
DOI: 10.1007/978-3-319-25789-1_20

when they are extracted from correct data (textual and/or manual transcriptions of audio) or from automatic transcriptions (generated by a speech recognition system). The studies related to automatic speech recognition systems have to additionally take into account the speech recognition errors which get more frequent for poor sound qualities and on spontaneous speech, and can highly decrease the classification performance.

Regarding the prosodic features, different studies on different languages consider different features computed over different parts of the speech signal. In [2], the prosodic features (pitch and energy) computed on the last 700 ms of speech were used for the detection of French questions. In [6], the energy and the fundamental frequency were the key features in the detection of Arabic questions. In [9] the English question asking behavior was designed in order to improve the intelligent tutoring systems; their study concluded that the most useful features were the pitch slope of the last 200 ms of a turn. Another detector of French questions (versus statements) made use of 12 prosodic features derived from the fundamental frequency of the entire utterance [4].

When dealing with correct data (e.g. manual transcriptions), considering both prosodic and lexical features proves very useful. In [5], the combined prosodic-lexical classifier considers lexical features relative to interrogative terms: the unigrams/bigrams preceding or succeeding interrogative terms and the presence (or absence) of interrogative terms. The use of web textual conversations to detect questions in conversational speech was analyzed in [7]. Their lexical features consider the presence or absence of unigrams through trigrams in the sentences (with respect to questions or statements).

When dealing with automatic transcriptions, the sentence modality detection becomes more challenging. In [1], 42 dialog acts were used to model and detect the discourse structure of natural English speech (human-to-human telephone conversations). They used three different types of information (linguistic, prosodic and a statistical discourse grammar) and achieved an accuracy of 65 % on ASR transcripts versus 72 % on reference manual transcripts. Combining recognized words with the discourse grammar was the most useful for this task. The detection of questions in English meetings was addressed in [10], using lexico-syntactic, turn related and pitch related information. They achieved an accuracy of 54 % on ASR transcripts versus 70 % on reference manual transcripts. The lexico-syntactic features were the most useful for this task. The automatic punctuation (comma, period, question mark) of French and English speech-to-text data was studied in [8]. Their boosting-based model uses linguistic (based on word n-grams) and prosodic information and was tested under real world conditions.

Based on the state of the art of question detection, we apply multiple feature combinations on our French data. Several approaches are analyzed: creating a classifier with only prosodic features or one with only linguistic features or one that combines both linguistic and prosodic features. Moreover classifier evaluations are carried out using linguistic features stemming out, on the one hand, from manual transcriptions, and on the other hand, from automatic speech-to-text transcriptions.

The work presented in this paper is part of the RAPSODIE project which aims at studying, deepening and enriching the extraction of relevant speech information, in order to support communication with deaf or hard of hearing people. The detection of sentence modality (questions versus statements) is therefore a key problem here, the deaf or hard of hearing people must be informed when a question is directed to them, as they should respond or ask for further clarifications.

The paper is organized as follows: Sect. 2 is devoted to the description of the data and tools used in our experiments, Sect. 3 provides a description of the features used for question detection, and Sect. 4 analyzes the results.

2 Experimental Setup

2.1 Textual Data for Training Language Models

Textual punctuated data is used for modeling the lexical and syntactic characteristics of questions and statements. The available data corresponds to more than 800 million words from the French Gigaword corpus [11]. Based on a vocabulary of 97 K words, 89K questions and 16M statements were extracted from this corpus by filtering the sentences ending with a question mark, respectively with a dot. The lexical data was also annotated with part-of-speech (POS) tags; this provided the syntactic data.

Based on the lexical (word-based) data we learned two language models, one for questions and one for statements, with a shared lexicon of 97 K words. These language models have the purpose of representing the main word sequences that occur in a question rather than a statement (like for example in French: "est-ce que ...", "qu'est-ce que ...", etc.).

Based on the syntactic (POS-based) data we learned two other language models, one for questions and one for statements, with a shared lexicon of 36 POS tags. These language models have the purpose of representing the main syntactic sequences that occur in a question rather than a statement (like for example in French the verb-pronoun inversions: "regardez vous ...", "pourrait on ...", "fallait il ...", etc.).

Table 1 describes the resulting 3-gram language models based on questions and statements, when using word-based sentences or POS-based sentences.

Table 1. Number of 3-grams in the language models computed over questions and statements

Language model	Word-based	POS-based
questions	718K	9K
statements	68M	16K

2.2   Speech and Textual Data for Modality Detection

The speech corpora used to train and evaluate the modality detection classifiers (questions versus statements) come from the ESTER2 [12] and ETAPE [13] evaluation campaigns, and from the EPAC [14] project. The ESTER2 and EPAC data are French broadcast news collected from various radio channels (prepared speech and interviews). The ETAPE data correspond to debates collected from various radio and TV channels (spontaneous speech). These corpora were manually transcribed and punctuated (the segmentation of speech into sentences is therefore already given).

The set of questions and statements were extracted from these corpora by filtering the sentences ending with a question mark and respectively with a dot. The training sets of ESTER2, EPAC and ETAPE corpora are used to train the question detection classifiers; the development and test sets of the ESTER2 and ETAPE corpora are used to evaluate them.

The speech training data set contains 10 K questions and 98 K statements. However, binary classifiers do not work well when trained with imbalanced data sets: new instances are likely to be classified as the class that has more training samples. In order to avoid this overfitting problem, we chose to resample the data set by keeping all questions and randomly extracting subsets of statements of the same size (ten different training data sets are considered based on the different random lists of statements). In the 'Experiments and results' section we present only the average performance (with the associated standard deviation) over all ten training data sets.

Table 2 gives more details on the number of questions and statements used in our experiments.

Table 2. Description of the data used in our experiments

Data	# Questions	# Statements
Training data	10077	10077
Evaluation data	831	7005

2.3   Configuration

The SRILM tools [15] were used to train the statistical language models. The TreeTagger software [16] was used to annotate the transcriptions with POS tags.

The WEKA software [17] was used to train and evaluate 5 question detection classifiers:

- logistic regression (LR) [18],
- C4.5 decision tree (J48) [19],
- rule learner (JRip - Repeated Incremental Pruning to Produce Error Reduction) [20],

- sequential minimal optimization algorithm for training a support vector classifier (SMO) [21],
- neural network using backpropagation to classify instances (MP - Multilayer Perceptron) [22].

The values of F0 in semitones and of the energy are computed every 10 ms from the speech signal using the ETSI/AURORA acoustic analysis [23].

The forced speech-text alignment is carried out with the Sphinx3 tools [24]. This provides the speech segmentation into phones and words, which is then used to compute the sound durations, as well as to obtain the location and the duration of pauses. As the speech signal quality is rather good, it can be assumed that the segmentation is obtained without major problems.

The pronunciation variants were extracted from the BDLEX lexicon [25] and from in-house pronunciation lexicons, when available. For the missing words, the pronunciation variants were automatically obtained using JMM-based and CRFbased Grapheme-to-Phoneme converters [26].

The Sphinx3 tools were also used to train the phonetic acoustic models and to decode the audio signals. More information on the large-vocabulary decoding system used in our experiments and its associated lexicon can be found in [27,28].

3 Features for Question Detection

3.1 Linguistic Features

Three linguistic features were used to distinguish questions from statements:

- Two log-likelihood ratios ($lexLLR$, $synLLR$)

Two of our linguistic features are represented by the difference between the log-likelihood of the sentence with respect to the 'question' language model and the log-likelihood of the sentence with respect to the 'statement' language model (as done in [3] for Chinese). Computed as:

$$LLR(sentence) = Log\left(\frac{P(sentence|questionLM)}{P(sentence|statementLM)}\right) \tag{1}$$

A sentence having a positive LLR value is likely to be a question. And vice-versa, a sentence having a negative LLR value is likely to be a statement.

To compute the lexical log-likelihood ratio ($lexLLR$) of a sentence we apply the lexical language models (of questions and statements) on its sequence of words.

To compute the syntactic log-likelihood ratio ($synLLR$) of a sentence we apply the syntactic language models (of questions and statements) on its sequence of POS tags.

- Presence of discriminative interrogative patterns (iP)

This feature indicates the presence (1) or absence (0) of some discriminative interrogative words or expressions. A sentence having an interrogative pattern is likely to be a question.

A list of sequential patterns was thus extracted from the Gigaword questions transcript with a modified version of the PrefixSpan software [29] that considers only consecutive patterns. Their frequencies were then compared between the Gigaword questions and statements transcripts: those with similar frequencies were removed. The patterns with no interrogative meaning were also removed.

The final list of discriminative interrogative patterns is: {quel, quelle, quels, quelles, comment, combien, pourquoi, est ce que, est ce qu', qu' est ce, qu' est ce que, qu' est ce qu'}, corresponding to {what, which, how, how much, why, ...}.

3.2   Prosodic Features

The prosodic features include duration, energy and pitch belonging to the last prosodic group of the sentence. Prosodic groups are determined according to linguistic information (for grouping grammatical words with corresponding lexical words) and further processing that relies on prosodic information as described in [30]. Ten prosodic features were considered in order to distinguish questions from statements. Five are associated to the last syllable of the sentence, and five other are computed on the ending part of the sentence.

The duration of the last vowel is computed from the phonetic segmentation that results from the forced alignment. Its energy corresponds to the mean value calculated over all the frames of the vowel segment. The vowel energy and the vowel duration are then normalized with respect to local mean values computed on non-stressed vowels of the current breath group (speech segment delimited by pauses). In practice we used the vowels that are not in a word final position. The F0 slope is calculated by linear regression on the speech frames corresponding to the vowel. In addition to the slope, we calculate also, for the vowel, the delta of F0 movement with respect to the preceding vowel. The fifth parameter is the product of the F0 slope by the square of the vowel duration (this is inspired from the glissando threshold). Other, more global, prosodic parameters are computed on the longest F0 slope that ends in the last syllable of the sentence. Starting from the last syllable, we go back in time up to detecting an inversion of the F0 slope. We then compute parameters on this longest final F0 slope: the F0 slope itself (determined by linear regression), the length of this longest slope, the total F0 variation between the beginning and the end of the slope, and also the product of the slope by the square of the duration. One last prosodic parameter is used, which corresponds to the F0 level at the end of the sentence, expressed as the percentage of the speaker F0 range (0 corresponding to the lowest F0 value for the speaker, 100 corresponding to the highest F0 value for the speaker).

4   Experiments and Results

The classifiers evaluated in our experiments (logistic regression, J48 decision tree, JRip rule learner, SMO sequential minimal optimization algorithm, neural network MP) gave similar results. Thus, only the results obtained with the classifier J48 are presented below.

The classifier evaluations are carried out using features stemming out from:

- automatic transcriptions (obtained with a large vocabulary speech recognizer) - to study the performance under real conditions
- manual transcriptions - to study the classifier's maximum performance, obtainable only in ideal conditions (i.e. with perfect transcripts).

The performance obtained on our imbalanced test data set (831 questions and 7005 statements) is evaluated by the harmonic mean between the ratio of correctly classified questions and the ratio of correctly classified statements, computed as:

$$H = 2 * \left(\frac{ccQuestions * ccStatements}{ccQuestions + ccStatements} \right) \tag{2}$$

where "cc" is an acronym for "correctly classified". This value allows us to estimate the global performance of our classifier, given that the performances achieved on questions and on statements are equally important.

4.1 Prosodic Features

The evaluated combinations of prosodic features are:

- the last F0 level (*lastF0level*)
- the 5 features computed over the last syllable (*lastSyl*),
- the 5 features computed over the last syllable plus the last F0 level (*lastSyl + lastF0level*),
- the 5 features computed over the ending part of the utterance (*lastPart*),
- the 6 features related to slope measurements (*slope*),
- all 10 features (*Prosodic*).

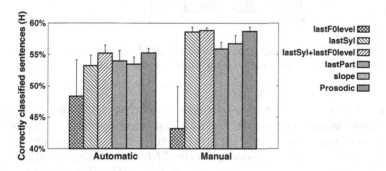

Fig. 1. Analysis of the average performance obtained when using different prosodic feature combinations on automatic and manual transcriptions

Figure 1 shows the average performance obtained with different prosodic feature combinations on automatic and manual transcriptions. The most important prosodic features are those computed over the last syllable of the utterance in combination with the last F0 level ($lastSyl + lastF0level$). Combining all 10 prosodic features (*Prosodic*) does not deteriorate this performance: they are all considered to be useful and kept in the following experiments. The performance loss between manual and automatic transcriptions (of about 3 %) is due to recognition errors and to the automatic word (phone) segmentation.

4.2 Linguistic Features

The evaluated combinations of linguistic features are:

- the lexical log-likelihood ratio ($lexLLR$)
- the syntactic log-likelihood ratio ($synLLR$),
- the lexical log-likelihood ratio plus the presence of discriminative interrogative patterns ($lexLLR + iP$),
- the syntactic log-likelihood ratio plus the presence of discriminative interrogative patterns ($synLLR + iP$),
- both log-likelihood ratios ($lexLLR + synLLR$),
- all 3 features (*Linguistic*).

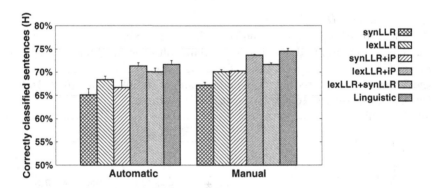

Fig. 2. Analysis of the average performance obtained when using different linguistic feature combinations on automatic and manual transcriptions

Figure 2 shows the average performance obtained with different linguistic feature combinations on automatic and manual transcriptions and it proves the importance of using and combining all of them. The most important linguistic feature is the lexical log-likelihood ratio ($lexLLR$). Combining it with the presence of discriminative interrogative patterns ($lexLLR + iP$) is more useful than combining it with the syntactic log-likelihood ratio ($lexLLR + synLLR$). The performance loss between manual and automatic transcriptions of the combined set of interrogative patterns and syntactic log-likelihood ratio ($synLLR + IP$) is bigger than the one achieved by the $lexLLR$ feature, which means that they are less tolerant to recognition errors. However, the combination of all three features (*Linguistic*) improves the classification performance, especially when dealing with correct transcriptions.

4.3   Combined Prosodic-Linguistic Features

Table 3 shows the average performance (harmonic mean H, along with the ratios of correctly classified questions and correctly classified statements) obtained with the prosodic, linguistic and combined features, when applied on automatic speech-to-text transcriptions and on manual transcriptions. It can be easily observed that the linguistic classifiers outperform the prosodic classifiers. The performance obtained with the linguistic classifiers when applied on the automatic transcriptions and on the manual transcriptions differs by about 3 % absolute, due to recognition errors (22 % word error rate on Ester and 28 % on Etape) and most likely to the misrecognition of the interrogative words. The combination of linguistic and prosodic features does not provide any improvement on manual transcripts and provides only a slight improvement on automatic transcription.

Table 3. Average performance (harmonic mean H, along with the ratios of correctly classified questions and correctly classified statements respectively) obtained on automatic and manual transcriptions, for prosodic features alone, linguistic features alone and with a combination of prosodic and linguistic features

Transcripts	Prosodic	Linguistic	Combined
automatic	55.24 (51.71; 60.23)	71.64 (66.62; 77.77)	72.21 (69.55; 75.25)
manual	58.69 (57.97; 59.55)	74.47 (71.57; 77.93)	74.26 (75.18; 73.46)

Table 4 gives more detailed results obtained with the combined prosodic-linguistic classifier on the manual transcriptions, when trained on a single random training set. 627 out of 831 questions were correctly classified as questions (ccQ=75.45 %) and 5047 out of 7005 statements were correctly classified as statements (ccS=72.05 %). The harmonic average performance is here H=73.71 %.

4.4   Combined Outputs

A final experiment consisted in combining the outputs of all five classifiers (when using all 13 prosodic-linguistic features). Each classifier makes a class prediction (question or statement) on each utterance of the test data set. The final decision is made by a majority vote: if most of the classifiers (in this case at least 3) assign the utterance to class "question", than the utterance is assigned to class "question"; if not, than the utterance is assigned to class "statement".

Table 4. Confusion matrix between questions and statements obtained on manual transcriptions with the combined prosodic-linguistic classifier, when trained on a single random training set

	Classified as question	Classified as statement	
Question	627	204	ccQuestions=75.45 %
Statement	1958	5047	ccStatements=72.05 %

Table 5 shows the average performance (H) obtained with all five classifiers separately, and with their combination (by majority vote). The majority vote and the 5 classifiers have similar performances, thus confirming that the 5 classifiers are likely to agree on the class predictions.

Table 5. Average performance (H) obtained with all 5 classifiers and with their combination (by majority vote) on manual and on automatic transcriptions

	LR	J48	JRip	SMO	MP	Combination
Automatic	72.04	72.21	72.81	69.56	72.07	72.66
Manual	73.34	74.26	74.12	72.09	74.33	74.91

5 Conclusions

This paper analyzed the impact of linguistic features, prosodic features and combined linguistic-prosodic features when developing an automatic question detector. The context of this work is to support the communication with deaf or hard of hearing people, which requires an automatic detection of questions in order to inform them when a question is directed to them. The experiments were carried out using three French speech corpora: ETAPE, EPAC and ESTER2.

Different types of classifiers (logistic regression, decision tree, rule learner, sequential minimal optimization algorithm, neural network) were evaluated, but they all give similar results.

The prosodic classifier (based on 10 prosodic features) has a poor performance: it hardly exceeds 55 % of correctly classified sentences. The most important prosodic features are those computed over the last syllable, in combination with the last F0 level.

The linguistic classifier (based on 3 linguistic features) provides by far better results: 72 % when it is applied on ASR transcriptions (with perfect sentence boundaries) versus 74 % when it is applied on reference manual transcripts. The most important linguistic feature is the lexical log-likelihood ratio (computed with respect to word-based language models).

The combination of prosodic and linguistic features does not provide any improvement on manual transcripts, but it provides a slight improvement on automatic transcription.

Future work will investigate further prosodic and linguistic features; confidence measures will also be considered in the computation of the linguistic features.

Acknowledgements. The work presented in this article is part of the RAPSODIE project, and has received support from the "Conseil Régional de Lorraine" and from the "Région Lorraine" (FEDER) (http://erocca.com/rapsodie).

References

1. Jurafsky, D., Bates, R., Coccaro, N., Martin, R., Meteer, M., Ries, K., Shriberg, E., Stolcke, A., Taylor, P., Van Ess-Dykema, C.: Automatic detection of discourse structure for speech recognition and understanding. In: IEEE Workshop on Automatic Speech Recognition and Understanding, pp. 88–95 (1997)
2. Kral, P., Kleckova, J., Cerisara, C.: Sentence modality recognition in french based on prosody. In: International Conference on Enformatika, Systems Sciences and Engineering - ESSE 2005. vol. 8, pp. 185–188 (2005)
3. Yuan, J., Jurafsky, D.: Detection of questions in chinese conversational speech. In: IEEE Workshop on Automatic Speech Recognition and Understanding, pp. 47–52 (2005)
4. Quang, V.M., Castelli, E., Yên, P.N.: A decision tree-based method for speech processing: question sentence detection. In: Wang, L., Jiao, L., Shi, G., Li, X., Liu, J. (eds.) FSKD 2006. LNCS (LNAI), vol. 4223, pp. 1205–1212. Springer, Heidelberg (2006)
5. Quang, V.M., Besacier, L., Castelli, E.: Automatic question detection: prosodic-lexical features and crosslingual experiments. In: Proceedings of Interspeech, pp. 2257–2260 (2007)
6. Khan, O., Al-Khatib, W.G., Cheded, L.: A preliminary study of prosody-based detection of questions in Arabic speech monologues. Arab. J. Sci. Eng. **35**(2C), 167–181 (2010)
7. Margolis, A., Ostendorf, M.: Question detection in spoken conversations using textual conversations. In: Association for Computational Linguistics, pp. 118–124 (2011)
8. Kolar, J., Lamel, L.: Development and evaluation of automatic punctuation for French and English speech-to-text. In: Proceedings of Interspeech (2012)
9. Liscombe, J., Venditti, J.J., Hirschberg, J.: Detecting question-bearing turns in spoken tutorial dialogues. In: Proceedings of Interspeech (2006)
10. Boakye, K., Favre, B., Hakkani-Tur, D.: Any questions? automatic question detection in meetings. In: IEEE Workshop on Automatic Speech Recognition and Understanding, pp. 485–489 (2009)
11. Mendonça, A., Graff, D., DiPersio, D.: French Gigaword third edition. In: Proceedings of the Linguistic Data Consortium (2011)
12. Galliano, S., Gravier, G., Chaubard, L.: The ESTER 2 evaluation campaign for rich transcription of French broadcasts. In: Proceedings of Interspeech (2009)
13. Gravier, G., Adda, G., Paulson, N., Carré, M., Giraudel, A., Galibert, O.: The ETAPE corpus for the evaluation of speech-based TV content processing in the French language. In: Proceedings of the International Conference on Language Resources, Evaluation and Corpora (LREC) (2012)

14. Estève, Y., Bazillon, T., Antoine, J.Y., Béchet, F., Farinas, J.: The EPAC corpus: manual and automatic annotations of conversational speech in French broadcast news. In: Proceedings of the Seventh International Conference on Language Resources and Evaluation (LREC) (2010)
15. Stolcke, A.: SRILM an extensible language modeling toolkit. In: Conference on Spoken Language Processing (2002)
16. Schmid, H.: Probabilistic part-of-speech tagging using decision trees. In: Proceedings of the International Conference on New Methods in Language Processing, pp. 44–49 (1994)
17. Hall, M., Frank, E., Holmes, G., Pfahringer, B., Reutemann, P., Witten, I.H.: The WEKA data mining software: an update. SIGKDD Explorations **11**(1), 10–18 (2009)
18. le Cessie, S., van Houwelingen, J.: Ridge estimators in logistic regression. Appl. Stat. **41**(1), 191–201 (1992)
19. Quinlan, J.R.: C4.5: Programs for Machine Learning. Morgan Kaufmann Publishers Inc, San Francisco (1993)
20. Cohen, W.: Fast effective rule induction. In: Proceedings of the Twelfth International Conference on Machine Learning, pp. 115–123 (1995)
21. Keerthi, S., Shevade, S., Bhattacharyya, C., Murthy, K.: Improvements to platt's SMO algorithm for SVM classifier design. Neural Comput. **13**(3), 637–649 (2001)
22. Ruck, D.W., Rogers, S.K., Kabrisky, M., Oxley, M.E., Suter, B.W.: The multilayer perceptron as an approximation to a Bayes optimal discriminant function. IEEE Trans. Neural Netw. **1**(4), 296–298 (1990)
23. ETSI ES 202 212: Speech processing, transmission and quality aspects (STQ); distributed speech recognition; extended advanced front-end feature extraction algorithm; compression algorithms. ETSI ES (2005)
24. Placeway, P., et al.: The 1996 Hub-4 Sphinx-3 System. In: DARPA Speech Recognition Workshop (1996)
25. de Calmès, M., Pérennou, G.: BDLEX : a lexicon for spoken and written French. In: Proceedings of the International Conference on Language Resources and Evaluation (LREC), pp. 1129–1136 (1998)
26. Jouvet, D., Fohr, D., Illina, I.: Evaluating grapheme-to-phoneme converters in automatic speech recognition context. In: Proceedings of the IEEE International Conference on Acoustics, Speech and Signal Processing (ICASSP), pp. 4821–4824 (2012)
27. Jouvet, D., Fohr, D.: Combining forward-based and backward-based decoders for improved speech recognition performance. In: Proceedings of Interspeech (2013)
28. Jouvet, D., Langlois, D.: A machine learning based approach for vocabulary selection for speech transcription. In: Habernal, I., Matoušek, V. (eds.) TSD. LNCS, vol. 8082, pp. 60–67. Springer, Heidelberg (2013)
29. Pei, J., Han, J., Mortazavi-asl, B., Pinto, H., Chen, Q., Dayal, U., Hsu, M.C.: PrefixSpan: Mining sequential patterns efficiently by prefix-projected pattern growth. In: International Conference on Data Engineering, pp. 215–224 (2001)
30. Bartkova, K., Jouvet, D.: Automatic detection of the prosodic structures of speech utterances. In: Železný, M., Habernal, I., Ronzhin, A. (eds.) SPECOM 2013. LNCS, vol. 8113, pp. 1–8. Springer, Heidelberg (2013)

Corpus Based Methods for Learning Models of Metaphor in Modern Greek

Konstantinos Pechlivanis[1,2](✉) and Stasinos Konstantopoulos[2]

[1] Technical University of Crete, Chania, Grecce
[2] Institute of Informatics and Telecommunications, NCSR 'Demokritos',
Athens, Greece
{kpechlivanis,konstant}@iit.demokritos.gr

Abstract. In this paper we propose a method for detecting metaphorical usage of content terms based on the hypothesis that metaphors can be detected by being characteristic of a different domain than the one they appear in. We formulate the problem as one of extracting knowledge from text classification models, where the latter have been created using standard text classification techniques without any knowledge of metaphor. We then extract from such models a measure of how characteristic of a domain a term is, providing us with a reliable method of identifying terms that are surprising for the context within which they are used. To empirically evaluate our method, we have compiled a corpus of Greek newspaper articles where the training set is only annotated with the broad thematic categories assigned by the newspapers. We have also manually annotated a test corpus with metaphorical word usage. In our experiment, we report results using tf-idf to identify the literal (characteristic) domain of terms and we also analyse the interaction between tf-idf and other typical word features, such as Part of Speech tags.

Keywords: Metaphor detection · Information extraction · Distributional semantics · Term extraction · Machine learning

1 Introduction

Metaphor is a figure of non-literal usage of words that can greatly impact the accuracy of automatically analysing and interpreting language. The computational treatment of metaphor aims at *detecting* metaphor and, more ambitiously, at *interpreting* it into literal semantics.

The first efforts to detect metaphor were based on subcategorization frames and similar semantic resources to detect violations of selectional restrictions in a certain context [1,2]. More recent systems are striving to be less demanding in the required linguistic resources and rely on more statistical approaches to semantics. The TroFi system [3], for example, assumes a user-provided set of seed sentences and detects metaphors by computing the similarity between a sentence and all of the seed sentences. Other systems rely on semantic hierarchies: Krishnakumaran and Zhu [4] for instance predict metaphorical phrases at the sentence level using

© Springer International Publishing Switzerland 2015
A.-H. Dediu et al. (Eds.): SLSP 2015, LNAI 9449, pp. 219–228, 2015.
DOI: 10.1007/978-3-319-25789-1_21

the hyponymy relation in WordNet. Also, Shutova [5] interprets metaphorical phrases as a paraphrasing task. So, for each metaphorical expression there is a literal paraphrase which is obtained by applying a probabilistic model in order to rank all the possible paraphrases of the certain metaphorical phrase at the certain context.

In the work described here we are interested in detecting *novel* metaphorical usage of content terms, excluding idiomatic metaphorical expressions. This task is motivated by *text categorization* applications, where metaphorical terminology can lead to misclassifications. Furthermore, we are interested in developing methods that can be applied to languages that lack rich semantic resources, such as subcaterization frame dictionaries or semantic network dictionaries. Although the ability to correctly interpret metaphors would be useful, in such a setting even detecting metaphors is a challenging task and can still be applied to exclude or reduce the weight of metaphorical terms in text categorization models.

Closer to our setting and methodology is the work by Schulder and Hovy [6] on detecting metaphors using a purely statistical approach to word semantics and the hypothesis that novel metaphoric language is *unusual* in a given context. In order to calculate whether a term is typical of its context, they use statistical metrics to identify words *commonly used in* and *characteristic of* a domain as opposed to words commonly used across all domains. They extract domain-specific document collections using term searching. The query terms are a set of seed terms that are considered typical for a domain. The evaluation of their results is based on manually chosen seed terms or the terms with the highest relevance for document search, generating a single governance domain.

In this paper we push further in the direction of minimizing the resources required in order to train the system, to present a method that only relies on having text placed in very broad thematic categories. Our use cases primarily stem in newspaper content categorization: newspaper content is organized in very broad thematic sections that can be used to detect out-of-topic, typically metaphorically used, terms. Metaphorically used terms can then be excluded or treated exceptionally in subsequent classification of the content in finer categories.

In the remainder of this paper, we first present our method (Sect. 2) and its implementation (Sect. 3). We then present the empirical evaluation of our method (Sect. 4) and our conclusions (Sect. 5).

2   Method

In the work described here, we pursue the same core hypothesis as Schulder and Hovy [6], namely, that metaphors can be detected by being characteristic of a different domain than the one they appear in. Unlike Schulder and Hovy, we do not rely on any manually chosen seed terms, neither do we formulate the problem as a classification problem where metaphor annotation is the output of a classifier.

Instead, we formulate the problem as one of *extracting knowledge* from text classification models, where the latter have been created using standard text classification techniques without any knowledge of metaphor. We then extract from such models a measure of how characteristic of a domain a term is, providing us with a reliable method of identifying terms that are *uncharacteristic* of the context within which they are used.

By doing this, we build upon the rich text classification literature and the robustness of its statistical methods in identifying domain-characteristic terms versus terms that are generally frequent across all domains. Furthermore, we provide a methodology that does not rely on any seeds or other semantic resources at the level of individual terms, but only on building a classifier that predicts very broad thematic annotations for complete articles, such as the 'politics', 'sports', and similar categories readily available in any newspaper corpus. This methodology does not require sentence structure information or semantic resources and can be applied to less-resourced languages, is robust to noisy data, and is efficient enough to be applied on large-scale corpora.

In the experiments presented here, we instantiate this generic methodology using the *Term Frequency – Inverse Document Frequency (tf-idf)* of the terms appearing in a document as features for *language models* that predict the domain based on the document terms. tf-idf balances between the frequency of a term in a particular context (tf) and its frequency across all contexts (idf) and is very well suited for identifying 'surprizing' words in a given context.

In our experiments we used a *Maximum Likelihood Classifier* that uses term weighting as its only metric. We, then, assume that the 'native' or literal-usage domain of a term is the domain where the term has the highest weight.

3   Implementation

3.1   Corpus Collection and Preprocessing

In order to investigate our research proposal we started with compiling a corpus of articles from three Greek newspapers that offer content on-line: 'Lefkaditika Nea', 'Thraki', and 'Avgi'. In order to have an initial classification, we mapped the sections of these three newspapers to domains from the top level of the relevant taxonomy of the *International Press Telecommunications Council (IPTC)*.[1] Table 1 lists the seven domains and the number of articles in each.

The articles of the corpus were downloaded from the web as HTML files and cleaned into plain text using the Boilerpipe library [7]. After tokenization and stopword removal, we dropped tokens that consist of single alphanumeric characters and several symbols, dropped stress marks and other diacritics, and stemmed the data. Stemming improves results because the presence of different word forms for the same term makes training harder, and this is more pronounced in morphologically rich languages such as Greek. Although there is a variety of stemmers, the unique morphological system of each language doesn't allow the creation of a global rule-based algorithm which would be able to find out the stem of each word. Especially, in some languages with a rich morphological system,

---

[1] Please cf. www.iptc.org for more details.

Table 1. Distribution of articles in topics.

IPTC code	Domain	Number	Percentage
01000000	Art, Culture and Entertainment	3178	20.2%
04000000	Economy,Business and Finance	3132	20.0%
06000000	Environment	693	4.4%
07000000	Health	771	4.9%
11000000	Politics	6618	42.2%
13000000	Science and Technology	210	1.3%
15000000	Sport	1100	7.0%
	All corpus	15702	

like Greek, it is even more difficult to find the word stem by reducing the suffix from inflected or derived words. It is useful to mention that a wide variety of suffixes exist in the Greek morphological system, some of them may appear in different parts of speech. For this reason, it is necessary to point out the part of speech of the certain word before trying to find out the root of the concrete word. Our stemmer is available on-line.[2]

Text Feature Extraction of Term Weighting in Corpus: Metric TF-IDF. Term weighting is an important aspect of IR systems. Terms are words, phrases or any other indexing units used to identify the contents of a text. Since different terms have different level of importance in a text, an important indicator, which is called *(term weight)*, is associated with each term. Three main components that affect the importance of a term in a text are the *term frequency factor (tf)*, the *inverse document frequency factor (idf)* and the *normalization factor*. More specifically:

– *term frequency factor (tf):* Long documents usually use the same terms repeatedly. As a result, the term frequency factors may be large for long documents, increasing the average contribution of its terms towards the query - document similarity.
– *inverse document frequency factor (idf):* Long documents also have numerous different terms. This increases the number of matches between a query and a long document, increasing the query - document similarity and the chances of retrieval of long documents in preference over short documents. Moreover, the high rate of appearance of a word doesn't imply that this word is directly related to the topic of the specific document. The word with the highest occurrence rate may be an auxiliary verb, like the Greek verb *eimai* ('to be'). For this reason, the *inverse document frequency (idf)* is used, which is based on counting the number of documents in the collection being searched which contain the term in question.

---

[2] Please see https://bitbucket.org/dataengineering/stemming.

– *Normalization factor:* is a way of penalizing the term weights for a document in accordance with its length. Normalization factor retain the number of featured terms, normalizing the weights and ensuring that all their values are between 0 and 1. The use of a logarithmic function in the TF-IDF equation constitutes the normalization factor.

As a consequence, in order to calculate the weight of a term of the domain, we use the formula:

$$\text{tf-idf}(t, d) = \text{tf}(t, d)\,\text{idf}(t, d)$$
$$= \frac{\text{freq}(t, d)}{|T_d|}\,\log\frac{|D|}{|D_t|}$$

where $\text{freq}(t, d)$ is the frequency of term t in domain d, T_d is the set of terms appearing in domain d, D is the set of domains, and D_t is the set of domains where t appears. At this point we should mention that we adapted this method by treating all texts of a domain as a single 'document'.

Terms with a great impact receive high scores, while low scores are assigned to words that are either not frequent in the document or otherwise are too frequent among documents.

3.2 Classification

The *Maximum Likelihood Classifier (MLC)* is one of the most popular methods of classification in text/word classification, in which a text/word with the maximum likelihood is classified into the corresponding class. We use MLC in order to determine probabilistically the domain of a word. Specifically, we have already estimated the TF-IDF value of words for each domain. Each word is classified in the domain where it appears with the highest TF-IDF value. More formally, given a term t and the set d_t of all the domains where t appears:

$$\text{MLC}(t, d_t) = \text{argmax}_{d \in d_t}\text{tf-idf}(t, d)$$

If a term has zero TF-IDF values for all domains, then this term remained unclassified and was not used in the procedure of metaphor detection. Also, if a term had almost the same probability to belong in more than one literal domains, that term was classified to all these possible domains (Fig. 1).

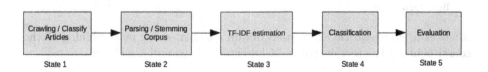

Fig. 1. Processing stages.

4 Empirical Evaluation

In order to evaluate our method, we have manually annotated 10 articles from the corpus of news articles presented in Sect. 3.1. The manual annotation was carried out by two initial annotators, with a third expert annotator resolving inconsistencies to create the golden corpus. Table 2 gives some statistics about this golden test corpus.

The annotation task was defined by extensive, written guidelines provided to the annotators, comprising the following steps:

1. Reading of the whole text in order to get a general understanding and assigning the text as a whole to one of the seven domains (IPTC top-level categories)
2. Establishing the contextual meaning of each lexical unit and determining if it has a more 'basic' meaning than the contextual meaning.
3. Marking the metaphor span in the text, as well as the providing the top-level category where the basic meaning is expected to be encountered.

The notion of a basic meaning was explained to the annotators as one that is more concrete, more closely related to the human body and its functions, or one that is historically older.

Besides the metaphor span, the annotators also marked each metaphor as being of one of the following types:

1. Multi-word metaphorical expression, explained below
2. Indirect metaphors, explained below
3. Direct or *is-a metaphors* where the copula connects the subject with a non-literal complement, such as 'time is money'
4. Idiomatic metaphorical expressions, such as 'kick the bucket'.

Indirect, lexical metaphors operate at the level of a single word where a term identifies metaphorically another term which is literal. Characteristic examples include:

(1) magika   nisia
    magic   islands
    magical islands

(2) Omirikoi kavgades
    Homeric quarrels
    fierce quarrels

Multi-word metaphorical expressions obtain a new meaning by combining all the constituents of the certain phrase. Characteristic examples include:

(3) Evale   to heri tou
    put-3ps the hand his-CLITIC
    He helped

(4) evale   freno
    put-3ps brake
    He slowed down

Metaphor spans can possibly overlap as in the example below:

(5) [[Rokanizontas   dramatika]   ton rolo] tou tupografou
    Shaving (carpentry) dramatically the role the typographer-GEN
    Dramatically gnawing at the typographer's role

where neither 'dramatically gnawing' nor 'gnawing at the role' are interpreted
literally.

Table 2. Annotated articles.

IPTC	All words	Content words	Metaphors
01000000	567	312	11
04000000	756	411	32
04000000	619	323	29
04000000	1158	650	34
07000000	321	169	13
11000000	760	414	51
11000000	961	518	52
11000000	715	404	15
11000000	985	558	50
11000000	987	546	53
All articles	7829	4305	340

Our method yields single-word binary decisions about metaphorical usage.
Following standard practice, we define *precision, recall* as follows:

- Precision is the percentage of positive decisions that were inside at least one
  span annotated as metaphor.
- Recall is the percentage of spans annotated as metaphors that include at least
  one positive decision.

F_1-score is defined in the usual manner over this precision and recall.

Moreover, we tested how metaphor detection interacts with lexical features,
and specifically with Part-of-Speech features. To this end, we evaluated our
model separately for nouns, verbs, and adjectives and compared that to the
overall evaluation results (Table 3). The drop in recall is explained by the fact
that as metaphors become sparser, each metaphorical phrase provides fewer

Table 3. Evaluation results.

	All PoS	Noun	Adjective	Verb
Precision	0.410	0.424	0.509	0.340
Recall	0.494	0.242	0.355	0.209
$F_{\beta=1}$	0.448	0.308	0.419	0.258

Table 4. Evaluation results for the 1/3 of all the terms with the highest TF-IDF value.

	All PoS	Noun	Adjective	Verb
Precision	0.443	0.421	0.597	0.555
Recall	0.285	0.150	0.209	0.066
$F_{\beta=1}$	0.347	0.221	0.300	0.119

Table 5. Evaluation results for 26,500 classified words.

	All PoS	Noun	Adjective	Verb
Precision	0.397	0.445	0.483	0.285
Recall	0.629	0.346	0.432	0.322
$F_{\beta=1}$	0.487	0.389	0.456	0.303

words, fewer 'hints' for the system to recognize it. The difference in precision is more interesting, indicating that metaphorical usage is harder to capture for verbal terms than for nominal terms. This is corroborates the intuitiion that nominal are more characteristic of a domain of discourse whereas verbs are more consistently used across domains.

Also, we evaluated our system only with the words which appear to have the strongest impact (Table 4, Fig. 2). For 24,463 classified words, we used only the 8,154 words with the highest TF-IDF value in order to detect the non literal phrases. The results are slightly different from the previous experiment. This is expected since making fewer classification decisions results in fewer detections (lower recall), but more accurate ones (higher precision). Naturally, there is a bigger gain in verbs, since these have a lower precision to start with.

The vocabulary of the model comprises 28,305 unique words. Of these words, 24,463 are classified using the MLC classifier. The remaining 3,842 words have zero tf-idf and are left unclassified. Trying to improve our system we estimate a alternative method to exploit these unclassified words. We set a rule which classify each word depending on the term frequency (calculated as in tf-idf above) and *document frequency (df)*. Document frequency of term t in document collection C is defined as the average number of occurrences of t in each document of C:

$$\mathrm{df}(t, C) = \frac{\mathrm{freq}(t, C)}{|C|}$$

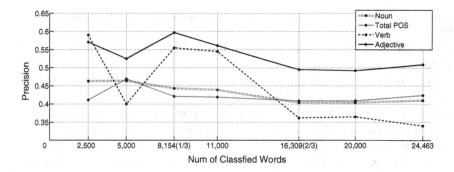

Fig. 2. Evaluation results for the terms with the highest TF-IDF value.

We then categorize terms with a low df (threshold determined empirically) using tf instead of tf-idf. Of the 3,842 unclassified words because if zero if-idf, we used this to classify 2,037 additional words (Table 5). Although precision is slightly worse, increased recall improves the overall F_1-score.

In order to investigate the influence of the number of classified words in model, we trained a model with larger vocabulary, and 40,723 classifies words. The improvement in Precision, Recall and F1 was tiny. As a result, we accept the previous model which retain approximately the same performance with lower complexity and cost.

A word with low TF-IDF might also indicate a word that is common among all domains. To filter out such candidates, we use document frequency as additional feature. We exclude word with high document frequency, but the results were not so promising.

There are a lot of cases that a word belongs to more than one literal domain. Thus, we try to build a more robust model, using alternative approach of classification. We classify each term in the two most probably domains, but our results did not improved.

5 Conclusion

In this paper we presented a statistical methodology for detecting metaphorical usage of content words. The main advantage of our methodology is that it only relies on a corpus of documents assigned to broad thematic categories and does not require any other semantic resources. This gives our method a very wide scope of application across less-resourced languages.

For our experiments we have used a newspaper corpus assuming the topics under which articles were posted as such thematic categories. We experimented with the F_1-score obtained by our method and found significant variation between the various Parts of Speech. Thus, we had the opportunity to study the structure of the Greek articles, to find out words which can be used as non literal indicates and to detect common words which are used to appear at more than one article's domains and so they don't have any value for contribution to differentiation between literal and non literal speech.

For future work we plan to revisit our experiments with more test data of different domains. The vast majority of articles currently belong to 'Art, Culture and Entertainment', 'Economy, Business and Finance' and 'Politics'. The latter two are the hardest domains in that they encompass several themes and might use words from different domains literally. Domains such as 'Sport' and 'Health' on the other hand are more amenable to our approach, but have considerably fewer articles (Table 1). We are continuing the article collection and pre-processing as more articles becomes available, in the hope that a larger dataset will include enough articles from all domains to allow a more thorough statistical investigation of the differences between domains.

Furthermore, a larger dataset will allow us to increase the order of our models to bigrams or trigrams without encountering severe data sparsity problems. This will allow us to experiment with detecting more complex metaphorical construction that cannot be captured by uni-term models.

Acknowledgments. The authors are grateful to the annotators for their contribution in preparing the test corpus. We would also like to thank 'Lefkaditika Nea' and 'Thraki' for granting us permission to use their articles for our research and 'Avgi' for offering its content under a creative commons license.

References

1. Fass, D.: met*: a method for discriminating metonymy and metaphor by computer. Comput. Linguist. **17**(1), 49–90 (1991)
2. Mason, Z.J.: CorMet: a computational, corpus-based conventional metaphor extraction system. Comput. Linguist. **30**(1), 23–44 (2004)
3. Birke, J., Sarkar, A.: A clustering approach for the nearly unsupervised recognition of nonliteral language. In: Proceedings of EACL-2006, Trento, pp. 329–336 (2006)
4. Krishnakumaran, S., Zhu, X.: Hunting elusive metaphors using lexical resources. In: Proceedings of the Workshop on Computational Approaches to Figurative Language, pp. 13–20, Rochester, Association for Computational Linguistics, April 2007
5. Shutova, E.: Metaphor identification as interpretation. In: Proceedings of the Second Joint Conference on Lexical and Computational Semantics (*SEM 2013), Atlanta, Georgia, USA, 13–14 June 2013
6. Schulder, M., Hovy, E.: Metaphor detection through term relevance. In: Proceedings of the 2nd Workshop on Metaphor in NLP, Baltimore, pp. 18–26, 26 June 2014
7. Kohlschütter, C., Fankhauser, P., Nejdl, W.: Boilerplate detection using shallow text features. In: Proceedings of The Third ACM International Conference on Web Search and Data Mining (WSDM 2010), New York City (2010)

Probabilistic Speaker Pronunciation Adaptation for Spontaneous Speech Synthesis Using Linguistic Features

Raheel Qader[1]([envelope]), Gwénolé Lecorvé[1], Damien Lolive[1], and Pascale Sébillot[2]

[1] IRISA/Université de Rennes 1, Lannion, France
{raheel.qader,gwenole.lecorve,damien.lolive,pascale.sebillot}@irisa.fr
[2] IRISA/INSA de Rennes, Rennes, France

Abstract. Pronunciation adaptation consists in predicting pronunciation variants of words and utterances based on their standard pronunciation and a target style. This is a key issue in text-to-speech as those variants bring expressiveness to synthetic speech, especially when considering a spontaneous style. This paper presents a new pronunciation adaptation method which adapts standard pronunciations to the style of individual speakers in a context of spontaneous speech. Its originality and strength are to solely rely on linguistic features and to consider a probabilistic machine learning framework, namely conditional random fields, to produce the adapted pronunciations. Features are first selected in a series of experiments, then combined to produce the final adaptation method. Backend experiments on the Buckeye conversational English speech corpus show that adapted pronunciations significantly better reflect spontaneous speech than standard ones, and that even better could be achieved if considering alternative predictions.

Keywords: Pronunciation adaptation · Linguistic features · Feature selection · Spontaneous speech synthesis · Conditional random fields

1 Introduction

Pronunciation variations are changes operated by speakers on standard pronunciations of words and phrases. These variations are valuable since they reflect the emotional state of a speaker, his/her intention, a specific accent or the context of the speech itself. However, pronunciation models and lexicons used by most current text-to-speech (TTS) systems still only rely on standard pronunciations, which limits the expressiveness of the resulting synthetic speech and prevents it from conveying a spontaneous style. A solution to this problem is to adapt standard pronunciations in order to fit this style. In a machine learning perspective, this task consists in predicting an adapted sequence of phonemes from an input sequence of canonical phonemes, i.e., deciding whether input phonemes should be deleted, substituted, simply kept as is, or if new phonemes should be inserted.

© Springer International Publishing Switzerland 2015
A.-H. Dediu et al. (Eds.): SLSP 2015, LNAI 9449, pp. 229–241, 2015.
DOI: 10.1007/978-3-319-25789-1_22

This paper proposes a new pronunciation adaptation method whose goal is to mimic the spontaneous style of individual speakers for the purpose of TTS. The strength of this method is to rely on linguistic information solely and on a probabilistic framework, namely conditional random fields (CRFs). Contrary to pronunciation adaptation in automatic speech recognition (ASR), the goal in TTS is to produce a unique adapted pronunciation rather than to cover all possible variants. In this scope, linguistic information is specifically important since no other type of information is available before synthesis. Nonetheless, this paper is wilfully limited to the generation of adapted pronunciations, the synthesis and evaluation of the corresponding speech signals being kept for future work. Finally, CRFs offer several advantages. They are widely used in grapheme-to-phoneme converters [9,14,19], thus enabling an easy integration of their outputs. And they also allow to explicitly consider and combine a large set of features.

Related work in pronunciation variant generation can be examined w.r.t. the nature of the proposed methods and the type of information they rely on. While early work has mostly concentrated on using phonological rules extracted from data to create alternative pronunciations [8,17], most recent techniques are machine learning approaches. Notably, decision trees [7,18], random forests [6], neural networks [5,10], hidden Markov models [16], and CRFs [10] have been investigated. In [18], decision trees and statistical contextual rules are even combined. Alternatively, [11] proposed to produce accented pronunciations by interpolating different grapheme-to-phoneme models. Unfortunately, these methods are hardly comparable as they are rarely used on the same data nor for the same exact task. Still, a picture of input information can be drawn. Acoustic features can be extracted from speech signals of the target style and considered as indicators for pronunciation adaptation (F0, energy, duration, speaking rate, etc.) [2–4], while linguistic features can be derived from textual data (distinction between content and function words, word predictability, syllable locations, lexical stress, etc.) [3,4,18]. Recently, [6] presented a deep study on the combination of both types of features, including even others like age and gender. This last work is close to the current paper, especially since the same corpus is used. However, machine learning techniques are different and [6] only focuses on making standard pronunciation shorter. Finally, [5] showed that predicting pronunciations transformations should not be carried out on each canonical phoneme independently, but including their neighborhood too. It is important to highlight that most of related work target ASR, whereas TTS approaches are still rare and none makes an extensive and exclusive use of linguistic features as achieved here. Hence, the presented pronunciation adaptation method is new and original.

The rest of this paper is structured as follows: Sect. 2 introduces the Buckeye speech corpus used in the experiments while Sect. 3 draws an overview of the method and describes the experimental setup. Section 4 provides the method details before Sect. 5 presents backend experiments and discusses the results.

2   The Buckeye Corpus

This work is conducted on a corpus of English conversational speech called the Buckeye corpus [15]. This corpus consists of 307,000 words collected through interviews with 40 speakers from central Ohio, USA, each interview lasting about 1 h. The proportions of gender and age of the speakers are equally balanced. The questions asked by the interviewers are of general topics to which the speakers have to answer based on their own opinion. Interviews are annotated with the orthographic transcription and each word is provided with two phonemic transcriptions: the standard pronunciation (*canonical phonemes*) and the one effectively uttered by the speaker (*realized phonemes*). Transcriptions have been automatically generated, manually checked and corrected.

In this work, 20 speakers from the Buckeye corpus are considered, other speakers being set aside for future work. They have been randomly selected under the constraint to maintain the age and gender proportions. Among selected speakers, the average number of phonemes per speaker is 22,789, and the average number of words is 7,354. As listed in Table 1, data has been enriched with additional information about utterances, words, stems, parts of speech (POS), syllables, and graphemes, forming a total of 22 linguistic features for each canonical phoneme. All frequencies have been grouped into three categories with equal probability masses (frequent/medium/rare), and stop words have been identified using a list of 500 words in English. Finally, canonical and realized phonemes have been automatically aligned using the Levenshtein distance. Alignments show that 30 % of the phonemes and 57 % of the words are pronounced differently from the standard pronunciation.

3   Method Overview

In this section, the proposed pronunciation adaptation method is described before introducing CRFs and the experimental setup.

3.1   Overall Method

The underlying problem in pronunciation adaptation is to predict the sequence of realized phonemes for a given utterance from an input sequence of canonical phonemes. As such, the adaptation accuracy is defined as an error rate between the realized and the predicted phonemes. Our method proposes to add information to canonical phonemes to improve this accuracy.

Mainly, every canonical phoneme has been augmented with a wide range of linguistic features and information about their neighborhood, i.e., surrounding canonical phonemes and their linguistic description. Linguistic features have been selected to discriminate those which enhance adaptation from useless and harmful others. Defining a neighborhood as a phoneme window centered on a canonical phoneme to be adapted, benefits of neighborhoods have been evaluated by adjusting the size of the window to optimize the adaptation accuracy.

Table 1. Features along with their number of votes for greedy backward, forward methods, and the sum of both.

Feature	Backward	Forward	Sum	
Canonical phoneme	20	20	40	
Word	20	20	40	
Is a stop word (true/false)	13	11	24	
Syllable lexical stress	11	13	24	
Syllable part (onset/nucleus/coda)	11	13	24	
Word frequency in English	7	15	22	
Reverse phoneme position in syllable	11	11	22	
Phoneme position in syllable	9	11	20	*50 %*
Syllable location (first/middle/last)	10	10	20	*strategy*
Stem frequency in the interview	11	8	19	
Word frequency in the interview	9	9	18	
Syllable type (open/close)	6	12	18	
POS	8	9	17	
Number of syllables of the word	8	9	17	
Stem frequency in English	7	9	16	
Grapheme	7	9	16	*Best 17*
Word length	8	5	13	*strategy*
Reverse utterance position	3	1	4	
Utterance position	2	1	3	
Word position	1	1	2	
Reverse word position	0	0	0	
Word occurrence count in interview	0	0	0	

In addition, contexts of the realized phonemes have been studied, the underlying question being whether predicting a given phoneme depends on the preceding predicted phoneme. The effect of cross-word information has also been investigated by comparing pronunciation adaptation when performed independently on each word of a given utterance or directly on all phonemes of all the words.

Linguistic feature selection and window size tuning have been performed in a first series of experiments, leading to the final method evaluated in backend experiments. Cross-dependencies over predicted phonemes and cross-word information have been evaluated in all experiments. Before presenting results of these experiments in Sects. 4 and 5, the remainder introduces the underlying probabilistic machine learning framework, CRFs, and the experimental setup.

3.2   Conditional Random Fields

CRFs are probabilistic models for labelling sequential data [12]. They model the conditional probability of a sequence of T labels $\mathbf{y} = (y_1, \ldots, y_T)$ given an input sequence of observations $\mathbf{x} = (x_1, \ldots, x_T)$ as follows:

$$\Pr(\mathbf{y}|\mathbf{x}) = \frac{1}{Z_\theta(\mathbf{x})} \exp\left(\sum_{t=1}^{T}\sum_{k=1}^{K} \theta_k f_k(y_{t-1}, y_t, x_t)\right), \tag{1}$$

where $Z_\theta(\mathbf{x})$ is a normalization factor, $\{f_k\}_{1 \leq k \leq K}$ are K so-called feature functions, $\{\theta_k\}_{1 \leq k \leq K}$ are their associated weights estimated on training data such that the error rate on a given development set is minimized.

Feature functions are a powerful mean to combine input information. They typically return 1 when the condition of the feature is met, 0 otherwise. An example of a condition in our case might be "the current input canonical phoneme x_t is /t/ *and* the output realized phoneme y_t is /?/". If desired, feature functions can also take advantage of the previous phoneme y_{t-1} to predict y_t. This configuration is referred to as *bigram* configuration, as opposed to *unigram* when only y_t is considered. Unigram and bigram features functions can be considered together (referred to as *uni+bigram* in the remainder). In order to test which configuration performs the best, all three have been tried for all the experiments in this work using the toolkit Wapiti [13].

Table 2. PERs and WERs (%) of canonical pronunciations (baseline error rates).

	PER on isolated words	PER on utterances	WER
Development set	30.4	30.3	57.0
Test set	30.5	30.3	57.2

3.3   Experimental Setup

Adaptation CRFs are trained and evaluated *independently for each speaker*. However, the objective is to determine a same set of features for any speaker adaptation. To do so, each interview is randomly divided into a training set (60 % of the utterances), a development set (20 %), and a test set (20 %). The linguistic feature selection and the window size tuning are performed on the development set while final experiments are conducted on the test set. For each speaker, the phoneme error rate (PER) and the word error rate (WER) are computed by comparing the realized phonemes with either the canonical phonemes (*baseline*) or those resulting from an adaptation. Mean error rates are then reported by averaging PERs and WERs over all the speakers. Baseline error rates on the development and test sets are shown in Table 2. PERs on isolated words and utterances are different since the computation on utterances is more tolerant of some errors. Typically, a deletion and an insertion may be merged into one substitution when computing cross-word error rates. These numbers can later be used to compare the different tested pronunciation adaptation configurations. These average error rates hide big differences across speakers: individual PERs range from 22.0 % to 39.8 %, and WERs from 45.0 % to 66.0 %. This disparity is a strong argument to perform pronunciation adaptation on a speaker basis rather than on all the speakers together, as capturing variations may be very difficult in the latter case. Finally, as stated in the introduction, no speech signal has been generated in this work, this for two reasons. First, the phoneme set used in the Buckeye corpus is more precise than the one supported by our TTS system, especially by including allophones. Synthesizing speech would then require to degrade the pronunciation precision, which could possibly erase some pronunciation variants. Second, results of perceptual tests may be biased by the style of the speech corpus on which the TTS system relies. Hence, this paper focuses on validating the approach through objective evaluations, leaving perceptual tests for future work.

4   Feature and Window Size Selection

In this section, the details of the linguistic feature and window size selections are given.

4.1   Linguistic Feature Selection

Training CRFs on too many features might result in overfitting the data. Therefore, it is important to reduce the number of features by removing less useful features and keeping only relevant ones. Moreover, it helps reducing the time and memory needed for training. For this purpose, a selection process has been applied on the development set. The basic idea of this process is to run an election over linguistic features by searching for the best feature set, i.e., the set with minimal PER, for each speaker. Features receive a vote each time they appear in the best set of some speaker. To make the selection process more robust, two selection schemes are considered, and votes for both schemes are finally added up. First, a greedy backward elimination was conducted where all features are considered at the beginning and features are eliminated one at a time until the best set is found. Second, a greedy forward selection was applied, i.e., the process starts with canonical phonemes as a unique feature and other features are added one at a time until the optimal set is found. This selection process has been carried out either disregarding or using cross-word information, i.e., on either isolated words or utterances, respectively.

Table 1 reports the total number of votes obtained by each feature over all speakers in the case of isolated words. Results are given for the backward and forward schemes, and when adding up votes from both methods. As a result, it appears that complementing canonical phonemes with information about the actual word is essential since this feature received the maximum number of votes (40). In the same trend, the status of the word in the language is also important. It can also be highlighted that syllable-based features are in the top of the list. These conclusions are consistent with previous studies [1,3,18].

Given the sorted list of features according to total votes, two strategies were tested to determine the feature set for the adaptation method. The first one considers all the features with at least 50 % of the speakers' votes, i.e., with 20 votes or more, while the second consists in choosing the threshold where a sharp decline is seen. Here, this strategy leads to select the best 17 features since the other worst 5 features received nearly no vote. Table 3 compares the PER and WER on the development set when ignoring linguistic features, i.e., only canonical phonemes are used, and when considering linguistic features selected with each strategy. Many conclusions can be drawn from these results. First, the selected features bring significant improvement over the baseline, whatever the CRF configuration (unigram, bigram or uni+bigram). This is all the more interesting since adaptation does not bring any improvement when CRFs are trained using unigrams and canonical phonemes only. Secondly, results on bigrams show that this configuration performs badly. This is probably due to data sparsity in the training set where only a limited number of realized phoneme bigrams can be observed. For all that, combining unigrams and bigrams leads to better results than the sole unigrams. Finally, the results of the two considered selection strategies are close. However the 50 % strategy leads to the lowest error rates and utilizes less features (9 against 17). Thus, the features selected by the 50 % strategy are chosen as the accepted set of features for the final experiments. In addition, as the bigram configuration does not provide any improvement, its results will not be reported in the following.

The same process was repeated on utterances after adding an explicit word boundary feature to keep track of the phoneme position inside their corresponding word. Very similar results were achieved, the only difference being the inclusion of the word boundary feature in the list of the best features. So the features used for training utterance-based models are the same as those selected by the 50 % strategy on isolated words plus this word boundary feature.

Table 3. PER and WER (%) without linguistic features and with features selected according to the 50 % and "Best 17" strategies. Absolute variations with the baseline are reported between brackets.

	No linguistic feature		50 % strategy		Best 17 strategy	
	PER	WER	PER	WER	PER	WER
Unigram	30.4 (0.0)	57.0 (-0.2)	24.7 (-5.7)	50.7 (-6.5)	24.4 (-6.0)	50.3 (-6.9)
Bigram	47.7 (+17.3)	82.3 (+25.1)	31.8 (+1.4)	59.3 (+2.1)	32.1 (+1.7)	59.7 (+2.5)
Uni+bigram	25.7 (-4.7)	50.1 (-7.1)	**24.1 (-6.3)**	**49.4 (-7.8)**	24.4 (-6.0)	50.2 (-7.0)

4.2 Window Size Selection

One important step apart from feature selection is to decide on the neighborhood scope around each canonical phoneme, that is determining the best suited size of canonical phoneme windows. These windows are centered on the canonical phoneme to be adapted. They are symmetrically[1] defined by the number W of the left and right hand surrounding phonemes. For instance, $W = \pm 2$ means that 2 neighbors from each side are considered along with the current canonical phoneme, hence considering 5 phonemes in total. The maximum value for W was set to ± 5.

Figure 1 presents PERs and WERs obtained without windows ($W = 0$) or with different window sizes, for both isolated words and utterances. CRFs were trained on unigram features, without any linguistic feature. First, results show that phoneme neighborhoods bring significant improvements. For both isolated words and utterances, results seem to converge after a given size is reached. However, convergence is slower for utterances and results are worse than for isolated words. The reason is probably that word boundaries are not known in this configuration, CRFs being only trained on cross-word canonical phonemes. As a conclusion, the window $W = \pm 2$ is considered in the backend experiments.

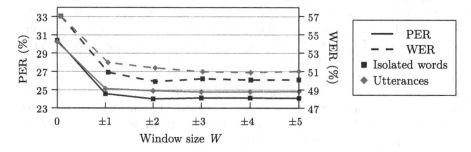

Fig. 1. PER and WER according to the window size, for isolated words and utterances.

5 Backend Experiments and Discussion

Experiments are carried out on the test set of all each speaker using canonical phonemes with or without linguistic features, with or without contextual windows, and on the basis of isolated words or utterances. Linguistic features and windows are those selected in Sect. 4. This section first presents the raw results before developing a deeper analysis.

Table 4 presents results for all combinations. First, configurations already evaluated on the development set lead to similar results, i.e., enriching canonical phonemes with either selected linguistic features or phoneme windows brings lower error rates; and the uni+bigrams configuration performs well in one or the other of the two settings. Then, new conclusions can be drawn when combining linguistic features and phoneme windows. On unigrams, it can be noticed that extra improvements are obtained, leading to the lowest PER for both isolated words and utterances. On isolated words, this improvement is small

---

[1] Asymmetric windows were also tested but they led to worse results.

(0.2 w.r.t. using the phoneme window only) but statistically significant[2]. On utterances, the improvement is large. A part of it is probably due to the inclusion of the word boundary information. Compared to canonical pronunciations, the PER relatively decreased by about 23 %, while the WER decreases by about 14 % in this case. On the contrary, when looking at the uni+bigram results, this combination degrades the results. We think that this is due to the too large number of parameters implied by this configuration in relation to the training set size, which leads to poor feature weight estimates. As a conclusion, overall results demonstrate that (i) the proposed pronunciation adaptation method clearly outperforms the baseline and results from the simplest CRFs, based on independent canonical phonemes, (ii) the inclusion of linguistic features is useful though the impact is small, and (iii) considering words in the context of their utterance does not lead to any improvement.

Table 4. PERs and WERs (%) on the test set for isolated words and utterances. Absolute variations with the baseline are reported between brackets.

			No window		$W = \pm2$	
			PER	WER	PER	WER
Isolated words						
Canonical phonemes (baseline)			30.5	57.2	–	–
Adapted phonemes based on	Canonical phonemes only	Unigram	30.4 (-0.1)	57.2 (0.0)	23.8 (-6.7)	49.5 (-7.7)
		Uni+bigram	25.5 (-5.0)	50.6 (-6.6)	24.0 (-6.5)	**48.8 (-8.4)**
	+ Ling. feat. (50% strat.)	Unigram	24.3 (-6.2)	50.0 (-7.2)	**23.6 (-6.9)**	49.2 (-8.0)
		Uni+bigram	24.1 (-6.4)	49.0 (-8.2)	24.2 (-6.3)	49.2 (-8.0)
Utterances						
Canonical phonemes (baseline)			30.3	57.2	–	–
Adapted phonemes based on	Canonical phonemes only	Unigram	30.2 (-0.1)	57.2 (0.0)	24.9 (-5.4)	51.8 (-5.4)
		Uni+bigram	25.9 (-4.4)	51.3 (-5.9)	24.2 (-6.1)	49.1 (-8.1)
	+ Ling. feat. (50% strat.)	Unigram	24.1 (-6.2)	50.0 (-7.2)	**23.4 (-6.9)**	48.9 (-8.3)
		Uni+bigram	23.9 (-6.4)	**48.7 (-8.5)**	24.4 (-5.9)	49.6 (-7.6)

To illustrate the results, Table 5 presents adapted pronunciation samples along with the realized and standard ones. First, it clearly appears that the standard pronunciation is very different from the realized one. Then, this example shows how adaptation only changes few phonemes, sometimes even none. Most of the time, these changes are deletions, substitutions with an allophone or simplifications of a diphthong into a monophthong. Nonetheless, adapted pronunciations are usually closer to spontaneous speech than the baseline, while still far from perfect. Overall, this example highlights one major difficulty in pronunciation adaptation: the way people speak is not deterministic and thus, in many cases, several pronunciations could be accepted for a same utterance and speaker. Consequently, error rates on single best hypotheses output by CRFs are probably not sufficient to measure how good is an adaptation model. In order to develop this analysis, extra measurements have been achieved.

---

[2] The p-values are 0.01037 and 0.008844 using a paired t-test and a paired Wilcoxon test, respectively, with a confidence level $\alpha = 0.05$.

Table 5. Pronunciations for the phrase "concentrated in Ohio". Presented adapted pronunciations have been generated on isolated words, using unigram features. Errors w.r.t. the realized pronunciations are marked in bold.

Realized phonemes		/kɑnsn̩ _tɹeɪ_ɪd·ɪř·oʊhɑ ʌ /
Canonical phonemes (baseline)		/kɑnsʌntɹeɪtʌd·ɪn·oʊhaɪoʊ/ (7 errors)
Adapted phonemes based on	canonical phonemes only	/kɑnsʌntɹeɪtʌd·ɪn·oʊhaɪoʊ/ (7 errors)
	+ linguistic feat.	/kɑnsʌntɹeɪtʌd·ɪn·oʊhaɪoʊ/ (7 errors)
	+ window	/kɑnsn̩n_ɹeɪtɪd·ɪn·oʊhaɪoʊ/ (6 errors)
	+ linguistic feat. + window	/kɑnsn̩n_ɹeɪɾɪd·ɪn·oʊhaɪoʊ/ (6 errors)

Table 6. Oracle PERs of n-best hypotheses on isolated words for n between 1 and 50.

n ▶		1	2	3	4	5	6	7	8	9	10	20	30	40	50
Canonical phonemes		30.4	23.0	19.0	16.3	14.8	13.5	12.4	11.7	11.0	10.5	7.8	6.6	6.0	5.6
+ ling. feat		24.3	17.1	13.8	11.8	10.4	9.5	8.8	8.3	7.8	7.3	5.3	4.5	4.0	3.6
	+ window	23.8	16.5	13.3	11.2	9.9	9.0	8.2	7.7	7.2	6.9	4.8	4.0	3.5	3.2
+ ling. feat	+ window	23.6	16.4	13.1	11.1	9.8	8.9	8.2	7.6	7.2	6.8	4.9	4.1	3.7	3.3

First, oracle PERs have been measured on the n-best hypotheses, i.e., only the best matching pronunciation is considered among the n generated by the CRF instead of the sole 1-best. Table 6 shows oracle PERs for different settings with n ranging from 1 to 50. As it can be seen, by only adding the 2-best hypothesis, the PER drops from 23.6 % to 16.4 % for the best performing configuration. The same trend is observed for all the other configurations. Then, results are improved as the number of hypotheses increases until apparently approaching a lower bound, which can be thought as "unpredictable", or at least very hardly predictable, pronunciation variants. Interestingly, this lower bound seems to be significantly higher for the most basic CRF (canonical phonemes only) than for others ($n = 50$). This tends to show that the latter models not only rerank phoneme probabilities but also introduce new adaptation possibilities.

Second, adaptation models have also been compared by measuring how well they can predict the realized pronunciations, that is how high is their probability. This can be achieved by computing the perplexities of the test set according to the different models, the lower perplexity the better. Perplexity is interesting since a model may assign a high probability to the realized pronunciations while not considering it as the most likely hypothesis though. As a consequence, there is no direct relation between perplexity and error rates. Perplexities over phonemes are presented in Table 7 for all the adaptation models. Evidently, the model based on canonical phonemes only and $W = 0$ achieves the highest perplexity. Other results on isolated words confirm the PER and WER results of Table 4 since the lowest perplexity is achieved by the combination of linguistic features and the window $W = \pm 2$. On utterances, the lowest perplexity is surprisingly achieved by the use of linguistic features without any window. This shows that linguistic features are relevant cues to predict pronunciation variants. Finally, perplexity can also be interpreted as a branching factor, i.e., the number of phonemes to be tested in the descending order of their probability before finding the realized

Table 7. Perplexity over phonemes of all the configurations. Relative variation w.r.t. the simplest CRFs (canonical phonemes only) are reported between brackets.

	Isolated words		Utterances	
	No window	$W = \pm 2$	No window	$W = \pm 2$
Canonical phonemes only	2.85	2.19 (-23%)	2.85	2.48 (-13%)
+ Linguistic features (50% strategy)	2.21 (-22%)	2.15 (-25%)	2.20 (-23%)	2.45 (-14%)

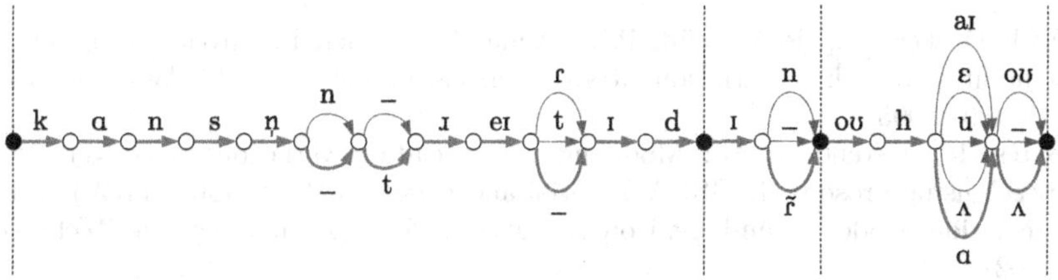

Fig. 2. Confusion network for "concentrated in Ohio". Black circles delimit words.

one. Numbers in Table 7 thus indicate that low oracle PERs could be achieved by only considering the few best phoneme predictions, typically the 2 or 3 best. For instance, Fig. 2 draws the minimal confusion network for the sample utterance "concentrated in Ohio" when using the best performing model. Edges are sorted according to their descending posterior probability from top to bottom. Realized phonemes are on the bold green edges. In this example, considering only the 3 best hypotheses for each phoneme would lead to an oracle PER of 5.6%. Such small confusion networks could be automatically using a very low static number of alternatives, and then post-processed or directly fed to a TTS system.

6   Conclusion and Future Work

This paper proposed a new CRF-based speaker pronunciation adaptation method for the purpose of spontaneous speech synthesis. While adapted pronunciations are significantly better than canonical ones, experiments on the Buckeye corpus demonstrate that including linguistic features contributes to achieving these good results. This work also shows that CRF features need to be selected since combining all possible features tends to decrease PER and WER gains produced by the adaptation. A deeper analysis of the results also showed that very low error rates could be achieved if considering alternative predictions.

Several tasks could be achieved in the future to improve the current work. First, results should be completed with speech synthesis experiments now that the current method has been validated by objective measures. The use of alternative hypotheses should be tested as well. Second, while this paper was focused on testing the relevance of the sole linguistic features, the proposed method could be enriched with phonetic features, e.g., phoneme aperture, manner, place of articulation, etc. These additional features could bring complementary information about how strong or weak a phoneme is. Finally, pronunciation adaptation should be used along with an automatic phonetizer. Especially, it would be interesting to couple the proposed approach with a CRF-based grapheme-to-phoneme converter and to interpolate probabilities returned by each of the components with the hope to outperform the quality of the finally adapted pronunciations.

References

1. Adda-Decker, M., de Mareüil, P.B., Adda, G., Lamel, L.: Investigating syllabic structures and their variation in spontaneous French. Speech Commun. **46**(2), 119–139 (2005)
2. Bates, R., Ostendorf, M.: Modeling pronunciation variation in conversational speech using prosody. In: ISCA Tutorial and Research Workshop (ITRW) on Pronunciation Modeling and Lexicon Adaptation for Spoken Language Technology (2002)
3. Bell, A., Brenier, J.M., Gregory, M., Girand, C., Jurafsky, D.: Predictability effects on durations of content and function words in conversational english. J. Mem. Lang. **60**(1), 92–111 (2009)
4. Bell, A., Jurafsky, D., Fosler-Lussier, E., Girand, C., Gregory, M., Gildea, D.: Effects of disfluencies, predictability, and utterance position on word form variation in english conversation. J. Acoust. Soc. Am. **113**(2), 1001–1024 (2003)
5. Chen, K., Hasegawa-Johnson, M.: Modeling pronunciation variation using artificial neural networks for English spontaneous speech. In: Proceedings of the Annual Conference of the International Speech Communication Association (Interspeech) (2004)
6. Dilts, P.C.: Modelling phonetic reduction in a corpus of spoken english using random forests and mixed-effects regression. Ph.D. thesis, University of Alberta (2013)
7. Fosler-Lussier, E., et al.: Multi-level decision trees for static and dynamic pronunciation models. In: Proceedings of the European Conference on Speech Communication and Technology (Eurospeech) (1999)
8. Giachin, E., Rosenberg, A., Lee, C.H.: Word juncture modeling using phonological rules for HMM-based continuous speech recognition. Proceedings of the IEEE International Conference on Acoustics, Speech and Signal Processing (ICASSP) **5**, 155–168 (1990)
9. Illina, I., Fohr, D., Jouvet, D.: Grapheme-to-phoneme conversion using conditional random fields. In: Proceedings of the Annual Conference of the International Speech Communication Association (Interspeech) (2011)
10. Karanasou, P., Yvon, F., Lavergne, T., Lamel, L.: Discriminative training of a phoneme confusion model for a dynamic lexicon in ASR. In: Proceedings of the Annual Conference of the International Speech Communication Association (Interspeech) (2013)
11. Kolluru, B., Wan, V., Latorre, J., Yanagisawa, K., Gales, M.J.F.: Generating multiple-accent pronunciations for TTS using joint sequence model interpolation. In: Proceedings of the Annual Conference of the International Speech Communication Association (Interspeech) (2014)
12. Lafferty, J., McCallum, A., Pereira, F.C.: Conditional random fields: probabilistic models for segmenting and labeling sequence data (2001)

13. Lavergne, T., Cappé, O., Yvon, F.: Practical very large scale CRFs. In: Proceedings of the 48th Annual Meeting of the Association for Computational Linguistics (2010)
14. Lecorvé, G., Lolive, D.: Adaptive statistical utterance phonetization for French. In: Proceedings of IEEE International Conference on Acoustics, Speech and Signal Processing (ICASSP) (2015)
15. Pitt, M.A., Johnson, K., Hume, E., Kiesling, S., Raymond, W.: The Buckeye corpus of conversational speech: labeling conventions and a test of transcriber reliability. Speech Commun. **45**(1), 89–95 (2005)
16. Prahallad, K., Black, A.W., Mosur, R.: Sub-phonetic modeling for capturing pronunciation variations for conversational speech synthesis. In: Proceedings of IEEE International Conference on Acoustics, Speech and Signal Processing (ICASSP), vol. **1** (2006)
17. Tajchman, G., Foster, E., Jurafsky, D.: Building multiple pronunciation models for novel words using exploratory computational phonology. In: Proceedings of the European Conference on Speech Communication and Technology (Eurospeech) (1995)
18. Vazirnezhad, B., Almasganj, F., Ahadi, S.M.: Hybrid statistical pronunciation models designed to be trained by a medium-size corpus. Comput. Speech Lang. **23**(1), 1–24 (2009)
19. Wang, D., King, S.: Letter-to-sound pronunciation prediction using conditional random fields. IEEE Signal Process. Lett. **18**(2), 122–125 (2011)

Weakly Supervised Discriminative Training of Linear Models for Natural Language Processing

Lina Maria Rojas-Barahona[1] and Christophe Cerisara[2]([⊠])

[1] Université de Lorraine/LORIA, Nancy, France
`lina.rojas@loria.fr`
[2] CNRS/LORIA, Nancy, France
`christophe.cerisara@loria.fr`

Abstract. This work explores weakly supervised training of discriminative linear classifiers. Such features-rich classifiers have been widely adopted by the Natural Language processing (NLP) community because of their powerful modeling capacity and their support for correlated features, which allow separating the expert task of designing features from the core learning method. However, unsupervised training of discriminative models is more challenging than with generative models. We adapt a recently proposed approximation of the classifier risk and derive a closed-form solution that greatly speeds-up its convergence time. This method is appealing because it provably converges towards the minimum risk without any labeled corpus, thanks to only two reasonable assumptions about the rank of class marginal and Gaussianity of class-conditional linear scores. We also show that the method is a viable, interesting alternative to achieve weakly supervised training of linear classifiers in two NLP tasks: predicate and entity recognition.

1 Introduction

Unsupervised training of discriminative models poses serious theoretical issues, which prevent such models from being widely adopted in tasks where annotated corpora do not exist. In such cases, generative models are thus often preferred. Nevertheless, discriminative models have various advantages that might be desirable even in these cases, for example their very interesting capacity to handle correlated features and to be commonly equipped with many rich features. Hence, many efforts have been deployed to address this issue, and some unsupervised training algorithms for discriminative models have been proposed in the Natural Language Processing (NLP) community, for instance Unsearn [3], Generalized Expectation [4] or Contrastive Training [15] amongst others.

Our approach[1] relies on a novel approximation of the risk of binary linear classifiers proposed in [1]. This approximation relies on only two assumptions:

---

[1] This work has been partly funded by the ANR ContNomina project.

© Springer International Publishing Switzerland 2015
A.-H. Dediu et al. (Eds.): SLSP 2015, LNAI 9449, pp. 242–254, 2015.
DOI: 10.1007/978-3-319-25789-1_23

the rank of class marginal is assumed to be known, and the class-conditional linear scores are assumed to follow a Gaussian distribution. Compared to previous applications of unsupervised discriminative training methods to NLP tasks, this approach presents several advantages: first, it is proven to converge towards the true optimal classifier risk; second, it does not require any constraint; third, it exploits a new type of knowledge about class marginal that may help convergence towards a relevant solution for the target task. But the original approach in [1] has a very high computational complexity, and we investigate in this work two options to reduce this issue: we first derive a closed-form expression of the objective function, which allows a much faster algorithmic implementation. We also propose to pre-train the classifier on a very small amount of data to speed-up convergence. Finally, we validate the proposed approach on two new binary NLP tasks: predicate identification and entity recognition.

2   Classifier Risk Approximation

We first briefly review the approximation of the risk proposed in [1]. A binary linear classifier associates a score $f_{\theta^0}(X) = \sum_i^{N_f} \theta_i X_i$ (resp $f_{\theta^1}(X) = -f_{\theta^0}(X)$) to the first class 0 (resp 1) for any input $X = (X_1, \cdots, X_{N_f})$ composed of N_f features X_i. The parameter $\theta_i \in \mathbb{R}$ represents the weight of the feature indexed by i for class 0. In the following, we may use both notations $f_{\theta^0}(X)$ or $f_\theta(X)$ equivalently. X is classified into class 0 iff $f_{\theta^0}(X) \geq 0$, otherwise X is classified into class 1. The objective of training is to minimize the classifier risk:

$$R(\theta) = E_{p(X,Y)}[\mathcal{L}(Y, f_\theta(X))] \tag{1}$$

where Y is the true label of the observation X, and $\mathcal{L}(Y, f_\theta(X))$ is the loss function, such as the hinge loss used in SVMs, or the log-loss used in CRFs. This risk is often approximated by the empirical risk that is computed on a labeled training corpus. In the absence of labeled corpus, an alternative consists in deriving the true risk as follows:

$$R(\theta) = \sum_{y \in \{0,1\}} P(y) \int_{-\infty}^{+\infty} P(f_\theta(X) = \alpha | y) \mathcal{L}(y, \alpha) d\alpha \tag{2}$$

We use next the following hinge loss:

$$\mathcal{L}(y, \alpha) = (1 + \alpha_{1-y} - \alpha_y)_+ \tag{3}$$

where $(x)_+ = \max(0, x)$, and $\alpha_y = f_{\theta^y}(X)$ is the linear score for the correct class y. Similarly, $\alpha_{1-y} = f_{\theta^{1-y}}(X)$ is the linear score for the wrong class.

Given y and α, the loss value in the integral can be computed easily. Two terms in Eq. 2 remain: $P(y)$ and $P(f_\theta(X) = \alpha | y)$. The former is the class marginal and is assumed to be known. The latter is the class-conditional distribution of the linear scores, which is assumed to be normally distributed. This implies that $P(f_\theta(X))$ is distributed as a mixture of two Gaussians (GMM):

$$P(f_\theta(X)) = \sum_{y \in \{0,1\}} P(y) \mathcal{N}(f_\theta(X); \mu_y, \sigma_y)$$

where $\mathcal{N}(z; \mu, \sigma)$ is the normal probability density function. The parameters $(\mu_0, \sigma_0, \mu_1, \sigma_1)$ can be estimated from an unlabeled corpus $\mathcal{U}$ using a standard Expectation-Maximization (EM) algorithm for GMM training. It is then possible to compute the integral in Eq. 2 and thus an estimate $\hat{R}(\theta)$ of the risk without relying on any labeled corpus. The authors of [1] prove that:

- $\hat{R}(\theta)$ converges towards the true risk $R(\theta)$;
- The estimated optimum converges towards the true optimal parameters, when the size of the unlabeled corpus $\mathcal{U}$ increases infinitely:

$$\lim_{|\mathcal{U}| \to +\infty} \arg\min_\theta \hat{R}(\theta) = \arg\min_\theta R(\theta)$$

They further prove that this is still true even when the class priors $P(y)$ are not known precisely, but only their relative order (rank) is known. These priors must also be different $P(y = 0) \neq P(y = 1)$.

3 Risk Minimization Algorithm

Given the estimated Gaussian parameters, the authors of [1] use numerical integration to compute Eq. 2. But numerical integration only gives an approximate integral and requires some tuning to balance between accuracy and computation time. We thus propose next a closed-form derivation of the risk that computes the exact integral at a lower cost.

3.1 Closed-Form Risk Estimate

Figure 1 summarizes the main steps of our proposed derivation of the risk. It exploits the following Gaussianity assumptions:

$$P(f_{\theta^i}(X)|Y = j) \sim \mathcal{N}(X; \mu_{j,i}, \sigma_{j,i}) \quad \forall i, j \in \{0, 1\}$$

The final risk for any binary linear classifier with the hinge loss in Eq. 3 is then:

$$\hat{R}(\theta) = P(Y = 0)\frac{1 - 2\mu_{0,0}}{4\sigma_{0,0}\sqrt{\pi}}\left(1 + \mathrm{erf}\left(\frac{\frac{1}{2} - \mu_{0,0}}{\sigma_{0,0}}\right)\right) + \frac{P(Y = 0)}{2\pi}\exp\left(-\frac{(\frac{1}{2} - \mu_{0,0})^2}{\sigma_{0,0}^2}\right)$$

$$+ P(Y = 1)\frac{1 + 2\mu_{1,0}}{4\sigma_{1,0}\sqrt{\pi}}\left(1 - \mathrm{erf}\left(\frac{-\frac{1}{2} - \mu_{1,0}}{\sigma_{1,0}}\right)\right) + \frac{P(Y = 1)}{2\pi}\exp\left(-\frac{(-\frac{1}{2} - \mu_{1,0})^2}{\sigma_{1,0}^2}\right)$$

$$(4)$$

In the following experiments, we evaluate the gain in computation time and accuracy resulting from the use of Eq. 4 by comparing it with a numerical integration algorithm, which is applied to compute the integrals in step 2 of Fig. 1.

Step 1 — Given $\alpha_1 = -\alpha_0$, exploit the constraints: $\mu_{0,1} = -\mu_{0,0}, \mu_{1,1} = -\mu_{1,0}, \sigma_{0,1} = \sigma_{0,0}$ to reduce double integrals to a single integral:

$$R(\theta) = P(Y = 0)\int_{-\infty}^{+\infty} N(\alpha_0; \mu_{0,0}, \sigma_{0,0})N(-\alpha_0; \mu_{0,1}, \sigma_{0,1})(1 - 2\alpha_0)_+d\alpha_0 +$$

$$P(Y = 1)\int_{-\infty}^{+\infty} N(\alpha_0; \mu_{1,0}, \sigma_{1,0})N(-\alpha_0; \mu_{1,1}, \sigma_{1,1})(1 + 2\alpha_0)_+d\alpha_0$$

Step 2 — Change integral boundaries to remove discontinuities:

$$R(\theta) = P(Y = 0)\int_{-\infty}^{\frac{1}{2}} N(\alpha_0; \mu_{0,0}, \sigma_{0,0})^2(1 - 2\alpha_0)d\alpha_0 +$$

$$P(Y = 1)\int_{-\frac{1}{2}}^{+\infty} N(\alpha_0; \mu_{1,0}, \sigma_{1,0})^2(1 + 2\alpha_0)d\alpha_0$$

Step 3 — Develop, simplify with $N(x; \mu, \sigma)^2 = \frac{1}{2\sigma\sqrt{\pi}}N(x; \mu, \frac{\sigma}{\sqrt{2}})$

and $x\exp\left(-\frac{(x-\mu)^2}{2\sigma^2}\right) = \mu\exp\left(-\frac{(x-\mu)^2}{2\sigma^2}\right) - \sigma^2\frac{\partial\exp\left(-\frac{(x-\mu)^2}{2\sigma^2}\right)}{\partial x}$

$$R(\theta) = P(Y = 0)\frac{1 - 2\mu_{0,0}}{2\sigma_{0,0}\sqrt{\pi}}\int_{-\infty}^{\frac{1}{2}} N(x; \mu_{0,0}, \frac{\sigma_{0,0}}{\sqrt{2}})dx + \frac{P(Y = 0)}{2\pi}\exp\left(-\frac{(\frac{1}{2} - \mu_{0,0})^2}{\sigma_{0,0}^2}\right) +$$

$$P(Y = 1)\frac{1 + 2\mu_{1,0}}{2\sigma_{1,0}\sqrt{\pi}}\int_{-\frac{1}{2}}^{+\infty} N(x; \mu_{1,0}, \frac{\sigma_{1,0}}{\sqrt{2}})dx + \frac{P(Y = 1)}{2\pi}\exp\left(-\frac{(-\frac{1}{2} - \mu_{1,0})^2}{\sigma_{1,0}^2}\right)$$

Step 4 — Integrate with the error function to obtain Eq-4

Fig. 1. Main steps of the proposed derivation; details available at authors' website.

1: Initialize the weights by training the linear classifier on N annotated sentences
2: **for** every iteration t **do**
3:     **for** every feature index i **do**
4:         Move temporary the weight $\theta_i(t) = \theta_i(t - 1) + \epsilon$
5:         Compute the linear scores on the corpus $\mathcal{U}$
6:         Apply EM to train the 4 Gaussian parameters
7:         Compute the risk $\hat{R}_+(\theta(t))$ with Eq. 4
8:         Similarly move temporary the weight in the other direction to compute $\hat{R}_-(\theta(t))$
9:         Compute the gradient with finite difference and update the weights according to this gradient
10:     **end for**
11: **end for**

Fig. 2. Overview of the training algorithm

3.2   Weakly Supervised Algorithm

Our weakly supervised training algorithm (Fig. 2) implements a coordinate gradient descent with finite difference.

n the training algorithm, only initialization exploits a few (we experimented with $N = 10$ and $N = 20$ annotated sentences) annotated data: the main iterations (from 2 to 11) do not use any supervised label at all. This initialization incurs no significant additional costs and provides a good enough starting point for the optimization algorithm, which can thus converge in a reasonable amount of time. Our preliminary experiments with random initialization actually did not converge even after several weeks on a single computer, which supports the principle of using a small supervised initialization in real NLP applications.

4   Target Tasks

We validate next our proposed weakly-supervised training algorithm on two NLP tasks: predicate identification and entity recognition.

4.1   Task 1: Predicate Identification

We consider the task of identifying the semantic predicates in a sentence, which constitutes the first stage in any semantic role labeling (SRL) system. We thus adapt the state-of-the-art supervised MATE SRL system [2], which exploits a linear classifier for predicate identification, by training this classifier with our proposed weakly-supervised algorithm.

We use the parallel Europarl corpus, CLASSIC [13], which contains 1000 French sentences from the European parliament that have been manually annotated with semantic roles. The MATE system relies on syntactic features, which are computed as follows:

- **Part of speech (POS) tags** are automatically computed with the Treetagger [14].
- **Dependency trees** are automatically produced by a parser trained on the French Treebank, as described in [13].

The initial weights are obtained after training the linear classifier on 10 manually annotated sentences (i.e., $N = 10$ in the algorithm presented in Sect. 3.2). We set the label priors $P(y)$ so that 80 % of the words are not predicates. This ratio is only based on our intuition, and is not very accurate: additional experiments suggest that the non-predicate class marginal should actually be larger. However, such approximation errors should not impact too much the performances, at least theoretically.

4.2   Task 2: Entity Recognition

The goal of this task is to detect whether any word form in a text refers to *an entity* or not, where an *entity* is defined as a mention of a person name, a place, an organization or a product. We use the ESTER2 corpus [5], which collects broadcast news transcriptions in French that are annotated with named entities. The following features are used to train the Stanford supervised linear classifier of the Stanford NLP toolkit[2]:

---

[2] http://nlp.stanford.edu/nlp.

- **Character n-grams** with $n = 4$.
- **Capitalization**: the pattern "Chris2useLC", as defined in Stanford NLP, describing lower, upper case and special characters in words [10].
- **POS tags**: the POS tag of every word as given by the Treetagger [14].

The POS tags as well as capitalization of words are common important features for entity recognition, while character n-grams constitute a smoother (less sparse) alternative to word forms and are also often used in this context. The label priors $P(y)$ are set so that 90 % of the words are not entities and only 10 % of the words are entities. The initial weights are obtained after training the linear classifier on 20 manually annotated sentences (i.e., $N = 20$ in the algorithm presented in Sect. 3.2). The same set of features is used both in the supervised initialization and the weakly supervised risk minimization iterations.

5 Results and Discussion

5.1 On the Gaussianity Assumption

The proposed approach assumes that the class-conditional linear scores are distributed normally. We invite the interested reader to read [1], where theoretical arguments are given that support the validity of this assumption in various contexts. However, this assumption can not always be taken for granted, and the authors suggest to verify it empirically.

Figure 3(a) thus shows the distribution of $f_\theta(X)$ for the first task with the initial weights on the CLASSIC corpus. We can observe that this distribution can be well approximated by two Gaussians (one for each Y), which confirms the validity of the Gaussianity assumption in this case. For this task, we have also tracked during the whole search the Kurtosis metric as a rough approximate measure of Gaussianity, but we have not observed any strong variation of this measure, which suggests that the distribution of the scores does not vary too much during the search.

Likewise for the second task, the distributions of $f_\theta(X)$ with the initial and final weights (i.e. the weights obtained after training) on the ESTER2 corpus are shown in Fig. 3(b) and (c) respectively. These distributions are clearly bi-normal on this corpus, which suggest that this assumption is reasonable in both our NLP tasks.

5.2 Study of the Risk Estimator

We now study the risk estimate for oracle parameters that have been trained with an increasing number of supervised labels. The objective is to check whether decreasing Eq. 4 may lead to parameters that are close to the supervised ones. This is an important question for every unsupervised model, where the resulting solution is optimal only with respect to the chosen features and given constraints. Because of their "implicit" definition, the resulting clusters may indeed differ from the expected ones.

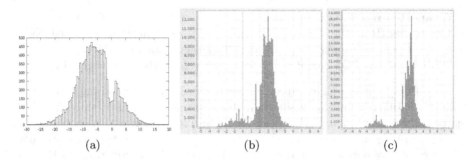

(a)   (b)   (c)

Fig. 3. Distribution of $f_\theta(X)$ on (a) the CLASSIC corpus, using the initial weights trained on 10 sentences; (b) the ESTER2 corpus, using the initial weights trained on 20 sentences; (c) the ESTER2 corpus, using the weights at iteration 6340 of the gradient descent algorithm. The largest mode is on the left in (a), because the *predicate* class is 0, while the largest mode is on the right in (b) and (c) because the *entity* class is 1.

Figure 4 shows the risk computed with Eq. 4 on parameters trained in a supervised way on a larger and larger labeled corpus, both with MATE for task 1 and the Stanford software for task 2.

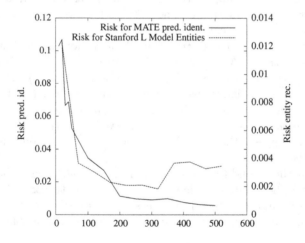

Fig. 4. $\hat{R}(\theta)$ for both supervised classifiers (MATE SRL and Stanford linear) as a function of the number of annotated sentences in the training corpus

We can observe that the risk estimate globally decreases when the linear classifier is trained on more annotated data, which confirms that the proposed risk correlates nicely with the expected clusters given the chosen features. However, the minimum risk on the curve of the second task is not located at the rightmost x-value, but rather at about 300 sentences. This suggests that, on the second task, the risk optimization algorithm will fail to return the best performing parameters. This limitation may be addressed by choosing input features that are better correlated to the target clustering into entities.

5.3 Experiments with Gradient Descent

We now apply the optimization algorithm described in Fig. 2.

Task 1: predicate identification. For task 1, the risk is estimated only with the closed-form Eq. 4. We start from initial weights trained on 10 sentences and then apply the gradient descent algorithm for 10,000 iterations. The results obtained are shown in Table 1 (top half).

Table 1. Performances of the proposed weakly supervised system in both tasks.

Task 1			
System	F1	precision	recall
MATE trained on 10 sent	64.8%	72.1%	58.9%
MATE trained on 500 sent	87.2%	92.0%	82.9%
Weakly supervised	**73.1%**	63.1%	**87.1%**
Task 2			
System	F1	precision	recall
Stanford trained on 20 sent	77.4%	89.8%	68%
Stanford trained on 520 sent	87.5%	90.3%	84.7%
Graph-based Semi.Sup.(MAD)	64.7%	2.37%	4.57%
Weakly sup. closed-form risk	**83.5%**	88.9%	**78.7%**
Weakly sup. numerical integration	**83.6%**	88.7%	**79%**

We can observe that, after the unsupervised optimization iterations, the F1-measure of predicate identification increases, but does not reach the F1 of the supervised model that is trained on 500 sentences. This is due to the existence of weights that give a lower risk than the supervised weights. These results may thus be improved by considering better features or including other constraints during the search.

An analysis of the resulting sentences shows that the inferred classifier has learnt a linguistic property that was not captured by the initial models trained on 10 sentences: verbs without subject can also be predicate. Hence, Fig. 5 shows an example of a sentence in the corpus with 2 predicates: the verbs *suffisent* (are enough) and *faire* (to do). The initial MATE model only identified the main verb, but missed its complement because it does not have any explicit subject.

We have noted that this is a typical mistake realized by the initial model, which results from the fact that the model has only been trained on 10 sentences without any occurrence of a similar pattern. But after unsupervised training, the same model is able to capture also the predicate without subject. This results from a better clustering of similar patterns when it is computed in an unsupervised way on a larger corpus.

Fig. 5. Example of sentence with 2 predicates that are identified by the weakly-supervised model but not by the initial model

Task 2: Entity Recognition. For task 2, experiments are realized both with the closed-form risk estimation in Eq. 4 and numerical integration. For numerical integration, we have made preliminary experiments with both the trapezoidal and Monte Carlo methods [7], and have chosen the former because it was more efficient in our experimental setup. The final performance figures are shown in Table 1 (bottom part), while Fig. 6 shows the convergence of optimization with the closed-form risk estimate. According to our experiments, both the closed-form risk and numerical integration reach the same performances (the differences shown in Table 1 are not statistically significant): after 2,000 iterations, the F1-measure is 83.5 %. Therefore, when evaluated on a test set of 167,249 words and 10,693 sentences, both methods outperform the supervised linear classifier trained on 20 sentences.

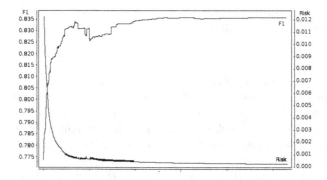

Fig. 6. F1 and $\hat{R}(\theta)$ (from Eq. 4) for entity detection, in function of the number of iterations (step 2 of Fig. 2), up to 6340 outer iterations.

In general the proposed model is prone to detect person names that are undetected by the baseline (i.e., the Stanford linear classifier trained on 20 sentences). Table 2 shows two examples of family names (e.g., Drouelle and Floch-Prigent) that are correctly recognized by our model but ignored by the baseline.

Table 2. Excerpt of examples correctly classified by the weakly supervised approach for entity recognition, improving the baseline (i.e. the Stanford linear classifier trained on 20 sentences). The last column shows the output probability of the winning class.

-	Baseline (Sup. on 20 sents)		Proposed model	
Word	Class	Prob.	Class	Prob.
Fabrice	Entity	0.94	Entity	**0.99**
Drouelle	NO	0.53	**Entity**	**0.79**
Floch-Prigent	NO	0.58	**Entity**	**0.69**
Iran	Entity	0.66	Entity	**0.82**
F16	NO	0.73	**Entity**	**0.91**

Our model also correctly detects entities other than person names, such as the aircraft F16, which are not captured by the initial model. Note also that for the first name *Fabrice* and the country *Iran*, the unsupervised model correctly augments their probabilities (where the probabilities correspond to the normalized scores $f_\theta(X)$ given by the model) to belong to the class entity.

5.4 Closed-Form vs. Numerical Integration

This comparison is realized for task 2 to assess the relative gain in terms of computational costs when using closed-form integration. Figure 7 shows three curves, in function of the chosen setting used for numerical integration (number of trapezoids):

- The downward curve is the error: $|\hat{R}(\theta) - \hat{R}_{num}(\theta)|^{1/2}$ between the risk estimated respectively with the closed-form Eq. 4 and trapezoidal integration. We use the root-square of the approximation error to better view the details, because the trapezoidal method is known to converge in $O(n^{-2})$.
- The horizontal line is the computation time (in seconds) required to compute the risk with Eq. 4; it obviously does not depend on the number of trapezoids.
- The rising line is the computation time required to compute the risk with numerical integration: it increases linearly as a function of the number of parameters used for numerical integration.

We can observe that increasing the number of trapezoids also increases the accuracy of numerical integration, and that the approximation error becomes smaller than 10 % of the risk value for 20 trapezoids and more. This corresponds to a computational cost that is about 6 times higher for numerical integration than for closed-form estimate.

We can conclude that the advantage of the proposed closed-form solution is twofold: it reduces the required computation time by at least a factor of 6 while providing a better estimate of the risk, as compared to the original algorithm. Note however that we have not observed in our experiments any impact of this exact solution in terms of F1-measure.

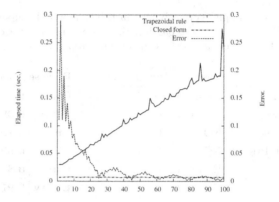

Fig. 7. Computational cost and approximated error of the trapezoidal rule with regard to the number of trapezoids (segments) used for approximating the integrals of the risk.

6   Related Work

A number of previous works have already proposed approaches to train discriminative models without or with few labels. Please refer, e.g., to [6,9] for a general and theoretical view on this topic. For NLP tasks several approaches have also been proposed. Hence, the traditional self- and co-training paradigm can be used to leverage supervised classifiers with unsupervised data [8,12]. In [4] exploit the Generalized Expectation objective function, which penalizes the mismatch between model predictions and linguistic expectation constraints. In contrast, our proposal does not use any manually defined prototype nor linguistic constraint.

Another interesting approach is *Unsearn* [3], which predicts the latent structure Y and then a corresponding "observation" $\hat{X}$, with a loss function that measures how well $\hat{X}$ predicts X. This method is very powerful and generalizes the EM algorithm, but its performances heavily depend on the quality of the chosen features set for discriminating between the target classes. A related principle is termed "Minimum Imputed Risk" in [11] and applied to machine translation. Our proposed approach also depends on the chosen features, but in a less crucial way thanks to both new assumptions, respectively the known label priors and discrimination of classes based on individual Gaussian distributions of scores. Another interesting generalization of EM used to train log-linear models without labels is *Contrastive Estimation*, where the objective function is modified to locally remove probability mass from implicit negative evidence in the neighborhood of the observations and transfer this mass onto the observed examples [15].

Comparatively, the main advantage of our proposed approach comes from the fact that the algorithm optimizes the standard classifier risk, without any modification nor constraint. The objective function (and related optimal parameters) is thus the same as in classical supervised training.

7 Conclusion

This work investigates the applicability of a novel framework to train linear classifiers without labels to the NLP domain. It is validated on two binary tasks, namely predicate and entity recognition. We show that convergence can only be obtained in a reasonable amount of time when initializing the classifier with good-enough values, which are obtained with supervised training on a very small amount of annotated data. We also show that the main assumption of the approach, i.e., gaussianity of the class-conditional distributions of the linear scores, is fulfilled in both our tasks. We finally propose and derive a closed-form expression of the risk estimator for a binary linear classifier, which reduces the algorithmic complexity of the proposed implementation. An interesting extension of the current approach would be to further consider some penalization term for non-Gaussian conditional distributions of the linear scores, in order to guarantee that this assumption is preserved during the whole optimization process. We also plan to work on the optimization algorithm to further reduce its complexity and thus make it applicable to a wider range of applications. This shall also involve generalizing the risk derivation to multiclass classifiers.

References

1. Balasubramanian, K., Donmez, P., Lebanon, G.: Unsupervised supervised learning II: margin-based classification without labels. J. Mach. Learn. Res. **12**, 3119–3145 (2011)
2. Björkelund, A., Hafdell, L., Nugues, P.: Multilingual semantic role labeling. In: Proceedings of CoNLL: Shared Task, pp. 43–48. Stroudsburg, PA, USA (2009)
3. Daumé III, H.: Unsupervised search-based structured prediction. In: Proceedings of ICML, Montreal, Canada (2009)
4. Druck, G., Mann, G., McCallum, A.: Semi-supervised learning of dependency parsers using generalized expectation criteria. In: Proceedings of ACL, pp. 360–368. Suntec, Singapore, August 2009
5. Galliano, S., Gravier, G., Chaubard, L.: The ester 2 evaluation campaign for the rich transcription of french radio broadcasts. In: Proceedings of INTERSPEECH, pp. 2583–2586 (2009)
6. Goldberg, A.B.: New directions in semi-supervised learning. Ph.D. thesis, University of Wisconsin-Madison (2010)
7. Gould, H., Tobochnik, J.: An Introduction to Computer Simulation Methods: Applications to Physical Systems. Addison-Wesley, Series in physics (1988)
8. Kaljahi, R.S.Z.: Adapting self-training for semantic role labeling. In: Proceedings Student Research Workshop, ACL, pp. 91–96. Uppsala, Sweden, July 2010
9. Kapoor, A.: Learning Discriminative Models with Incomplete Data. Ph.D. thesis, Massachusetts Institute of Technology, February 2006
10. Klein, D., Smarr, J., Nguyen, H., Manning, C.: Named entity recognition with character-level models. In: Proceedings of CoNLL, pp. 180–183. Stroudsburg, USA (2003)
11. Li, Z., Wang, Z., Eisner, J., Khudanpur, S., Roark, B.: Minimum imputed-risk: unsupervised discriminative training for machine translation. In: Proceedings of EMNLP, pp. 920–929 (2011)

12. Liu, X., Li, K., Zhou, M., Xiong, Z.: Enhancing semantic role labeling for tweets using self-training. In: Proceedings of AAAI, pp. 896–901 (2011)
13. van der Plas, L., Samardžić, T., Merlo, P.: Cross-lingual validity of propbank in the manual annotation of french. In: Proceedings of the Fourth Linguistic Annotation Workshop, ACL. pp. 113–117. Uppsala, Sweden, July 2010
14. Schmid, H.: Improvements in part-of-speech tagging with an application to german. In: Proceedings of the Workshop EACL SIGDAT, Dublin (1995)
15. Smith, N.A., Eisner, J.: Unsupervised search-based structured prediction. In: Proceedings of ACL (2005)

Merging of Native and Non-native Speech for Low-resource Accented ASR

Sarah Samson Juan[1]([✉]), Laurent Besacier[2], Benjamin Lecouteux[2], and Tien-Ping Tan[3]

[1] Faculty of Computer Science and Information Technology,
Universiti Malaysia Sarawak, Kota Samarahan, Sarawak, Malaysia
sjsflora@unimas.my
[2] Grenoble Informatics Laboratory (LIG),
University Grenoble-Alpes, Grenoble, France
{laurent.besacier,benjamin.lecouteux}@imag.fr
[3] School of Computer Science, Universiti Sains Malaysia, Gelugor, Penang, Malaysia
tienping@cs.usm.my

Abstract. This paper presents our recent study on low-resource automatic speech recognition (ASR) system with accented speech. We propose multi-accent Subspace Gaussian Mixture Models (SGMM) and accent-specific Deep Neural Networks (DNN) for improving non-native ASR performance. In the SGMM framework, we present an original language weighting strategy to merge the globally shared parameters of two models based on native and non-native speech respectively. In the DNN framework, a native deep neural net is fine-tuned to non-native speech. Over the non-native baseline, we achieved relative improvement of 15 % for multi-accent SGMM and 34 % for accent-specific DNN with speaker adaptation.

Keywords: Automatic speech recognition · Cross-lingual acoustic modelling · Non-native speech · Low-resource system · Multi-accent SGMM · Accent-specific DNN

1 Introduction

Performance of non-native automatic speech recognition (ASR) is poor when few (or no) non-native speech is available for training / adaptation. Many approaches have been suggested for handling accented-speech in ASR, such as acoustic model merging [2,16,22,23], applying maximum likelihood linear regression (MLLR) for adapting models to each non-native speaker [8], or adapting lexicon [1,4].

Lately, Subspace Gaussian Mixture Models (SGMMs) [17,18] have shown to be very promising for ASR in limited training conditions (see [11,13]). In SGMM modelling, the acoustic units are all derived from a common GMM called the Universal Background Model (UBM). This UBM, which in some way represents the acoustic space of the training data, can be estimated on large amount of

A.-H. Dediu et al. (Eds.): SLSP 2015, LNAI 9449, pp. 255–266, 2015.
DOI: 10.1007/978-3-319-25789-1_24

untranscribed data from one or several languages. The globally shared parameters do not need the knowledge about the phone units used in the source language(s). Without this constraint of source-target mapping of acoustic units, the UBM can be well used in cross-lingual or multilingual (multi-accent) settings.

In the mean time, Deep Neural Networks (DNNs) have been increasingly employed for building efficient ASR systems. HMM/DNN hybrid systems clearly outperform HMM/(S)GMM systems for many ASR tasks [6] which include dealing with low-resource systems [9,14,25]. Several studies have shown that multilingual DNNs can be achieved by utilizing multilingual data for conducting unsupervised pretraining [21] or training the whole network simultaneously [5,9,25].

In the above techniques, acoustic model merging can easily be conducted through sharing the UBMs (for SGMM) and hidden layers (for DNN) with other systems. But what is the optimal way to do so? Can we merge a large amount of native speech with a small quantity of non-native data? This paper tries to respond to these questions using both SGMM (less efficient than DNNs but more compact for embedded applications) and DNN (state-of-the-art) frameworks. We apply our methods to Malaysian English ASR, where a large amount of native (English) data is available (TED-LIUM corpus [20]), while only 2h of non-native speech is available. More precisely, we propose one strategy for each framework: (1) language weighting for multi-accent SGMMs and (2) accent-specific top layer for DNN. The first strategy is novel and involves manipulating the number of Gaussians of each native / non-native model for (multi-accent) UBM merging. In the second approach, we build accent-specific DNN similarly to last year's work of [10] but we make it work for a very low-resource setting and with speaker adaptation on top of it.

The rest of the paper is organized as follows. In Sect. 2 we describe the background of SGMM and DNN as well as their application to multilingual and multi-accent ASR. Section 3 presents the experimental setup for building native and non-native systems as well as the results of our baselines. In Sects. 4 and 5, we describe the proposed strategies and show their benefits to low-resource accented ASR. Last but not least, Sect. 6 concludes this paper.

2   Background of Acoustic Modelling for Cross-Lingual or Accented ASR

2.1   Subspace Gaussian Mixture Models

The GMM and SGMM acoustic models are similar since each emission probability of each HMM state is modelled with a Gaussian mixture model. However, in the SGMM approach, the Gaussian means and mixture component weights are generated from the phonetic and speaker subspaces along with a set of weight projections. For SGMM, the state probabilities are defined following the equations below [18]:

$$p(\mathbf{x}|j) = \sum_{m=1}^{M_j} c_{jm} \sum_{i=1}^{I} w_{jmi}\mathcal{N}(\mathbf{x}; \mu_{jmi}, \mathbf{\Sigma}_i) \tag{1}$$

$$\mu_{jmi} = \mathbf{M}_i \mathbf{v}_{jm}, \tag{2}$$

$$w_{jmi} = \frac{\exp \mathbf{w}_i^T \mathbf{v}_{jm}}{\sum_{i'=1}^{I} \exp \mathbf{w}_{i'}^T \mathbf{v}_{jm}} \tag{3}$$

where $\mathbf{x} \in \mathbb{R}^D$ denotes the D-dimensional feature vector, $j \in \{1..J\}$ is the HMM state, i is the Gaussian index, m is the substate and c_{jm} is the substate weight. Each state j is associated to a vector $\mathbf{v}_{jm} \in \mathbb{R}^S$ (S is the phonetic subspace dimension) which derives the means, μ_{jmi} and mixture weights, w_{jmi}, I is the number of Gaussians for each state. The phonetic subspace $\mathbf{M}_i$, weight projections $\mathbf{w}_i^T$ and covariance matrices $\mathbf{\Sigma}_i$, i.e., the globally shared parameters $\mathbf{\Phi}_i = \{\mathbf{M}_i, \mathbf{w}_i^T, \mathbf{\Sigma}_i\}$ are common across all states.

These parameters can be shared and estimated over multiple language data. [11,13] presented cross-lingual and multilingual work using SGMM for improving ASR with very limited training data. In both studies, the cross-lingual approach was carried out by porting the UBM which was trained using source language data, to SGMM training of target language. Basically, the SGMM model was derived from the UBM of source language. For the second approach, the strategy involved training UBM using more than one language data and then employed the multilingual UBM for SGMM training of a specific language. Applying both methods improved ASR performance of monolingual system.

This idea motivates us to investigate a multi-accent approach using this framework. We propose to build SGMM models which are derived from *merged* UBMs, rather than carrying out the SGMM training in a multilingual fashion (see studies on non-native SGMM in [15,24]). The method is particularly appealing if one wishes to consider borrowing UBMs of other systems. Our strategy and experiments are described in Sect. 4.

2.2   Deep Neural Networks

Deep Neural Network (DNN) for ASR is a feedforward neural network with hidden layers. Mathematically, each output of the l-th layer of a DNN can be defined as

$$\mathbf{x}_l = \sigma(\mathbf{b}_l + \mathbf{W}_l \mathbf{x}_{l-1}), \; for \; 1 \leq l < L \tag{4}$$

where $\mathbf{W}_l$ is the connection weight from $\mathbf{x}_l$ and $\mathbf{x}_{l-1}$, the output of the $(l-1)$-th layer, while $\mathbf{b}_l$ is the bias. The hidden output $\mathbf{x}_l$ is a sigmoid function defined as $\sigma(x) = (1 + \exp(-x))^{-1}$. The last ($L$-th) layer of the DNN uses a softmax function to obtain the posterior probability of each HMM state j given the acoustic observation $\mathbf{o}_t$ at time t:

$$p(j|\mathbf{o}_t) = \frac{\exp(\mathbf{x}_L)}{\Sigma_{j'} \exp(\mathbf{x}_L)}. \tag{5}$$

Optimizing hidden layers can be done by pretraining the network using Restricted Boltzmann Machines (RBM) [7]. The generative pretraining strategy builds stacks of RBMs corresponding to the number of desired hidden layers and provides better starting point (weights) for DNN fine-tuning through back-propagation algorithm. Pretraining a DNN can be carried out in a unsupervised manner because it does not involve specific knowledge (labels, phone set) of a target language[1]. Only the softmax layer is sensitive to the target language. It is added on top of the hidden layers during fine-tuning and its ouput corresponds to the HMM states of the target language.

As shown in [21], using untranscribed data for RBM pretraining as a multilingual strategy has little effect on improving monolingual ASR performance. The *transfer learning* [5] approach has shown large recognition accuracy improvements. The method involves removing the top layer of a multilingual DNN and fine-tuning the hidden layers to a specific language.

Recently, a multi-accent DNN with accent specific softmax layer has been proposed for improving decoding performance of English ASR for British and Indian accents [10]. The accent adaptation approach yielded better decoding results compared to non-adapted DNNs. Another attempt to improve ASR performance on non-native task was done by [3] for Mandarin language. They also proved the interest of adapting non-native accents over the baseline DNN model.

In this paper, we investigate a method similar to [10] to build the accent-specific network models, but we apply it in a very low-resource setting. Previously, the method has been tested with larger amount of non-native speech (x10 or x100 compared to our experimental conditions). Hence, we try to measure the effectiveness of the approach when the ratio between non-native data and native data is largely unbalanced. In addition, we develop a strategy to handle cross-lingual DNNs with different feature transforms for speaker adaptation.

3   Experimental Setup

The ASR experiments were conducted on Kaldi speech recognition toolkit [19]. This section reports non-native and native speech databases used in our investigation. Besides that, we present the baseline results for non-native ASR based on GMM, SGMM and DNN.

3.1   Data

The non-native speech corpus contains 15 h of English speech spoken by 24 Malaysians (of Malay, Chinese and Indian origin). The data were collected by Universiti Sains Malaysia for conducting research on acoustic model merging for ASR (see [23] for more details). Table 1 shows the amount of data used to train and evaluate the non-native ASR. We employed 2 h of transcribed data for training the system and evaluate its performance on 4 h of transcribed speech. For SGMM training, 9 h of untranscribed data were added to the 2 h of transcribed speech to build the UBM. Our system used the CMU pronunciation dictionary

---

[1] In that sense, RBM pretraining (for DNN) and UBM training (for SGMM) are both unsupervised methods to get an initial representation of the acoustic space before modelling the speech units.

(no non-native adaptation of the lexicon) which has more than 100k words. Furthermore, we used a trigram language model for decoding. The model was trained on news data, taken from a local English news website[2]. After evaluating the LM on the test transcription data, the LM perplexity is 189 while the OOV rate is 2.5 %.

Table 1. Statistics of the non-native speech data for ASR.

Train		Test
Untranscribed	Transcribed	
9 h	2 h	4 h

To obtain a baseline for native ASR, we used the first release of TED-LIUM [20] corpus[3]. The transcriptions of this corpus were generated by the Laboratoire d'Informatique at Université du Maine (LIUM) for the International Workshop on Spoken Language Translation (IWSLT) evaluation campaign in 2011. The corpus contains speeches that were excerpted from video talks of the TED website. We used 118 h to train the system and 4 h for evaluation. Besides that, we used a pronunciation dictionary which was included in the package. For decoding, we used a trigram language model which was built on TED and WMT11 (Workshop on Machine Translation 2011) data. The model perplexity is 220 after estimation on the test data.

3.2 Baseline Systems

For the non-native ASR system, we trained a triphone acoustic model (39 MFCC with deltas and deltas deltas) using 776 states and 10K Gaussians. Then, we trained SGMM using the same decision trees as in the previous system. The SGMM was derived from a UBM with 500 Gaussians and phonetic subspace dimension was $S = 40$. The UBM was trained on 11h data. We built a DNN based on state-level minimum Bayes risk [12] (sMBR) and the network had 7 layers, each of the 6 hidden layers had 1024 hidden units. The network was trained from 11 consecutive frames (5 preceding and 5 following frames) of the same MFFCs as in the GMM system. Besides that, the same HMM states were used as targets of the DNN. The initial weights for the network were obtained using Restricted Boltzmann Machines (RBMs) that resulted in a deep belief network with 6 stacks of RBMs. Fine tuning was done using Stochastic Gradient Descent with per-utterance updates, and learning rate 0.00001 which was kept

---

[2] http://www.thestar.com.my/.

[3] We are aware that TED-LIUM is not a truly native English corpus (non-native speakers of multiple origins) but we consider here that the corpus permit to build an efficient system to decode native English ASR. Thus, in this paper we call it "excessively" a native corpus.

constant for 4 epochs. To run our DNN experiments, we utilized a GPU machine and CUDA toolkit to speed up the computations.

For the native ASR system, we built a triphone acoustic model with 3304 states and 40K Gaussians. Subsequently, we built SGMM system using the same decision trees and 500 UBM Gaussians. Lastly, we trained a DNN with 7 layers using the same setting for building non-native DNN. The three systems were evaluated on native speech (TED task) and we achieved the following WER results: 30.55 % for GMM, 28.05 % for SGMM and 19.10 % for DNN.

Table 2. Word error rates (WER %) of ASR with non-native (2 h) and native (118 h) acoustic models on the non-native evaluation data (4 h test) - same pronunciation dictionary and language model for both system.

Acoustic models	Non-native	Native
GMM	41.47	57.09
SGMM	40.41	45.84
DNN	32.52	40.70

Table 2 presents the baseline results of systems that used non-native and native acoustic models, evaluated on accented speech. For non-native acoustic modelling, SGMM and DNN systems outperformed the GMM system. The systems gave 3 % and 22 % relative improvement, respectively. Using these non-native models (trained on 2 h only!) to decode non-native speech resulted lower word error rate (WER) compared to the pure native ASR systems (trained on 118 h). In the following sections, we try to take advantage of both corpora (*large* native and *small* non-native) by merging acoustic models (or data) efficiently.

4   Language Weighting for Multi-accent Subspace Gaussian Mixture Models

4.1   Proposed Method

In SGMM, the system is initialized by a Universal Background Model (UBM) which is a mixture of full-covariance Gaussians. This single GMM is trained on all speech classes that are pooled together. The advantage of this model is that it can be trained on large amount of untranscribed data or multiple languages, as shown in [13] for cross-lingual SGMM in low-resource conditions. The authors showed that the SGMM global parameters are transferable between languages, especially when the parameters are trained in multilingual fashion. Thus, this gives an opportunity for low-resource systems to borrow UBM trained from other sources.

Figure 1 illustrates the process of UBM merging through language weighting. The first step is to choose a language weight, α to L_1 in order to determine the

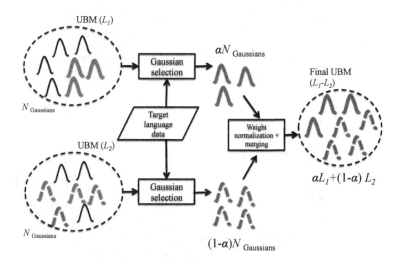

Fig. 1. An Illustration of UBM merging through language weighting

number of Gaussians to be kept for merging ($(1 - \alpha)$ is given to L_2). Intuitively, a larger α should be given to the less represented source data. Then, we use data that are representative of the ASR task in order to find the top αN Gaussians in L_1 UBM using maximum likelihood criterion. The same process is done for the L_2 UBM but only $(1-\alpha)N$ Gaussians are selected. The final step applies weight normalization before merging all the Gaussians in a single GMM. The final UBM should have the same number of Gaussians if both initial UBMs are the same size.

For experiments, we built a multi-accent UBM by merging native and non-native models using our language weighting strategy. To implement this, we used UBM of native speech (trained on 118 h) and UBM of non-native speech (trained on 11 h). Each of the UBMs has 500 Gaussians. Using the two models, we employed the language weighting approach for obtaining several multi-accent UBMs. Thereafter, these UBMs were used to estimate the parameters of non-native SGMM systems. Subsequently, we trained multi-accent SGMMs with different numbers of substates, ranging from 800 to 8750. By doing this, we obtained several SGMMs for each multi-accent UBM applied. We summarize our results by reporting only the highest (maximum), average and best (minimum) SGMM results, as shown in Fig. 2.

4.2   Results

Our findings show that using the proposed strategy resulted in significant improvement from the SGMM baseline. We reach the lowest WER when the SGMM system was obtained from a multi-accent UBM with 250 Gaussians from native and 250 Gaussians from non-native ($\alpha = 0.5$, WER=37.71 %). This result proves that carefully controlling the contribution of two (unbalanced) data as sources

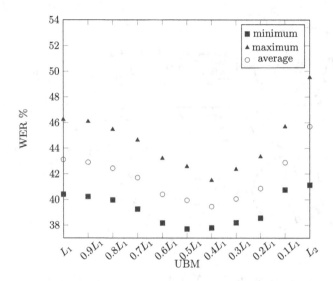

Fig. 2. Min, max and average performance (WER (%)) of multi-accent SGMM based on language weighting strategy for non-native ASR (4 h test). Note: non-native (L_1) native (L_2) and $\alpha = 0.1, ..., 0.9$.

for UBM training is a way to optimize ASR performance. In this experiment, the optimal α obtained tells us that non-native (Malaysian) data (in small quantity but very representative of the ASR task) and native (TED-LIUM) data (bigger corpus with speaker diversity) contribute equally to the acoustic space representation.

Furthermore, we did not gain WER improvements when the amount of substates increased. The minimum WERs shown in the figure are results for SGMMs with 800 substates. We extended our investigation to evaluate ASR performance for very compact (smaller number of Gaussians) UBM. The UBM was built with only 50 Gaussians using native/non-native data and then we applied the same language weighting strategy to obtain multi-accent UBMs.

Table 3. A summary of results from the SGMM experiments on the non-native ASR (4 h test). Different UBMs were employed for building SGMM with 2 h of non-native training data.

SGMM	WER (%)
Non-native UBM500	40.41 (baseline)
Native UBM500	41.13
For $\alpha = 0.5$,	
a. Multi-accent UBM500	37.71
b. Multi-accent UBM50	34.24

The non-native system significantly improved after applying this method. Table 3 shows the comparison between multi-accent SGMM with UBM=500 and UBM=50. For $\alpha = 0.5$, the new multi-accent SGMM outperformed the one with more UBM Gaussians by 9 % relative improvement on the WER. The result shows that deriving SGMM from a compact UBM gives better performance in very low-resource conditions. We also tried even smaller UBM but the WERs started to go back up (39.85 % for UBM with 5 Gaussians!).

5 Accent-Specific Top Layer for DNN

5.1 Proposed Method

Figure 3 illustrates the training process for obtaining an accent-specific DNN. We began with a network that was fine-tuned on native speech (last line and last column in Table 2). Then, we removed the softmax layer of native (source) DNN. Subsequently, a new softmax layer was added through fine-tuning the whole network on the non-native (target) training data. For this condition, we built the DNN on the GMM baseline for non-native.

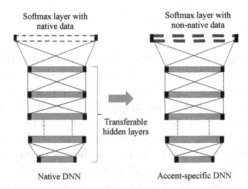

Fig. 3. Process of obtaining accent-specific DNN (right) using hidden layers of native DNN

We also built a second system for evaluating this approach. The system was speaker adapted and built upon new HMM/GMM acoustic models. First, we trained new native and non-native triphone models on new feature vectors using linear discriminant analysis (LDA) and maximum likelihood linear transform (MLLT), as well as speaker adaptive training using feature-space maximum likelihood linear regression (fMLLR). One important trick is to use feature transforms that were acquired from the native corpus (with large number of speakers), during LDA+MLLT training of non-native system. If not done this way, we observed no improvement with speaker adaptation (merging non-native and native DNNs with different feature transforms is not good). Then, we trained DNN for native and later we removed the top layer of the model. Subsequently, we fine-tuned the remaining DNN layers on the non-native data.

Table 4. WERs of accent-specific DNN on the non-native ASR task (4 h test).

DNN with accent-specific top layer	WER (%)
a. No speaker adaptation	24.89
b. Speaker adaptation	21.48

5.2 Results

We tested the DNNs on the same non-native evaluation data (4 h test). Table 4 presents our findings. Both results are significantly better than the pure non-native DNN baseline (last line in Table 2). For example, we achieved 24 % and 34 % relative improvement respectively over the non-native DNN baseline (32.52 %). Thus, the hidden layers of the native DNN proved to be useful for improving the low-resource non-native ASR. Besides that, our approach for building DNN with speaker adaptation and accent-specific top layer provided the best result. We obtained 14 % relative improvement over the accent-specific DNN without speaker adaptation.

6 Conclusions

We have proposed two approaches for optimal merging of native and non-native data in order to improve accented ASR with limited training data. The first approach introduced a language weighting strategy for constructing multi-accent compact SGMM acoustic models. In this approach, we used language weights to control the number of Gaussians of each UBM involved in the merging process. Improvement of the ASR performance was observed with language weighting. The second approach involved fine-tuning the hidden layers of native DNN on the non-native training data. We applied this approach for obtaining accent-specific DNN with and without speaker adaptation. For the former, we trained the DNN on HMM/GMMs that had feature transforms of the native speech data. Both DNNs outperformed the DNN baseline. Overall, the approaches used in this study resulted in encouraging improvement in WER. Over the non-native baseline, we achieved relative improvement of 15 % for SGMM (multi-accent UBM50) and 34 % for DNN (accent-specific with speaker adaptation).

References

1. Arslan, M.J., Hansen, J.L.: A study of the temporal features and frequency characteristics in american english foreign accent. J. Acoust. Soc. **102**(1), 28–40 (1996)
2. Bouselmi, G., Fohr, D., Haton, J.P.: Fully automated non-native speech recognition using confusion-based acoustic model intergration. In: Proceedings of Eurospeech, Lisboa, pp. 1369–1372 (2005)
3. Chen, X., Cheng, J.: Deep neural network acoustic modeling for native and non-native mandarin speech recognition. In: Proceedings of International Symposium on Chinese Spoken Language Processing (2014)

4. Goronzy, S. (ed.): Robust Adaptation to Non-Native Accents in Automatic Speech Recognition. LNCS (LNAI), vol. 2560. Springer, Heidelberg (2002)
5. Heigold, G., Vanhoucke, V., Senior, A., Nguyen, P., Ranzato, M., Devin, M., Dean, J.: Multilingual acoustic models using distributed deep neural networks. In: Proceedings of ICASSP (2013)
6. Hinton, G., Deng, L., Yu, D., Mohamed, A.R., Jaitly, N., Senior, A., Vanhoucke, V., Nguyen, P., Dahl, T.S.G., Kingsbury, B.: Deep neural networks for acoustic modeling in speech recognition. IEEE Sig. Process. Mag. **29**(6), 82–97 (2012)
7. Hinton, G.E.: A practical guide to training restricted boltzmann machines. UTML Technical report 2010–003, Department of Computer Science, University of Toronto (2010)
8. Huang, C., Chang, E., Zhou, J., Lee, K.F.: Accent modeling based on pronunciation dictionary adaptation for large vocabulary mandarin speech recognition. In: Proceedings of the ICLSP, vol. 2, pp. 818–821 (2000)
9. Huang, J.T., Li, J., Yu, D., Deng, L., Gong, Y.: Cross-language knowledge transfer using multilingual deep neural network with shared hidden layers. In: Proceedings of ICASSP (2013)
10. Huang, Y., Yu, D., Liu, C., Gong, Y.: Multi-accent deep neural network acoustic model with accent-specific top layer using the KLD-regularized model adaptation. In: Proceedings of Interspeech (2014)
11. Imseng, D., Motlicek, P., Bourlard, H., Garner, P.N.: Using out-of-language data to improve under-resourced speech recognizer. Speech Commun. **56**, 142–151 (2014)
12. Kingsbury, B.: Lattice-based optimization of sequence classification criteria for neural network acoustic modeling. In: Proceedings of IEEE International Conference on Acoustics, Speech and Signal Processing, pp. 3761–3764, April 2009
13. Lu, L., Ghoshal, A., Renals, S.: Cross-lingual subspace gaussian mixture models for low-resource speech recognition. IEEE/ACM Trans. Audio Speech Lang. Process. **22**, 17–27 (2014)
14. Miao, Y., Metze, F.: Improving low-resource CD-DNN-HMM using dropout and multilingual DNN training. In: Proceedings of INTERSPEECH, pp. 2237–2241 (2013)
15. Mohan, A., Ghalehjegh, S.H., Rose, R.C.: Dealing with acoustic mismatch for training multlingual subspace gaussian mixture models for speech recognition. In: Proceedings of ICASSP, pp. 4893–4896. IEEE, Kyoto, March 2012
16. Morgan, J.J.: Making a speech recognizer tolerate non-native speech through gaussian mixture merging. In: Proceedings of ICALL 2004, Venice (2004)
17. Povey, D., Burget, L., Agarwal, M., Akyazi, P., Feng, K., Ghoshal, A., Glembek, O., Goel, N., Karafiat, M., Rastrow, A., Rose, R.C., Schwarz, P., Thomas, S.: Subspace gaussian mixture models for speech recognition. In: Proceedings of ICASSP (2010)
18. Povey, D., Burget, L., Agarwal, M., Akyazi, P., Kai, F., Ghoshal, A., Glembek, O., Karafiàt, N.G.M., Rastrow, A., Rose, R.C., Schwartz, P., Thomas, S.: The subspace gaussian mixture model - a structured model for speech recognition. Comput. Speech Lang. **25**, 404–439 (2011)
19. Povey, D., Ghoshal, A., Boulianne, G., Burget, L., Glembek, O., Goel, N., Hannemann, M., Motlíček, P., Schwarz, P., Silovský, J., Stemmer, G., Veselý, K.: The kaldi speech recognition toolkit. In: Society, I.S.P. (ed.) Proceedings of Workshop on Automatic Speech Recognition and Understanding, IEEE Catalog No.: CFP11SRW-USB, December 2011
20. Rousseau, A., Deléglise, P., Estève, Y.: TED-LIUM: an automatic speech recognition dedicated corpus. In: Proceedings of LREC, pp. 125–129. European Language Resources Association (ELRA) (2012)

21. Swietojanski, P., Ghoshal, A., Renals, S.: Unsupervised cross-lingual knowledge transfer in DNN-based LVCSR. In: Proceedings of ICASSP (2013)
22. Tan, T.P., Besacier, L.: Acoustic model interpolation for non-native speech recognition. In: Proceedings of ICASSP (2007)
23. Tan, T.P., Besacier, L., Lecouteux, B.: Acoustic model merging using acoustic models from multilingual speakers for automatic speech recognition. In: Proceedings of International Conference on Asian Language Processing (IALP) (2014)
24. Tong, R., Lim, B.P., Chen, N.F., Ma, B., Li, H.: Subspace gaussian mixture models for computer-assisted language learning. In: Proceedings of ICASSP, pp. 5347–5351. IEEE (2014)
25. Vu, N.T., Imseng, D., Povey, D., Motlíček, P., Schultz, T., Bourlard, H.: Multilingual deep neural network based acoustic modeling for rapid language adaptation. In: Proceedings of ICASSP (2014)

On Continuous Space Word Representations as Input of LSTM Language Model

Daniel Soutner[(✉)] and Luděk Müller

NTIS - New Technologies for the Information Society, Faculty of Applied Science,
University of West Bohemia, Pilsen, Czech Republic
{dsoutner,muller}@ntis.zcu.cz

Abstract. Artificial neural networks have become the state-of-the-art in the task of language modelling whereas Long-Short Term Memory (LSTM) networks seem to be an efficient architecture. The continuous *skip-gram* and the *continuous bag of words* (CBOW) are algorithms for learning quality distributed vector representations that are able to capture a large number of syntactic and semantic word relationships. In this paper, we carried out experiments with a combination of these powerful models: the continuous representations of words trained with *skip-gram/CBOW/GloVe* method, word cache expressed as a vector using *latent Dirichlet allocation* (LDA). These all are used on the input of LSTM network instead of *1-of-N* coding traditionally used in language models. The proposed models are tested on Penn Treebank and MALACH corpus.

Keywords: Language modelling · Neural networks · LSTM · Skip-gram · CBOW · GloVe · word2vec · LDA

1 Introduction

In last years, recurrent neural networks (RNN) have attracted attention among other types of language models (LM) caused by their better performance [5] and their ability to learn on a smaller corpus than conventional n-gram models. Nowadays, they are considered as a state-of-the-art, especially the Long-short Term Memory (LSTM) variant. Skip-gram and continuous bag of words (CBOW) [6] are recently developed technique for building a neural network that maps words to real number vectors, with the desideratum that words with similar meanings will be mapped to the similar vectors. In this paper, we propose using these vectors on the input of LSTM language model and we observe an effectivity of this model.

There is described our proposed language model architecture in Sect. 2, including the description of sub-parts of the model such as continuously distributed representations of words. Section 3 deals with experiments and results and Sect. 4 summarizes the results and draws conclusions.

© Springer International Publishing Switzerland 2015
A.-H. Dediu et al. (Eds.): SLSP 2015, LNAI 9449, pp. 267–274, 2015.
DOI: 10.1007/978-3-319-25789-1_25

2   Model Architecture

2.1   Recurrent Neural Network Language Model

In standard back-off n-gram language models, words are represented in a discrete space – in the vocabulary. This prevents better interpolation of the probabilities of unseen n-grams because a change in this word space can result in an arbitrary change of the n-gram probability. The basic architecture of neural network model was proposed by Y. Bengio in [1]. The main idea is to understand a word not as a separate entity with no specific relation to other words, but to see words as points in a finite dimensional (separable) metric space. The recurrent neural networks – using the similar principle – were successfully introduced to the field of language modelling by T. Mikolov [5] and have become widely used language modelling technique.

In our work, we aimed to discover whether we are able to step further and move from words that are projected into the continuous space by network itself to the words projected to vectors with some more powerful techniques (as it is shown in [7]) and to learn the RNN model on this vectors afterwards.

2.2   Skip-Gram and CBOW

Mikolov et al. in [6] introduced a new efficient architecture for training distributed word representations which belongs to the class of methods called "neural language models". Authors proposed two architectures: continuous skip-gram that tries to predict the context words given the input word and continuous bag-of-words (CBOW) that predicts the current word given the context (Fig. 1). The input words are encoded in 1-of-N coding, the model is trained with hierarchical softmax, a context in interval 5–10 word is usually considered. We used publicly available *word2vec*[1] tool in our experiments.

2.3   Log-Bilinear Variant of Skip-Gram and CBOW

The log-bilinear variant (LBL) of both previously described architectures (CBOW-LBL and skip-gram-LBL) learned by noise-contrastive estimation (more could be found in [9]) seemed to perform slightly better on word analogy tasks. Thus, we decided to compare word vectors obtained with these architectures with the previous ones. We used the publicly available *LBL4word2vec* tool in experiments[2].

2.4   GloVe (Global Vectors for Word Representation)

In essence, GloVe (Global Vectors for Word Representation) [10] is a log-bilinear model with a weighted least-squares objective. The main intuition which underlies this model is the simple observation that ratios of word-word co-occurrence

---

[1] https://code.google.com/p/word2vec.
[2] https://github.com/qunluo/LBL4word2vec.

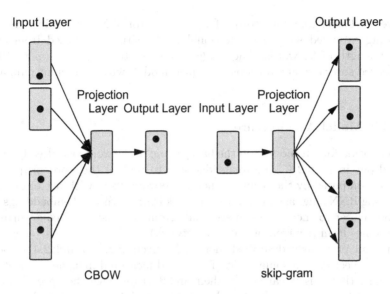

Fig. 1. The scheme of CBOW and skip-gram architecture. CBOW predicts the current word given the context and skip-gram that predicts the context words given the input word.

probabilities have the potential for encoding some form of meaning. The training criterion of GloVe is to estimate word vectors such that their dot product equals the logarithm of the words' probability of co-occurrence. Owing to the fact that the logarithm of a ratio equals the difference of logarithms, this objective associates (the logarithm of) ratios of co-occurrence probabilities with vector differences in the word vector space. Because these ratios can encode some form of meaning, this information gets encoded to the vector differences as well. Details about this architecture and implementation are published in [10]. We used the publicly available *GloVe* tool in our experiments[3].

2.5   LDA

In addition to the word embeddings – to exploit more information from the long span context – we decided to use the latent Dirichlet allocation (LDA) [2] in our experiments as we already did in our previous work [13].

The LDA process converts word representation of a document to a low-dimensional vector which represents a probability of the topic. It represents documents as mixtures of topics that split out words with certain probabilities. In our experiments, we fixed the length of the word cache while computing the topic distribution.

We divided text to documents of 10 non-overlapping sentences for Penn Treebank corpus. For MALACH corpus, we stacked sentences to documents so that

---

[3] http://nlp.stanford.edu/projects/glove.

they are not longer than 200 words. The input vector of NN is modified as word embedding extended with this additional LDA feature computed from cache with the length of 50 words. The models were created with *gensim* tool [12]. We explored several configurations of trained models with a different number of topics.

2.6   Our Model Architecture

The motivation for our model and the background of the work was already lightly sketched out in Sect. 2.1. The main idea was to replace the word projections, which are computed by the recurrent neural network itself, with better ones. By the training RNN language model, the network learns own word embeddings, but if we look closer to they are estimated only from the past word co-occurrences (which comes from recurrentness of the network).

While all previously described models (in Sects. 2.2, 2.3 and 2.4) produce better word vectors – in some point of view and measured in some applications. We assume, that this is caused by their architecture since they are all taking into account both words in past and words in the future context. This, in our opinion, makes the resulting vectors more accurate. The projections perform better while they are trained on more text or trained in more iterations, thus we run experiments with a various number of iterations over the training text.

We chose for our experiments the Long-short Term Memory (LSTM) [4] networks architecture due its better ability of learning [14]. The scheme of one LSTM cell is shown in Fig. 2. The simpler "vanilla" RNN network showed to be unstable while we were changing the input of the model in experiments.

To be able to catch the longer context, which is also crucial for accurate language model, we added LDA context features. This architecture, as we expect, should allow to discover more regularities in a language.

More technically, first the word vectors are computed for every word in vocabulary from training text (by different techniques mentioned above). Afterwards, in the training phase of the language model we give to the input of LSTM network these vectors instead of 1-of-N encoding of words. The input vector could be moreover extend simply by appending with the LDA vector – computed for every context.

3   Experiments and Results

In this section, we describe experiments which we did with proposed models and presenting the obtained results. The data are described in first part, the details about the training and the results follow.

3.1   Data

Penn Treebank. To maintain comparability with the other experiments in literature, we chose the well-known and widely used Penn Treebank (PTB) [3]

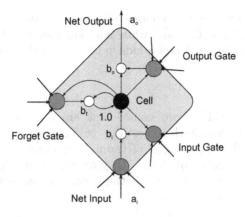

Fig. 2. The scheme of LSTM cell.

portion of the Wall Street Journal corpus for testing our models. This is quite rare in the language modelling field and allows us to compare direct performance of different techniques and their combinations. Following common preprocessing was applied to the corpora:

- words outside the 10 K vocabulary were mapped to a special token (unknown word)
- all numbers were unified into $\langle N \rangle$ tag
- punctuation was removed.

The corpus was divided into tree parts: Sections 0–20 were used as the training data, Sections 21–22 as the validation data and Sections 23–24 as the test data.

MALACH. We also wanted to carry out results on some real-world problem, so we decided to evaluate our models on the MALACH corpus [11]. Steven Spielberg (inspired by his experience making *Schindlers List*) established the Survivors of the Shoah Visual History Foundation (1994) to gather video testimonies from survivors and other witnesses of the Holocaust. It contains approximately 375 h of interviews with 784 interviewees along with transcripts and other documentation. The original release includes transcripts of the first 15 min of each interview, which makes in a textual form circa 2M tokens, the vocabulary consists of 21.7k words. We split this data to *train* (70 %), *development* and *test* folds (13 % and 17 %).

3.2 Training

For obtaining word embeddings, we used context window size of 10 words and the other parameters we anchored to default settings because tuning more parameters would be exhausting. For the language model – as remarked above – we used LSTM network, to speed up experiments the simplified version. During the training phase, the gradients were clipped into the interval $< -1, 1 >$; starting with learning rate at $\alpha = 0.16$ and while not achieving perplexity decrease we halved learning rate. The width of a hidden layer is fixed to 100 neurons.

3.3   Results

First, we evaluated, which word embeddings are suitable for this task; we employed algorithms described above. The performance of word embeddings also strongly depends on a number of training iterations, hence we produced results with various number of them – they are shown in Fig. 3. If we employ the LDA extension, we obtain a bit better results, as we supposed. The best results on PTB are shown in Table 1.

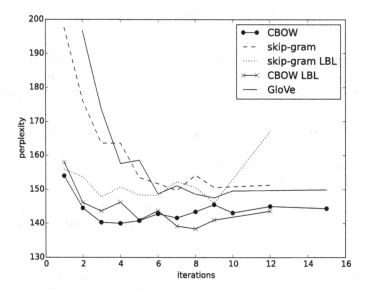

Fig. 3. Various types of word vectors, performance measured on PTB while using them in language model

Table 1. Perplexity results on Penn Treebank corpus.

Model	PPL
LSTM-100	144.0
LSTM-100&CBOW LBL-100	138.4
LSTM-100&CBOW LBL-100 + LDA-50	133.6

The results for the MALACH task are in Table 2, where we compared our model (LSTM-100&CBOW LBL-100) with RNN models (RNN-100 and RNN-400 with 100 respectively 400 neurons in hidden layer). The CBOW LBL embeddings were used as the best option from experiments before. The RNN models were trained with RNNLM Toolkit[4], number denotes a size of a hidden layer. We also added results achieved with a linear combination of neural models with more conventional models Knesser-Ney 5-gram (KN5) and maximum entropy on 5-gram features (ME5). For the completeness the result with LDA (with 50) extension is added and also the model mixed together with n-gram models (with linear interpolation, which coefficients were tuned on development data with EM algorithm).

Table 2. Perplexity results on MALACH corpus.

Model	PPL
RNN-100	107
RNN-400	100
LSTM-100	99
RNN-100 + KN5 + ME5	94
LSTM-100&CBOW LBL-100	**94**
LSTM-100&CBOW LBL-100 + LDA-50	91
LSTM-100&CBOW LBL-100 + LDA-50 + KN5 + ME5	83

4   Conclusion and Future Work

We proposed language model using continuous word representation as input and extended this input with information from context vectors. We employed techniques that are assumed as a state-of-the-art such as LDA, skip gram and a log-bilinear continuous bag of words. The experiments showed the improvement of 4–5 % in perplexity over standard LSTM/RNN models measured on Penn Treebank and MALACH corpus and with LDA extension circa 7 %–8 %.

We shown, that using continuous word vectors computed outside of neural network could improve LSTM-LM performance and even more if we add additional context information. Nevertheless it is good to notice, that the type and the quality matter i.e. number of train iterations of word vectors and algorithm. In our next work, we would like to further verify our approach on other corpora (such as Wikipedia Text8 or "One billion word benchmark") and on the speech recognition or automatic translation.

Acknowledgements. Access to computing and storage facilities owned by parties and projects contributing to the National Grid Infrastructure MetaCentrum, provided

---

[4] www.rnnlm.org.

under the programme "Projects of Large Infrastructure for Research, Development, and Innovations" (LM2010005), is greatly appreciated.

This research was supported by the Ministry of Culture Czech Republic, project No. DF12P01OVV022.

References

1. Bengio, Y., Ducharme, R., Vincent, P., Janvin, C.: A neural probabilistic language model. J. Mach. Learn. Res. **3**, 1137–1155 (2003)
2. Blei, D.M., Ng, Y.A., Jordan, M.I., Lafferty, J.: Latent dirichlet allocation. J. Mach. Learn. Res. **3**, 993–1022 (2003)
3. Charniak, E., et al.: BLLIP 1987–89 WSJ Corpus Release 1. Linguistic Data Consortium, Philadelphia (2000)
4. Hochreiter, S., Schmidhuber, J.: Long Short-term memory. Neural Comput. **9**(8), 1735–1780 (1997)
5. Mikolov, T., Kombrink, S., Deoras, A., Burget, L., Černocký, J.: RNNLM - Recurrent Neural Network Language Modeling Toolkit (2011)
6. Mikolov, T., Chen, K., Corrado, G., Dean, J.: Efficient estimation of word representations in vector space. In: Proceedings of Workshop at ICLR (2013)
7. Mikolov, T., Sutskever, I., Chen, K., Corrado G., Dean, J.: Distributed representations of words and phrases and their compositionality. In: Proceedings of NIPS (2013)
8. Mikolov, T., Yih, W., Zweig, G.: Linguistic regularities in continuous space word representations. In: Proceedings of NAACL HLT (2013)
9. Mnih, A., Kavukcuoglu, K.: Learning word embeddings efficiently with noise-contrastive estimation. Adv. Neural Inf. Process. Syst. **26**, 2265–2273 (2013)
10. Pennington, J., Socher, J., Manning., C.D.: GloVe: global vectors for word representation. In: Empricial Methods in Natural Language Processing (EMNLP) (2014)
11. Ramabhadran, B., et al.: USC-SFI MALACH Interviews and transcripts english LDC2012S05. Web Download. Philadelphia: Linguistic Data Consortium (2012)
12. Řehůřek, R., Sojka, P.: Software framework for topic modelling with large corpora. In: Proceedings of LREC 2010 Workshop New Challenges for NLP Frameworks. University of Malta, Valletta, Malta, pp. 4650–4655 (2010). ISBN 2-9517408-6-7
13. Soutner, D., Müller, L.: Application of LSTM neural networks in language modelling. In: Habernal, I. (ed.) TSD 2013. LNCS, vol. 8082, pp. 105–112. Springer, Heidelberg (2013)
14. Sundermeyer, M., Schlüter, R., Ney, H.: LSTM neural networks for language modeling. In: Interspeech (2012)

An Improved Hierarchical Word Sequence Language Model Using Word Association

Xiaoyi Wu[✉], Yuji Matsumoto, Kevin Duh, and Hiroyuki Shindo

Nara Institute of Science and Technology, Ikoma, Japan
{xiaoyi-w,matsu,kevinduh,shindo}@is.naist.jp

Abstract. Language modeling is a fundamental research problem that has applications for many NLP tasks. For estimating probabilities, most research on language modeling uses n-gram approach to factor sentence probabilities. However, the assumption of n-gram is too simple to cope with the data sparseness problem, which affects the final performance of language models. At the point, Hierarchical Word Sequence (abbreviated as HWS) language model, which uses word frequency information to convert raw sentences into special n-gram sequences, can be viewed as an effective alternative to normal n-gram method.

In this paper, we improve upon the basic HWS approach by generalizing it to exploit not only word frequencies but word association.

For evaluation, we compare word association based HWS models to normal HWS models and normal n-gram models. Both intrinsic and extrinsic experiments verify that word association based HWS models can achieve better performance.

Keywords: Language modeling · Modified Kneser-Ney (MKN) · Generalized Language Model (GLM) · Hierarchical Word Sequence (HWS) · Word association

1 Introduction

Probabilistic Language Modeling is a fundamental research direction of Natural Language Processing. It is widely used in many applications such as machine translation [1], spelling correction [2], speech recognition [3], word prediction [4] and so on.

Most research about Probabilistic Language Modeling, such as Katz back-off [5], Kneser-Ney [6], and modified Kneser-Ney [7], only focus on smoothing methods because they all take n-gram approach [8] as a default setting for extracting word sequences from a sentence. Yet even with 30 years worth of newswire text, more than one third of all trigrams are still unseen [9], which cannot be distinguished accurately even using a high-performance smoothing method such as modified Kneser-Ney (abbreviated as MKN). It is better to make these unseen sequences actually be observed rather than to leave them to smoothing method directly.

© Springer International Publishing Switzerland 2015
A.-H. Dediu et al. (Eds.): SLSP 2015, LNAI 9449, pp. 275–287, 2015.
DOI: 10.1007/978-3-319-25789-1_26

For the purpose of extracting more valid word sequences and relieving data sparsity problem, in [10], a heuristic approach is proposed to convert a sentence into a hierarchical word sequence (abbreviated as HWS) structure, by which special n-grams (HWS-n-grams) can be achieved. According to the results of experiments, HWS method helps to keep better balance between coverage and usage than the normal n-gram models and skip-gram models [11,12].

However, the frequency-based heuristic rule used for HWS approach is still too simple to construct high-quality hierarchical pattern structures. In this paper, we propose an improved HWS model by which the information of word association can be used for constructing HWS structures to achieve higher performance.

The paper makes the following scientific contributions:

(1) We present ways to construct HWS-n-grams using word association rather than word frequency.
(2) We propose two extra techniques to improve the performance of HWS models.
(3) We empirically observe how our models reduce perplexity and how real world applications benefit from them.

This paper is organized as follows. In Sect. 2, we give a complete review of the original HWS language model. Then we present our improved HWS model in Sect. 3. In Sects. 4 and 5, we show the effectiveness of our model by both intrinsic experiments and extrinsic experiments. Finally, we summarize our findings in Sect. 6.

2   Review of HWS Language Model

In [10], the *HWS structure* is constructed from training data in an unsupervised way as follows:

Step 1. Calculate word frequencies from training data and sort all these words by frequency. Then we can get a frequency-sorted list $V = \{v_1, v_2, ..., v_m\}$.

Step 2. According to V, for each sentence $s = w_1, w_2, ..., w_n$, the most frequently used word $w_i \in s(1 \leq i \leq n)$ is determined[1]. Then use w_i to split s into two substrings $s_l = w_1, ..., w_{i-1}$ and $s_r = w_{i+1}, ..., w_n$.

Step 3. Set $s' = s_l$ and $s'' = s_r$, then repeat Step 2 separately. The most frequently used words $w_j \in s_l(1 \leq j \leq i-1)$ and $w_k \in s_r(i+1 \leq k \leq n)$ can also be determined, by which s_l and s_r are splitted into two smaller substrings separately.

Executing Step 2 and Step 3 recursively until all the substrings become empty strings, then a binary tree $T = (\{w_i, w_j, w_k, ...\}, \{(w_i, w_j), (w_i, w_k), ...\})$ can be constructed, which is defined as an *HWS structure*.

---

[1] If w_i appears multiple times in s, then select the first one.

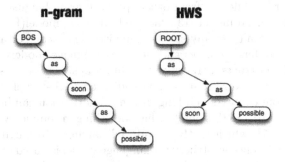

Fig. 1. A comparison of structures between HWS and n-gram

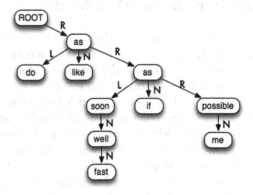

Fig. 2. An example of node selection tree

In an HWS structure T, assuming that each node depends on its preceding n-1 parent nodes, then special n-grams can be trained. Such kind of n-grams are defined as *HWS-n-grams*.

The advantage of HWS models can be considered as *discontinuity*. Taking Fig. 1 as an example, since n-gram model is a continuous language model, in its structure, the second 'as' depends on 'soon', while in the HWS structure, the second 'as' depends on the first 'as', forming a discontinuous pattern to generate the word 'soon', which is closer to our linguistic intuition. Rather than 'as soon ...', taking 'as ... as' as a pattern is more reasonable because 'soon' is quite easy to be replaced by other words, such as 'fast', 'high', 'much' and so on. Consequently, even using 4-gram or 5-gram, sequences consisting of 'soon' and its nearby words tend to be low-frequency because the connection of 'as...as' is still interrupted. On the contrary, the HWS model extracts sequences in a discontinuous way, even 'soon' is replaced by another word, the expression 'as...as' won't be affected. This is how the HWS models relieve the data sparseness problem.

The HWS model is essentially an n-gram language model based on a different assumption that a word depends upon its nearby high-frequency words instead of its preceding words. Different from other special n-gram language models, such as class-based language model [13], factored language model(FLM) [14], HWS language model doesn't use any specific linguistic knowledge or any abstracted categories. Also, differs from dependency tree language models [15,16], HWS language model constructs a tree structure in an unsupervised fashion.

In HWS structure, word sequences are adjusted so that irrelevant words can be filtered out from contexts and long distance information can be used for predicting the next word. On this point, it has something in common with structured language model [17], which firstly introduced parsing into language modeling. The significant difference is, structured language model is based on CFG parsing structures, while HWS model is based on pattern-oriented structures.

3 Word Association Based HWS Model

In the previous section, we discussed that the main idea of HWS is to arrange patterns in a hierarchical structure. However, for the purpose of extracting patterns, only using frequency information is not sufficient. More valid patterns can be extracted if word associations are taken into account.

Taking sentence pattern '... too ... to ...' as an example, suppose that 'to' is firstly selected as a node of HWS structure, it is preferable that 'too' is selected as the next node so that a hierarchical pattern can be constructed, however, if we select nodes only by frequency, it is highly possible that a more frequently occurring word would be selected prior and the connection between 'too' and 'to' are interrupted. Instead, if we use a word association score to selected nodes, then a higher-quality hierarchical pattern structure can be constructed.

In this section, we propose an improved HWS model by which the information of word associations is also incorporated.

3.1 Basic Idea

As we reviewed in Sect. 2, HWS-n-grams are extracted from HWS structures, while HWS structures are constructed by a frequency-sorted vocabulary list $V = \{v_1, v_2, ..., v_m\}$. Once V is given, the HWS structures and HWS-n-grams are deterministic. Thus, the performance of HWS model is determined by V.

V is actually a priority list for constructing HWS structures. Given a sentence S, we first judge whether $v_1 \in S$, if not, then we check whether $v_2 \in S$, until we find $v_i \in S(1 \leq i \leq m)$ and use it to divide S into two substrings.

In the original HWS model, the priorities of V are only arranged by word frequency, which is independent. For the purpose of constructing higher-quality HWS structures, we arrange priorities under certain context. We represent the priority list under context c as V_c, where $c = \{a_1^T, a_2^T, ..., a_n^T\}$ $(T \in \{L, R\})$. L(left) and R(right) represent the direction of each $v \in V_c$ relative to each $a \in c$.

For example, according to the priority list $V_{\{to^L, too^R\}} = \{good, bad, big, ...\}$, given a substring between 'too' and 'to', which correspond to context $\{to^L, too^R\}$, we first judge whether 'good' is in it, if not, then we check 'bad' in turn.

Thus, for each $v_i \in V_c$, $\{v_1, v_2, ..., v_{i-1}\}$ can be appended as part of its context by adding a special tag 'N'. Then in the above example, the whole context of 'big' can be written as $\{to^L, too^R, good^N, bad^N\}$.

By combining all priority list under all contexts in this form, a tree structure is constructed (as shown in Fig. 2). We call this tree as *Node Selection Tree* (abbreviated as NST).

Instead of frequency-sorted word list, we use the whole NST as the 'priority list' so that the information of word association can be incorporated in the HWS structures. We will discuss how to use word association to train an NST from a raw corpus in the next section.

3.2 Training of NST

Suppose we are given a corpus S for training, NST is trained by the following steps.

Step 1: Adding '⟨ROOT⟩' to the beginning of each sentence and set it as the default context c (also the root node of NST), for each word $w \in S$, we calculate the word association score by using Dice coefficient [18] (Eq. (1)).

$$score_{Dice}(c, w) = \frac{2 \times C(c, w)}{C(c) + C(w)} \tag{1}$$

Step 2: According to these scores, we choose the word with maximum score $\hat{w}$ as the child node of the root node '⟨ROOT⟩'. For each sentence $s \in S$, if $\hat{w} \notin s$, then we put s under the 'N' arc of $\hat{w}$, otherwise, we use the first $\hat{w}$ appeared in s to split s into 2 substrings s_l and s_r, then put s_l and s_r under the 'L' arc and 'R' arc of $\hat{w}$ separately. Finally, all strings (or substrings) of S are classified by three types of arcs of $\hat{w}$.

Step 3: For each arc of $\hat{w}$, $\hat{w}$ can be considered as the new context c', strings (or substrings) can be considered as a new corpus S'. Repeating Step 1 and Step 2 recursively, an NST is constructed.

3.3 Converting Sentences by NST

With the NST trained from corpus, any sentence can be converted into the HWS structure.

Suppose we use the NST shown in Fig. 2 to convert the sentence 'as soon as possible' into an HWS structure. Firstly, we start from the node 'as' and check whether it exists in this sentence. Although two 'as's are observed in this sentence, we use the first one to divide it and take the right substring 'soon as possible' to the 'R' arc. Since the child node 'as' of 'R' arc can also be observed in this substring, 'soon' and 'possible' are classified to its 'L' arc and 'R' arc respectively. Finally, 'as soon as possible' is converted to the same HWS structure as shown in Fig. 1. Similarly, sentence 'as fast as possible', 'as well as me' can also be converted into HWS structures by this NST.

Notice that even a well-trained NST cannot cover all possible situations. For instance, suppose we use the above NST to convert sentence 'as far as I could remember', although 'I could remember' is correctly classified as 'R' arc of the second 'as' node, it cannot be further analyzed because this NST doesn't offer more details in this arc. In this case, we use the original HWS approach (by word frequency) to select nodes from substring 'I could remember' as a covering of our method.

3.4   Two Extra Techniques for HWS Model

Besides using word association, we also use two extra techniques to improve the performance of our model.

1. Directionalization: Directionalization is about the extraction of HWS-n-grams.

In the normal n-gram models, since all the generation of words is one-sided (from left to right), the directional information of contexts can be omitted. However, HWS structures are essentially binary trees, which also generate words on the left side, a syntactical problem can be caused if we don't consider the direction of contexts.

Taking Fig. 1 as an example. According to the structure of HWS, HWS-3-grams are trained as {(ROOT, as, as), (as, as, soon), (as, as, possible)}, where 'soon' and 'possible' are generated from the context (as, as) without any distinction, which means, an illegal sentence such like 'as possible as soon' can be also generated from this HWS-3-gram model.

We can add directional information to distinguish the contexts of HWS-n-grams. As shown in Fig. 3, after a HWS structure (binary tree) being constructed, directional information can be easily attached to this tree. Then, assuming that each node depends on its n-1 preceding parent nodes with their directional information, we can train a special n-gram from this binary tree. For instance, 3-grams trained from this tree are {(ROOT-R, as-R, as), (as-R, as-L, soon), (as-R, as-R, possible)}.

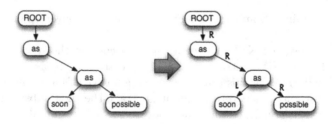

Fig. 3. An example of HWS structure with directional information

2. Unification: Unification is about the counting of word frequency.

Word frequency is a main factor that affects the quality of HWS structures. Even we use word association score to construct an NST, $C(w)$ still plays an important role for arranging the priority. Especially for top levels of NST, the more frequent a word is, the more priority it tends to have. Since constructing NST is a top-down approach, the inappropriate priority setting of top levels can bring negative effects to the whole NST and consequently to the HWS structures and HWS-n-grams.

For example, as an appropriate hierarchical pattern structure, the period should be the top level because it represents the most typical **sentence** pattern. However, in English, definite article 'the' is the most frequent word, consequently, the word 'the' tends to be the top of NST and be selected as the top of HWS structure.

For relieving this problem, no matter how many times a word appears **in one sentence**, we can count it **only once**. Then words actually appear in more sentences can be arranged as higher level of HWS structures, which improves the quality of hierarchical pattern structures.

3.5   Smoothing Methods for HWS

As for smoothing methods for HWS models, [10] only used an additive smoothing. Although HWS-n-grams are trained in a special way, they are essentially n-grams because each trained sequence is reserved as a $(n-1\ length\ context,\ word)$ tuple as normal n-grams, which makes it possible to apply MKN smoothing to HWS models. The main difference is that HWS models are trained by tree structures while n-gram models in a continuous way, which affects the counting of contexts $C(w_{i-n+1}^{i-1})$.

Taking Fig. 1 as an example. According to the structure of HWS, HWS-3-grams are trained as {(ROOT, as, as), (as, as, soon), (as, as, possible)}, while the HWS-2-grams are trained as {(ROOT, as), (as, as), (as, soon), (as, possible)}. In the HWS-3-gram model, as the context of 'soon' and 'possible', 'as ... as' appears twice, however, in the HWS-2-gram model, $C(as, as)$ is counted only once. In normal n-gram models, $C(w_{i-n+1}^{i-1})$ can be directly achieved from its lower model because they are continuous, but in HWS models, $C(w_{i-n+1}^{i-1})$ should be counted as $\sum_{w_j \in \{w_i : C(w_{i-n+1}^{i}) > 0\}} C(w_{i-n+1}^{i-1}, w_j)$, which means that the frequencies of contexts should be counted in the model with the same order. Taking this into account, MKN smoothing method can be also applied to HWS models.

As an alternative of MKN smoothing method, we can also use GLM [19]. GLM (Generalized Language Model) is a combination of skipped n-grams and MKN, which performs well on overcoming data sparseness.

Since GLM is a generalized version of MKN smoothing, it can also be applied to HWS models. In the following experiments, we will use MKN and GLM as smoothing methods.

4   Intrinsic Evaluation

4.1   Settings

We use two different corpus: **British National Corpus** and **English Giga-word Corpus**.

British National Corpus (BNC)[2] is a 100 million word collection of samples of written and spoken language from a wide range of sources. We randomly choose 449,755 sentences (10 million words) as training data.

English Gigaword Corpus[3] consists of over 1.7 billion words of English newswire from 4 distinct international sources. We randomly choose 44,702 sentences (1 million words) as test data.

As preprocessing of training data and test data, we use the tokenizer of NLTK (Natural Language Toolkit)[4] to split raw English sentences into words. We also converted all words to lowercase.

To ensure the openness of our research, the source code used for following experiments can be downloaded.[5]

As intrinsic evaluation of language modeling, perplexity [20] is the most common metric used for measuring the usefulness of a language model. In [10], coverage and usage are also proposed to evaluate efficiency of language models. The sequences of training data are defined as TR, while unique sequences of test data as TE, then the coverage is calculated by Eq. (2).

$$coverage = \frac{|TR \bigcap TE|}{|TE|} \qquad (2)$$

Usage (Eq. (3)) is used to estimate how much redundancy contained in a model and a balanced measure is calculated by Eq. (4).

$$usage = \frac{|TR \bigcap TE|}{|TR|} \qquad (3)$$

$$F\text{-}Score = \frac{2 \times coverage \times usage}{coverage + usage} \qquad (4)$$

4.2   Results

We compare our models (refer to 'DB-HWS' in the following tables) with normal n-gram models and the original HWS models (refer to 'FB-HWS' in the following tables). According to the results shown in Table 1, DB-HWS models using extra techniques (Sect. 3.4) get more stable and better performance on perplexity than other models.

---

[2] http://www.natcorp.ox.ac.uk.
[3] https://catalog.ldc.upenn.edu/LDC2011T07.
[4] http://www.nltk.org.
[5] https://github.com/aisophie/HWS.

On the other hand, since almost each word is distinguished as 'two words(-L and -R)' by using directionalization technique, the coverage and usage tend to be relatively lower. But we believe it is worth because the mechanism of natural language is modeled more precisely in this way.

We also noticed that for each model (n>2), perplexity is greatly reduced after applying GLM smoothing, which is consistent with the results reported in [19].

Table 1. Performance of language models

Models	PP(MKN)	PP(GLM)	C	U	F
2gram	1257.374	-	**0.475**	**0.086**	**0.145**
FB-HWS-2 (without extra)	1122.878	-	0.455	0.081	0.138
DB-HWS-2 (without extra)	1261.526	-	0.451	0.061	0.107
FB-HWS-2 (with extra)	**901.656**	-	0.445	0.078	0.133
DB-HWS-2 (with extra)	909.873	-	0.443	0.070	0.122
3gram	1119.764	929.773	0.227	0.030	0.052
FB-HWS-3 (without extra)	1053.359	862.821	**0.317**	**0.047**	**0.081**
DB-HWS-3 (without extra)	1152.181	923.108	0.289	0.033	0.059
FB-HWS-3 (with extra)	816.545	670.477	0.299	0.043	0.075
DB-HWS-3 (with extra)	**802.419**	**659.298**	0.305	0.041	0.072
4gram	1104.837	862.956	0.084	0.009	0.017
FB-HWS-4 (without extra)	1047.151	742.188	**0.242**	**0.032**	**0.057**
DB-HWS-4 (without extra)	1146.382	783.998	0.196	0.022	0.039
FB-HWS-4 (with extra)	809.877	583.237	0.217	0.028	0.049
DB-HWS-4 (with extra)	**798.930**	**570.317**	0.225	0.028	0.050

5 Extrinsic Evaluation

Perplexity is not a definite way of determining the usefulness of a language model since a language model with low perplexity may not work equally well in a real world application. Thus, we also performed extrinsic experiments to evaluate our model. In this paper, we use the reranking of n-best translation candidates to examining how language models work in a statistical machine translation task.

5.1 Settings

We use the French-English part of TED talks parallel corpus as the experiment dataset. The training data contains 139,761 sentence pairs, while the test data contains 1,617 sentence pairs. For training language models, we set English as the target language.

As for statistical machine translation toolkit, we use Moses system[6] to train the translation model and output 50-best translation candidates for each french sentence of the test data. Then we use the 139,761 English sentences to train language models. With these models, 50-best translation candidates can be reranked. According to these reranking results, the performance of machine translation system can be evaluated, which also means, the language models can be evaluated indirectly. In this paper, we use **BLEU** [21], **METEOR** [22] and **TER** [23] as the measures for evaluating reranking results.

5.2   Results

We use HWS models with/without extra techniques to perform experiments and compared them to normal n-gram models. As shown in Table 2, since the results performed by our implementation (3-gram+MKN) is almost the same as that performed by existing languge model toolkits IRSTLM[7] and SRILM[8,9], we believe that our implementation is correct. Based on the results, DB-HWS (with extra) performs best on each score. We also notice that for an SMT task, the performance doesn't benefit much by applying GLM smoothing, in some case, it get worse than applying MKN smoothing.

Table 2. Performance of SMT system using different language models

Models	Smoothing	BLEU	METEOR	TER
IRSTLM-3	MKN	31.2	33.5	49.1
SRILM-3	MKN	31.3	33.5	48.9
3-gram	MKN	31.3	33.5	49.1
	GLM	31.3	33.5	49.2
FB-HWS-3 (without extra)	MKN	31.3	33.4	48.6
	GLM	31.2	33.4	48.7
FB-HWS-3 (with extra)	MKN	31.3	33.5	48.6
	GLM	31.4	33.5	**48.5**
DB-HWS-3 (with extra)	MKN	31.5	**33.6**	**48.5**
	GLM	**31.6**	**33.6**	**48.5**

Since as training data, 139,761 english sentences isn't big enough and is in-domain, we mix them with the 449,755 sentences (10 million words) that we used in Sect. 4.1 as a new training data and performe this experiment again.

---

[6] http://www.statmt.org/moses/.

[7] http://sourceforge.net/projects/irstlm/.

[8] http://www.speech.sri.com/projects/srilm/.

[9] For the settings of IRSTLM and SRILM, we use default settings except for using modified Kneser-Ney as the smoothing method.

As the results shown in Table 3, previous conclusions still stand. Considering all these experiment results, DB-HWS models using directionalization and unification get the best and the most stable performance for an SMT task.

Table 3. Performance of SMT system using different language models with mixed training data

Models	Smoothing	BLEU	METEOR	TER
IRSTLM-3	MKN	31.4	**33.6**	48.9
SRILM-3	MKN	31.5	**33.6**	48.8
3-gram	MKN	31.5	**33.6**	49.0
	GLM	31.3	**33.6**	49.1
FB-HWS-3 (without extra)	MKN	31.5	33.5	48.6
	GLM	31.3	33.5	48.6
FB-HWS-3 (with extra)	MKN	31.4	**33.6**	48.5
	GLM	31.3	**33.6**	48.6
DB-HWS-3 (with extra)	MKN	**31.6**	**33.6**	**48.4**
	GLM	**31.6**	33.5	**48.4**

6   Conclusion

In this paper, we presented an improved HWS language model, by which word association can be used for establishing HWS structure and achieving HWS-n-grams instead of only using word frequency.

We also proposed two extra techniques for improving the performance of HWS models. Furthermore, we discussed how to apply the current state-of-the-art smoothing methods to HWS models for higher performance.

For evaluation, we compared word association based HWS models to normal HWS models and normal n-gram models, and performed intrinsic and extrinsic experiments separately. The intrinsic experiment proved that the perplexity can be greatly reduced by using word association based HWS models, while the extrinsic experiments proved that SMT tasks can benefit from word association based HWS models. Both verified the effectiveness of our model.

In this paper, we only used Dice coefficient as the word association measure, there are many other alternative word association measures, such as mutual information, T-score and so on. We will evaluate our model using different word association measures in the future.

References

1. Brown, P.F., Cocke, J., Pietra, S.A., Pietra, V.J., Jelinek, F., Lafferty, J.D., Mercer, R.L., Roossin, P.S.: A statistical approach to machine translation. Comput. Linguist. **16**(2), 79–85 (1990)

2. Mays, E., Damerau, F.J., Mercer, R.L.: Context based spelling correction. Inf. Process. Manage. **27**(5), 517–522 (1991)
3. Rabiner, L., Juang, B.H.: Fundamentals of Speech Recognition. Prentice Hall, Englewood Cliffs (1993)
4. Bickel, S., Haider, P., Scheffer, T.: Predicting sentences using n-gram language models. In: Proceedings of the Conference on Human Language Technology and Empirical Methods in Natural Language Processing, HLT'05, pp. 193–200, Stroudsburg, PA, USA. Association for Computational Linguistics (2005)
5. Katz, S.: Estimation of probabilities from sparse data for the language model component of a speech recognizer. IEEE Trans. Acoust. Speech Signal Process. **35**(3), 400–401 (1987)
6. Kneser, R., Ney, H.: Improved backing-off for m-gram language modeling. In: 1995 International Conference on Acoustics, Speech, and Signal Processing. ICASSP-95, vol. 1, pp. 181–184. IEEE (1995)
7. Chen, S.F., Goodman, J.: An empirical study of smoothing techniques for language modeling. Comput. Speech Lang. **13**(4), 359–393 (1999)
8. Shannon, C.E.: A mathematical theory of communication. Bell Syst. Tech. J. **27**, 379–423 (1948)
9. Allison, B., Guthrie, D., Guthrie, L., Liu, W., Wilks, Y.: Quantifying the Likelihood of Unseen Events: A Further Look at the Data Sparsity Problem. Awaiting publication (2005)
10. Wu, X., Matsumoto, Y.: A hierarchical word sequence language model. In: Proceedings of The 28th Pacific Asia Conference on Language, Information and Computation (PACLIC), pp. 489–494 (2014)
11. Huang, X., Alleva, F., Hon, H.W., Hwang, M.Y., Lee, K.F.: The SPHINX-II speech recognition system: an overview. Comput. Speech Lang. **7**(2), 137–148 (1993)
12. Guthrie, D., Allison, B., Liu, W., Guthrie, L.: A Closer Look at Skip-gram Modeling. In: Proceedings of the 5th International Conference on Language Resources and Evaluation, pp. 1–4 (2006)
13. Brown, P.F., Desouza, P.V., Mercer, R.L., Pietra, V.J.D., Lai, J.C.: Class-based n-gram models of natural language. Comput. Linguist. **18**(4), 467–479 (1992)
14. Bilmes, J.A., Kirchhoff, K.: Factored language models and generalized parallel backoff. In: Proceedings of the 2003 Conference of the North American Chapter of the Association for Computational Linguistics on Human Language Technology, vol. 2, pp. 4–6 (2003)
15. Shen, L., Xu, J., Weischedel, R.M.: A new string-to-dependency machine translation algorithm with a target dependency language model. In: ACL, pp. 577–585 (2008)
16. Chen, W., Zhang, M., Li, H.: Utilizing dependency language models for graph-based dependency parsing models. In: Proceedings of the 50th Annual Meeting of the Association for Computational Linguistics, Long Papers, vol. 1. Association for Computational Linguistics, pp. 213–222 (2012)
17. Chelba, C.: A structured language model. In: Proceedings of ACL-EACL, Madrid, Spain, pp. 498–500 (1997)
18. Dice, L.R.: Measures of the amount of ecologic association between species. Ecology **26**, 297–302 (1945)
19. Pickhardt, R., Gottron, T., Körner, M., Staab, S.: A generalized language model as the combination of skipped n-grams and modified kneser-ney smoothing. In: Proceedings of the 52nd Annual Meeting of the Association for Computational Linguistics, pp. 1145–1154 (2014)

20. Manning, C.D., Schütze, H.: Foundations of Statistical Natural Language Processing. MIT Press, Cambridge (1999)
21. Papineni, K., Roukos, S., Ward, T., Zhu, W.J.: BLEU: a method for automatic evaluation of machine translation. In: Proceedings of the 40th Annual Meeting on Association for Computational Linguistics, pp. 311–318. Association for Computational Linguistics (2002)
22. Banerjee, S., Lavie, A.: METEOR: An automatic metric for MT evaluation with improved correlation with human judgments. In: Proceedings of the ACL Workshop on Intrinsic and Extrinsic Evaluation Measures for Machine Translation and/or Summarization, pp. 65–72 (2005)
23. Snover, M., Dorr, B., Schwartz, R., Micciulla, L., Makhoul, J.: A study of translation edit rate with targeted human annotation. In: Proceedings of Association for Machine Translation in the Americas, pp. 223–231 (2006)

Neural-Network-Based Spectrum Processing for Speech Recognition and Speaker Verification

Jan Zelinka[1(✉)], Jan Vaněk[2], and Luděk Müller[2]

[1] Department of Cybernetics, Faculty of Applied Sciences,
University of West Bohemia, Univerzitní 8, 306 14 Plzeň, Czech Republic
zelinka@kky.zcu.cz
[2] Faculty of Applied Sciences, New Technologies for the Information Society,
University of West Bohemia, Univerzitní 8, 306 14 Plzeň, Czech Republic
{vanekyj,muller}@kky.zcu.cz

Abstract. In this paper, neural networks are applied as a feature extractors for a speech recognition system and a speaker verification system. A long-temporal features with delta coefficients, mean and variance normalization are applied when a neural-network-based feature extraction is trained together with a neural-network-based voice activity detector and with a neural-network-based acoustic model for speech recognition. In speaker verification, the acoustic model is replaced with a score computation. The performance of our speech recognition system was evaluated on the British English speech corpus WSJCAM0 and the performance of our speech verification system was evaluated on our Czech speech corpus.

1 Introduction

This paper describes an application of neural networks (NNs) for the construction of a feature extractor, a voice activity detector, and an acoustic model for a speech recognition system. An application of the NN-based feature extraction is also presented for speaker verification. Nowadays, there is a tendency in speech processing to substitute a NN for standard feature extraction [3,4,8,12–14]. In a common NN-based acoustic model, several neighboring feature vectors are joined into one vector called long-temporal features as the input of NN. When a "raw" signal or a spectrum is the input of NN, long-temporal features can be generated in similar way and the NN can be a feed-forward NN or a convolutional NN [1]. But this approach brings some difficulties [17]. To make the long-temporal features more efficient in the sense of recognition error, we decided to place some operations between a NN-based feature extraction and a NN-based acoustic model. In this paper, the NN-based feature extraction substitutes the Perceptual Linear Prediction [6] (PLP) or Mel Frequency Cepstral Coefficients (MFCC) feature computation. Our experiments showed that a relatively small NN could sufficiently substitute and even overcome the PLP feature computation. A relatively small number of features generated by our NN-based feature extraction and the use of the time-shifting operations allows also to use standard signal processing methods. The novelty of this paper lies in the replacement of the PLP feature

© Springer International Publishing Switzerland 2015
A.-H. Dediu et al. (Eds.): SLSP 2015, LNAI 9449, pp. 288–299, 2015.
DOI: 10.1007/978-3-319-25789-1_27

computation, the use of long-temporal features with standard signal processing methods, namely delta coefficients computation, mean and variance normalization [7], and training a NN-based feature extraction, a NN-based voice activity detector and a NN-based acoustic model together.

NN-based feature extractions were tested in speech recognition and also in speaker verification scenario. Our experiments showed that our approach is beneficial in the latter case as well. The performance of our speech recognition system was evaluated on the British English speech corpus WSJCAM0 and the performance of our speech verification system was evaluated on our Czech speech corpus.

This paper is organized as follows: Sect. 2 describes our NN-based acoustic models. Our NN-based feature extraction and the role of a mean and variance normalization are discussed in Sect. 3. Section 4 deals with speaker verification. Results of our experiments where the NNs are combined are shown in Sect. 5. Some conclusions and description of the future work are presented in Sect. 6.

2   Neural-Network-Based Acoustic Models

A NN-based acoustic model computes posterior probabilities from its input features. Figure 1 shows a schema of a basic NN-based acoustic model. In our experiments, instead a shallow NN drawn in Fig. 1, a deeper NN was tested. The activation functions in the hidden layer (or layers) are the usual sigmoidal functions. The activation functions in the output layer are soft-max functions. A criterion for backpropagation is the cross-entropy (XENT).

Several time-shifting operations are applied on the input of the NN. The time-shifting operations, mean normalization, variance normalization could be computed for all features independently. Thus, the equations in this and the following section are for the i-th feature. The time-shifting operation $ts_i(t, s)$ with a shift s transforms an input sequence $x_i(1)$, ..., $x_i(T)$ into an output sequence $ts_i(1, s)$, ..., $ts_i(T, s)$ according to the formula

$$ts_i(t, s) = \begin{cases} x_i(t + s) & 1 \leq t + s \leq T \\ x_i(1) & 1 > t + s \\ x_i(T) & t + s > T \end{cases} . \tag{1}$$

The resultant block "time shiftings" (shown in Fig. 1 and following figures) consists of eleven time-shifting operations. The shifts in these operations are $-5, -4, \ldots, 4, 5$.

In non-hybrid speech recognition, using mean (cepstral) normalization (substraction) significantly decreases the recognition error. Our experiments have shown that using these methods decreases the recognition error as well in the case of NN/HMM hybrids. There are many ways how these methods can be applied. One of the most general ones was chosen for our experiments. Its schema is shown in Fig. 2. The schema in the figure shows that features with normalized means are computed from PLP features. Features with normalized variances are computed from these mean-normalized features. There are four delta coefficients

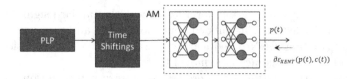

Fig. 1. A schema of standard NN-based acoustic model.

computation blocks in the schema. The first pair of delta coefficient blocks computes standard delta and delta-delta coefficients from the PLP features. Since the mean normalization does not change the value of delta coefficients, the second pair of delta coefficients blocks computes delta and delta-delta coefficients from the features with normalized variances.

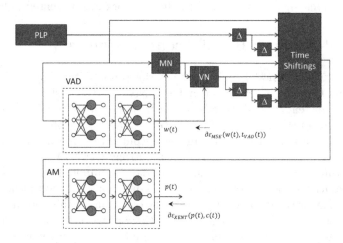

Fig. 2. A schema of standard NN-based acoustic model with mean normalization, variance normalization, delta features and delta-delta features

It is assumed that a Voice Activity Detector (VAD) is employed when mean or variance normalization is applied in a speech processing system. Figure 2 shows an application of a NN-based VAD. The NN has only a single neuron in its output layer. The neuron has the sigmoidal activation function. Thus, the VAD in a time t produces a weight $0 < w(t) < 1$. In backpropagation, not only one gradient is computed but some additional gradients are computed too [16]. The additional gradient which is denoted as $\partial \varepsilon_{MSE}(w(t), t_{VAD}(t))$ where ε_{MSE} means that the used criterion is the Mean Square Error (MSE) and $t_{VAD}(t) \in \{0, 1\}$ is a target for VAD in a time t. The resultant gradient is a sum of the additional gradient and "the main" gradient, i.e. $\partial \varepsilon_{XENT}$ for all layers where both gradients are computed.

3   Neural-Network-Based Feature Extraction

All NN-based feature extractions described in this paper process the absolute spectrum ($|$FFT$|$). The absolute spectrum processing is reasonable compromise between "raw" signal processing and using some more "lossy" method such as PLP. A schema of a basic posteriors computation that use a NN-based feature extraction is shown in Fig. 3.

In this paper, the NNs for the NN-based feature extraction have at least one hidden layer with sigmoidal activation functions. Activation functions in outputs layers were linear (diagonal) functions. The number of output neurons was always equal to number of the PLP features (i.e. 12). In our experiments, a NN trained using only $\partial \varepsilon_{XENT}$ and $\partial \varepsilon_{MSE}(w(t), t_{VAD}(t))$ is compared with a NN trained using these gradients and another additional gradient. This additional gradient is denoted as $\partial \varepsilon_{MSE}(x(t), PLP(t))$ and the additional gradient uses the MSE criterion. This additional gradient makes features produced by the NN-based feature extraction $x(t)$ close to the PLP features $PLP(t)$.

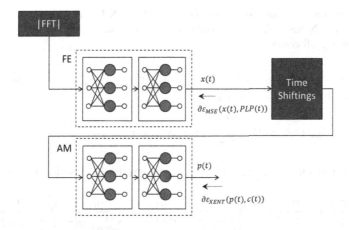

Fig. 3. A schema of a basic posteriors computation with a NN-based feature extraction and a NN-based acoustic model.

The NN-based feature extraction only replaces standard PLP feature computation that is shown in Fig. 1. In the same way, the PLP feature computation is replaced in the proposed more complex schema that is shown in Fig. 4. Besides the NN-based feature extraction and the NN-based acoustic model, there is also a NN-based VAD and its additional gradient $\partial \varepsilon(w(t), t_{VAD}(t))$ is computed.

In backpropagation, gradients for the normalizations, delta coefficient computation the long-temporal feature computation and the VAD must be computed. The time-shift operation and the delta coefficients computation are linear operations. Thus, its gradients can be computed by means of linear operation.

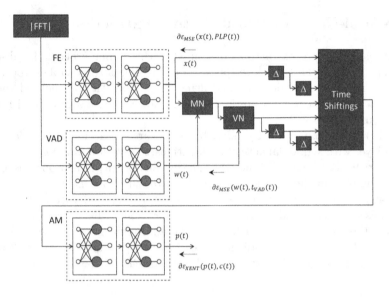

Fig. 4. A schema of a posteriors computation with a NN-based acoustic model, a NN-based VAD, a NN-based feature extraction, mean normalization, variance normalization and delta coefficients.

The mean normalization normalizes inputs $x_i(1), \ldots, x_i(T)$ using weights $w(1), \ldots, w(T)$ in the following way:

$$y_i(t) = x_i(t) - \sum_{\tau=1}^{T} \hat{w}(\tau) x_i(\tau), \ \hat{w}(t) = \frac{w(t)}{\sum_{\tau=1}^{T} w(\tau)}. \tag{2}$$

The required gradients $\frac{\partial \varepsilon}{\partial x_i(t)}$ and $\frac{\partial \varepsilon}{\partial w(t)}$ are computed according to the formula

$$\frac{\partial \varepsilon}{\partial x_i(t)} = \frac{\partial \varepsilon}{\partial y_i(t)} - \hat{w} \sum_{\tau=1}^{T} \frac{\partial \varepsilon}{\partial y_i(\tau)}, \ \frac{\partial \varepsilon}{\partial w(t)} = \frac{-y_i(t)}{\sum_{\tau=1}^{T} w(\tau)} \sum_{\tau=1}^{T} \frac{\partial \varepsilon}{\partial y_i(\tau)}. \tag{3}$$

Variance normalization is computed in the following way

$$y_i(t) = \frac{x_i(t)}{\sigma}, \ \sigma^2 = \sum_{\tau=1}^{T} \hat{w}(\tau) \left(x_i(\tau) - \mu \right)^2, \ \mu = \sum_{\tau=1}^{T} \hat{w}(\tau) x_i(\tau). \tag{4}$$

The gradient $\frac{\partial \varepsilon}{\partial x_i(t)}$ and the gradient $\frac{\partial \varepsilon}{\partial w(t)}$ are computed by means the following formula:

$$\frac{\partial \varepsilon}{\partial x_i(t)} = \frac{1}{\sigma} \frac{\partial \varepsilon}{\partial y_i(t)} - \frac{x_i(t) - \mu}{\sigma^3} \hat{w}(t) \sum_{\tau=1}^{T} \frac{\partial \varepsilon}{\partial y_i(\tau)} x_i(\tau). \tag{5}$$

The gradient for weights is computed using the following equation:

$$\frac{\partial \varepsilon}{\partial w(t)} = \frac{\sigma^2 - (x_i(t) - \mu)^2}{2\sigma^3 \sum_{\tau=1}^{T} w(\tau)} \sum_{\tau=1}^{T} \frac{\partial \varepsilon}{\partial y_i(\tau)} x_i(\tau). \tag{6}$$

4 Neural-Network-Based Feature Extraction for Speaker Verification

Our speaker verification system is a text-dependent speaker verification system based on Dynamic Time Warping (DTW) [2,5,18]. Our speaker verification consists of the following four steps:

1. Firstly, two recordings are recorded: recording of "suspicious" speaker and referential recording. Both recordings contain utterances of the same sentence. The goal of our verification system is to make a decision if the suspicious speaker is the same speaker as the speaker in the referential recording.
2. Afterwards, both recordings are synchronized by means of well-known DTW. The synchronized suspicious recording is denoted as $X_1 = (\mathbf{x}_1(1), \ldots, \mathbf{x}_1(T))$ where $\mathbf{x}_1(t) = (x_{1,1}(t), \ldots, x_{1,n}(t))$ for $t = 1, \ldots, T$. The synchronized referential recording is denoted as $X_2 = (\mathbf{x}_2(1), \ldots, \mathbf{x}_2(T))$ where $\mathbf{x}_2(t) = (x_{2,1}(t), \ldots, x_{2,n}(t))$ for $t = 1, \ldots, T$.
3. In the next step, both recordings are processed by means of some feature extraction method. In our experiments, MFCC, PLP and a NN-based feature extraction were used and tested.
4. Finally, distances between each pair $\mathbf{x}_1(t)$ and $\mathbf{x}_2(t)$ are computed. For $t = 1, \ldots, T$ and $i = 1, \ldots, n$ where n is the number of features, an i-th partial distance $\delta_i(t)$ is computed as $\delta_i(t) = (x_{1,i}(t) - x_{2,i}(t))^2$. A resultant distance in a time t is computed as $d(t) = \sigma\left(\sum_{i=1}^{n} \delta_i(t) + b\right)$ where b is a chosen bias and σ is a sigmoid function. A score of dissimilarity s, $0 \leq s \leq 1$, is computed for the pair of recordings. We did not want to compute a resultant score from all vectors but we wanted to use only suitable vectors, i.e. vectors where speakers are really speaking. Therefore, the resultant score is computed according to the formula:

$$ s = \sum_{\tau=1}^{T} \hat{w}(\tau)d(\tau), \tag{7} $$

where a weight $\hat{w}(t)$ is computed from $x_2(t)$ using a VAD.

The whole process for MFCC is shown in Fig. 5. The delta coefficients were computed before the synchronization. Because it is questionable if mean and variance normalization are beneficial for speaker verification, a feature vector contains all three options. One NN layer with linear activation functions reduces a number of features to forty. One neuron computes the resultant distance $d(t)$ as $d(t) = \sigma\left(\sum_{i=1}^{n} w_i\delta_i(t) + b\right)$. Its weights w_i were fixed and equal to one. Hence, our training algorithms changed only the bias.

Figure 6 shows a schema of the first proposed application of the NN-based feature extraction. This application does not employ any normalization.

The second proposed speaker verification system of a NN-based feature extraction does employ mean and variance normalization. A schema of the speaker verification system is shown in Fig. 7.

The figures show that two gradients are computed. The first gradient is computed for the score s using MSE criterion. The target t_s equals to 0 when

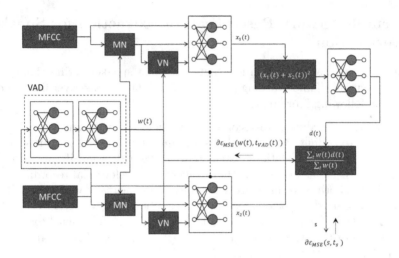

Fig. 5. A schema of a speaker verification system with standard MFCC feature computation.

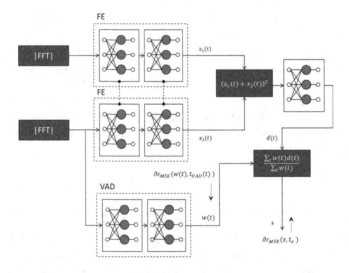

Fig. 6. A schema of a speaker verification system with a NN-based feature extraction.

the speaker who recorded the suspicious recording is the same speaker who recorded the referential recording. Otherwise, t_s equals to 1. The second gradient is an additional gradient for training the NN-based VAD.

Computation of gradients $\frac{\partial \varepsilon}{\partial x_{1,i}(t)}$ and $\frac{\partial \varepsilon}{\partial x_{2,i}(t)}$ is trivial. Thus, we present only the gradients

$$\frac{\partial \varepsilon}{\partial d(t)} = \hat{w}(t)\frac{\partial \varepsilon}{\partial s}, \quad \frac{\partial \varepsilon}{\partial w(t)} = \frac{\partial \varepsilon}{\partial s}\frac{d(t) - s}{\sum_{\tau=1}^{T} w(\tau)}. \tag{8}$$

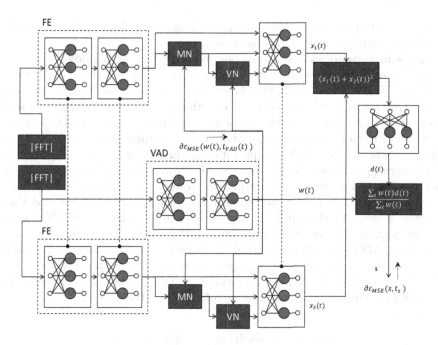

Fig. 7. A schema of a speaker verification system with a NN-based feature extraction and normalizations.

Due to recordings synchronization, some neighboring frames could be equal. Therefore, the delta coefficients computation might be distorted. One possible solution is to include the synchronization process into the backpropagation. Instead the delta coefficient computation, feature vectors were extended with several time-shift operations placed before the feature computation.

5   Experiments and Results

This section describes our experiments with speech recognition and speaker recognition.

5.1   Speech Recognition

The British English speech corpus WSJCAM0 [11] was used for the system performance evaluation in the following experiments. This corpus includes 7,861 utterances (i.e. approximately 15 h of speech) in the training set. Phonetic alphabet consists of 43 phones (including silence and inhale). The experiments were performed on the development sets si_dt5a and si_dt5b. In the corpus, a particular trigram language model for both sets is prescribed. The set si_dt5a was used to find the optimal word insertion penalty and language model weight.

The set si_dt5b was used strictly as an evaluation set. Our real-time LVSCR decoder [9] was applied in our experiments.

In the first experiment, standard non-hybrid speech recognition was tested. In the second experiment, the performance of standard NN/HMM hybrid was evaluated. The NN has one hidden layer with 1,024 neurons. A NN-based feature extraction is tested in the third experiment. The NN for the NN-based feature extraction has one hidden layer with 256 neurons. In the fourth experiment, the PLP features with mean and variance normalization and NN/HMM hybrid were tested. No VAD was applied in this and the next experiment (i.e. all weight were equal to one). In the fifth experiment a NN-based feature extraction with mean and variance normalization was tested. The sixth and the seventh experiment follow the fourth and the fifth experiment, except a NN-based VAD detector is used. In the third, fifth, and seventh experiments, the NN-based features were trained to approximate the PLP features. The results, i.e. word error rates (WER), are shown in Table 1. These results show that the proposed NN based feature extraction is significantly beneficial for speech recognition. The VAD utilization is not provably beneficial. This fact is probably caused by a low and relatively constant percentage of silence in the database.

Table 1. The results for speech recognition.

No.	Feat. Ext.	Signal processing	VAD	Ac. Model	si_dt5a	si_dt5b
1	PLP	$\Delta + \Delta\Delta$	None	GMM	16.3 %	17.8 %
2	PLP	$\Delta + \Delta\Delta$	None	NN	13.0 %	14.1 %
3	NN	$\Delta + \Delta\Delta$	None	NN	11.3 %	12.4 %
4	PLP	$\Delta + \Delta\Delta$ + MN + VN	None	NN	10.5 %	11.3 %
5	NN	$\Delta + \Delta\Delta$ + MN + VN	None	NN	8.4 %	9.1 %
6	PLP	$\Delta + \Delta\Delta$	NN	NN	10.2 %	11.2 %
7	NN	$\Delta + \Delta\Delta$ + MN + VN	NN	NN	**8.3 %**	**9.1 %**

In the next (from 8 to 10) experiments, the additional gradient application was investigated. Because a deep-NN training may be difficult, a potential benefit of the additional gradient manifests in a deep-NN training more likely. In the previous (from 1 to 7) experiments, shallow NNs were trained. In the next three experiments, deep NNs with four layers were trained. A connection of a NN-based feature extraction and a NN-based acoustic model creates a deep NN with eight layers. No VAD was applied in these experiments. In the eighth experiment, the PLP features with mean and variance normalization and with NN/HMM hybrid was tested. In the ninth experiment, a NN-based feature extraction and the additional gradient application was tested. In the tenth experiment, a NN-based feature extraction was also tested but no additional gradient was applied. The results are shown in Table 2. The additional gradient is slightly beneficial but it is not crucial for training a NN-based feature extraction.

Table 2. The results for speech recognition with deep NNs.

No.	Feat. Extraction	Additional gradient	si_dt5a	si_dt5b
8	PLP	N/A	8.7%	9.2%
9	NN	Yes	**7.1%**	**8.3%**
10	NN	No	7.8%	8.4%

5.2 Speaker Verification

Our Czech telephone database consists of recordings by 337 speakers. Each speaker recorded twelve sentences. Speakers repeated their recording after a couple of weeks.

Several hundred thousand pairs were randomly generated, MFCCs were computed and each pair was evaluated [10]. 15,000 true pairs that gives the highest score and 15,000 false pairs that give the lowest score were selected. Both selections were equally randomly divided into training and testing part. Equal error rate (EER) was chosen as the evaluation metric. Hence, our training and test set contain only pairs that are difficult to distinguish. Therefore, all EERs are significantly higher than they are in a usual case.

Due to a relatively small corpus, a so called dropout technique was employed [15]. In the first four experiments, standard feature extraction techniques MFCC and PLP were investigated. The system also permits to use |FFT| computation as a feature extraction method. A NN was applied for feature vector reduction according to the schema shown in Fig. 5. In these experiments, a NN with one single layer and a NN with two layers were tested. There were 256 neurons in each hidden layer. An EER-boosted MSE cost is used to train all NNs. The EER boost is based on weighting of the MSE gradient. The weights are derived from actual output score and distributions of true or false pair scores. These distributions were approximated by a normal distribution. The results are shown in Table 3.

Table 3. The results for speaker recognition with standard MFCC, PLP and |FFT| feature extraction.

Feat. Ext.	One layer	Two layers
MFCC	**10.79%**	**4.73%**
PLP	12.03%	5.83%
|FFT|	13.79%	8.28%

In the last four experiments, a NN-based feature extraction was investigated. Two experiments were done without any normalization according the schema shown in Fig. 6 and two experiments were done with mean and variance normalization according the schema in Fig. 7. In two experiments, additional

Table 4. The results for NN-based feature extractions.

	No add. time-shifting	Additional time-shifting
No normalization	6.07 %	**3.92 %**
MN + VN	6.93 %	5.64 %

time-shifting was performed as a substitution of delta coefficients computation. The NN for feature vector reduction had one layer. The results are shown in Table 4. These results show that the proposed NN-based feature extraction is beneficial in this case too. Surprisingly, mean normalization and variance normalization are not beneficial for speaker recognition.

6 Conclusions and Future Work

In this paper, simultaneous training of the NN-based feature extraction, the NN-based VAD and the NN based acoustic model for the automatic speech recognition system were investigated. Our experiments showed that the proposed NN-based feature extraction in a speech recognition system can sufficiently replace methods such as the PLP. Moreover, this paper demonstrates that mean or variance normalization and delta coefficients could be successfully utilized in training process directly. The benefit of the proposed NN-based feature extraction in speaker verification was proven too.

In the future, some experiments described in this paper will be done for context-dependent units. Also the role of NN-based feature extraction in text-independent speaker verification will be investigated.

Acknowledgments. This research was supported by the Ministry of Culture Czech Republic, project No. DF12P01OVV022.

References

1. Chang, S., Morgan, N.: Robust CNN-based speech recognition with Gabor filter kernels. In: Interspeech 2014, 15th Annual Conference of the International Speech Communication Association, Singapore, 14–18 September 2014, pp. 905–909 (2014)
2. Das, A., Tapaswi, M.: Direct modeling of spoken passwords for text-dependent speaker recognition by compressed time-feature representations. In: ICASSP, IEEE, March 2010
3. Astudillo, R.F., Abad, A., Trancoso, I.: Accounting for the residual uncertainty of multi-layer perceptron based features. In: 2014 IEEE International Conference on Acoustics, Speech and Signal Processing (ICASSP), pp. 6859–6863, May 2014
4. Grézl, F., Karafiát, M.: Semi-supervised bootstrapping approach for neural network feature extractor training. In: ASRU, pp. 470–475. IEEE (2013)
5. Hbert, M.: Text-dependent speaker recognition. In: Benesty, J., Sondhi, M., Huang, Y. (eds.) Springer Handbook of Speech Processing, pp. 743–762. Springer, Berlin Heidelberg (2008)

6. Heřmanský, H.: Perceptual linear predictive (PLP) analysis of speech. J. Acoust. Soc. Am. **57**(4), 1738–1752 (1990)
7. Ioffe, S., Szegedy, C.: Batch normalization: accelerating deep network training by reducing internal covariate shift. CoRR abs/1502.03167 (2015)
8. Narayanan, A., Wang, D.: Ideal ratio mask estimation using deep neural networks for robust speech recognition. In: 2013 IEEE International Conference on Acoustics, Speech and Signal Processing (ICASSP), pp. 7092–7096, May 2013
9. Pražák, A., Psutka, J.V., Psutka, J., Loose, Z.: Towards live subtitling of TV ice-hockey commentary. In: Cabello, E., Virvou, M., Obaidat, M.S., Ji, H., Nicopolitidis, P., Vergados, D.D. (eds.) SIGMAP, pp. 151–155. SciTePress (2013)
10. Ramasubramanian, V., Das, A., Praveen, K.V.: Text-dependent speaker-recognition using one-pass dynamic programming algorithm. In: 2006 IEEE International Conference on Acoustics Speech and Signal Processing, ICASSP 2006, Toulouse, France, 14–19 May 2006, pp. 901–904 (2006)
11. Robinson, T., Fransen, J., Pye, D., Foote, J., Renals, S.: Wsjcam0: a british english speech corpus for large vocabulary continuous speech recognition. In: Proceedings of ICASSP 1995, pp. 81–84. IEEE (1995)
12. Sainath, T.N., Kingsbury, B., Rahman Mohamed, A., Ramabhadran, B.: Learning filter banks within a deep neural network framework. In: ASRU, pp. 297–302. IEEE (2013)
13. Sainath, T.N., Peddinti, V., Kingsbury, B., Fousek, P., Ramabhadran, B., Nahamoo, D.: Deep scattering spectra with deep neural networks for LVCSR tasks. In: Interspeech 2014, 15th Annual Conference of the International Speech Communication Association, Singapore, 14–18 September 2014, pp. 900–904 (2014)
14. Seps, L., Málek, J., Cerva, P., Nouza, J.: Investigation of deep neural networks for robust recognition of nonlinearly distorted speech. In: Interspeech, pp. 363–367 (2014)
15. Srivastava, N., Hinton, G.E., Krizhevsky, A., Sutskever, I., Salakhutdinov, R.: Dropout: a simple way to prevent neural networks from overfitting. J. Mach. Learn. Res. **15**(1), 1929–1958 (2014)
16. Szegedy, C., Liu, W., Jia, Y., Sermanet, P., Reed, S., Anguelov, D., Erhan, D., Vanhoucke, V., Rabinovich, A.: Going deeper with convolutions. CoRR abs/1409.4842 (2014)
17. Tüske, Z., Golik, P., Schlüter, R., Ney, H.: Acoustic modeling with deep neural networks using raw time signal for lvcsr. In: Interspeech, Singapore, pp. 890–894. September 2014
18. Yegnanarayana, B., Prasanna, S., Zachariah, J., Gupta, C.: Combining evidence from source, suprasegmental and spectral features for a fixed-text speaker verification system. IEEE Trans. Speech Audio Process. **13**(4), 575–582 (2005)